Structure: from Physics to General Systems

Festschrift volume in honour of
E. R. Caianiello
on his seventieth birthday

Volume 1

Dipartimento di Fisica Teorica – Università di Salerno
Istituto Italiano per gli Studi Filosofici
International Institute for Advanced Scientific Studies

Structure: from Physics to General Systems

Festschrift volume in honour of
E. R. Caianiello
on his seventieth birthday

Volume 1

Amalfi, Salerno, Italy   20 – 24 October 1991

Edited by

M. Marinaro & G. Scarpetta

Dipartimento di Fisica Teorica, Universita di Salerno
I.I.A.S.S., Vietri sul mare, Salerno
Italia

World Scientific
Singapore • New Jersey • London • Hong Kong

Published by

World Scientific Publishing Co. Pte. Ltd.

P O Box 128, Farrer Road, Singapore 9128

USA office: Suite 1B, 1060 Main Street, River Edge, NJ 07661

UK office: 73 Lynton Mead, Totteridge, London N20 8DH

STRUCTURE: FROM PHYSICS TO GENERAL SYSTEMS

ISBN 981-02-1291-7 (Set)
ISBN 981-02-1964-4 (Vol. 1)

Printed in Singapore by Stamford Press Pte Ltd

Professor Eduardo R. Caianiello

PREFACE

These two volumes contain the invited lectures presented at the International Conference on "Structure: from Physics to General Systems" held in Amalfi and Naples, Italy, from 22 to 24 October 1991, to honour the teacher, the friend and the colleague, Professor Eduardo R. Caianiello. It also contains a collection of invited articles by scientists who have collaborated with him, or worked in fields related to his wide scientific interests.

Eduardo has contributed to the development of research in Theoretical Physics and in Cybernetics on different levels: with his research activity, that embraces wide sectors of Theoretical Physics, of the General Theory of Complex Systems, and of Neural Nets; by creating and directing until full growth the "Istituto di Fisica Teorica" at the Mostra d'Oltremare in Napoli, the "Laboratorio Internazionale di Cibernetica" in Arco Felice, the "Facoltá di Scienze" in Salerno, the "International Institute for Advanced Scientific Studies" in Vietri sul Mare; and with his deep dedication to teaching in the Universities of Naples and Salerno many courses at graduate and postgraduate level. There are many researchers who have drawn inspiration and support in many ways, and in diverse fields from his relentless activity.

The name chosen for the Conference reflects the unifying interdisciplinary mental approach to Science of Eduardo; from the very inception his research activity was dedicated to "physics", "neural nets", "self-organizing systems", "linguistics" and what is behind, at a time in which the very association of these words sounded unthinkable. The division in two volumes of the Proceedings of the Conference, the first dedicated to "Physics" and the second to "General Systems", does not therefore contradict the unity of the Conference.

During the Closing Ceremony, held in Naples at Palazzo Serra di Cassano on October 24th, Eduardo received from Academician Professor N. N. Bogolubov the nomination to full member "Akademia Tvorchestvo" and a Certificate of Appreciation of the V. A. Steklov Mathematical Institute of Moscow; similar Certificates were given to him by Professor Kurdyumov and Professor A. M. Aizerman, on behalf respectively of the M. V. Keldish Institute of Applied Mathematics and of the Institute of Problems of Control of Moscow.

The Proceedings open with a recollection by Bruno Preziosi of the birth of the Istituto di Fisica Teorica, the first among those founded by Eduardo.

We would like to thank the members of the Scientific Committee for their assistance and advice: Professor A. Barone, Professor A. Covello, Professor F. Lauria, Professor G. Marmo, Professor B. Preziosi and Professor L. M. Ricciardi of the Naples University; Professor L. De Cesare and Professor F. Mancini of the Salerno University; Professor R. Capocelli and Professor A. de Luca of the Rome University; Professor S. Fubini and A. Giovannini of the Torino University; Professor G. Germano of Pisa University; Professor A. Restivo of Palermo University and Professor A. Rimini of Pavia University.

We gratefully acknowledge the financial support we received from the following organizations: Universitá di Salerno; Dipartimento di Scienze Fisiche, Universitá di Napoli; Istituto Nazionale di Fisica Nucleare, Sezione di Napoli; Istituto di Cibernetica — CNR; Istituto Italiano per gli Studi Filosofici; International Institute for Advanced Scientific Studies; ICSC — World Laboratory.

Very deep are our appreciation and gratitude towards the eminent scientists who have promptly and friendly accepted to participate in this Conference to honour Eduardo R. Caianiello; all together they have created a stimulating scientific and warm human atmosphere, whose memory will last in all the participants.

Professor Maria Marinaro
Professor Gaetano Scarpetta

CONTENTS

АКАДЕМИЯ ТВОРЧЕСТВА

СССР

НА ОСНОВАНИИ УСТАВА
ИЗБРАЛА

Professor

Eduardo Caramello

ДЕЙСТВИТЕЛЬНЫМ ЧЛЕНОМ
— АКАДЕМИКОМ —
АКАДЕМИИ
ТВОРЧЕСТВА

ПРЕЗИДЕНТ Гладышев Г.П.ГЛАДЫШЕВ

ЧЛЕНЫ ПРЕЗИДИУМА

Dear Professor Caianiello !

Today we are glad to congratulate You, a prominent member of our scientific society, with a remarkable jubilee - a seventieth birthday.

For many years You have been a leader and a source of innovative and inspiring ideas in the field of quantum field theory and statistical physics. Your understanding of the importance of the renormalization group formalism in the field theory, generalization of the entropy conception for chaotic processes, usefulness of differential equations for neural networks dynamical investigations and other achievements have become cornerstones in the corresponding disciplines of theoretical physics.

Your important scientific contributions are naturally coupled with hospitality, openness, intellectual honesty, broad humanistic views and Christian integrity.

On the occasion of seventieth birthday we wish **You** many more years of happy creative life full of scientific and social activity.

Sincerely Yours colleagues from V.A. Steklov Mathematical Institute and Moscow State University with profound admiration and best regards.

За выдающийся вклад в физико — математические науки и в развитие русско — итальянского научного сотрудничества, а также в связи с 70 — летием Институт прикладной математики им. М.В.Келдыша награждает

профессора

ЭДУАРДО Р. КАЙАНЬЕЛЛО

Медалью М.В.Келдыша

15 октября 1991 года
Москва
Директор Института прикладной
математики им.М.В.Келдыша
профессор

С.П.Курдюмов

Considering an outstanding contribution to physico —mathematical sciences and to progress in the Italian — Russian scientific collaboration as well as in the connection of the 70 — th birthday the M.V.Keldysh Institute of Applied Mathematics awards

Professor

EDUARDO R. CAIANIELLO

with the Keldyshmedal

October 15 — 1991
Moscow
Director of M.V.Keldysh Institute
Professor

S.P.Kurdyumov

DEAR PROFESSOR CAIANIELLO

The Directorate and the scientists of the Institute of Control Sciences of Academy of Sciences are sending you to the day of your seventieth anniversary their warm greetings and sincere wishes of many happy years of successful creative activity.

In our country we have a very high estimate of your contribution to Theoretic Physics. You were the first in Italy to initiate studies in the domain of Theoretical Foundations of Cybernetics, you founded the Institute of Cybernetics in Naples, which became the first center of italian studies in different branches of cybernetics. All your activity in Salerno university was directed to the progress and effective development of this science and it met a live interest and high appreciation in our country.

More then 20 years is our Institute working in a close contact with you. We are certain that this collaboration will progress and strengthen in the forthcoming years to the mutual advantage both of italian and our own and thus to all the world scientists active in various fields of cybernetics.

We are sending you our friendly wishes for many years of happy and successful life.

Director of the Institute
of Control Sciences
academician

I.V. Prangishvily

M.A. Aizerman

L.I. Rozonoer

I.M. Smirnova

E.A. Andreeva

F.T. Aleskerov

V.I. Volskiy

E.S. Piatnitskiy

1956-1958

IL PERIODO EROICO

Bruno Preziosi

Dipartimento di Scienze Fisiche, Università di Napoli

Nel 1937, in occasione di un celebre concorso presieduto da Enrico Fermi, la Commissione chiese al Ministro "di nominare Ettore Majorana professore di Fisica Teorica per alta e meritata fama in una Università del Regno". Il Ministro aderì alla richiesta ed assegnò allo scopo una cattedra all'Università di Napoli. Tale cattedra oltre che da Ettore Majorana, che l'occupò solo per qualche mese, fu successivamente tenuta, per meno di un anno ciascuno, da Ezio Clementel e Luigi Radicati di Brozolo e fu infine messa a concorso nel 1955.

La commissione esaminatrice, composta da Bruno Ferretti, Nicoló Dallaporta e Giampiero Puppi, concluse i propri lavori all'inizio dell'autunno 1955 formulando la seguente terna ordinata di vincitori: Eduardo Renato Caianiello, Marcello Cini e Fausto Fumi. Al momento della nomina a professore di Fisica Teorica Eduardo si trovava a Princeton, dove, in qualità di Higgins visiting professor, era impegnato ad illustrare i propri lavori in un corso su "Advanced topics in quantum theory". Egli chiese, ed ottenne, proprio per rispettare gli impegni assunti con l'Institute of Physics, di posticipare la presa di servizio, che avvenne formalmente ai primi del febbraio 1956; si trattenne tuttavia a Napoli lo stretto indispensabile, perché dovette rientrare a Princeton per completare il suo lavoro scientifico, e di fatto prese possesso della sua carica accademica nel novembre 1956.

Per comprendere lo straordinario impulso dato da Eduardo alla fisica napoletana é bene tenerne presente lo scenario nel momento in cui Eduardo tornava a

Napoli da professore: vi erano, presso l'Istituto Fisico di via Tari, un solo professore, Antonio Carrelli, direttore dell'Istituto e professore di Fisica Sperimentale, e tre assistenti la cui attivitá scientifica era indirizzata allo studio sperimentale di proprietá macroscopiche di liquidi e solidi; a questo corpo docente erano affidate tutte le attivitá didattiche e formative e, anche a causa della mancanza della presenza stabile di un fisico teorico, era possibile per uno studente napoletano laurearsi in fisica senza aver seguito un corso di meccanica quantistica.

Per consentirgli di iniziare la sua attivitá didattica e scientifica, Carrelli concesse ad Eduardo tre stanze, adibite a magazzino, all'ultimo piano di via Tari, che furono arredate utilizzando il contributo speciale di un milione che l'Universitá di Napoli dava a quel tempo come dote ai nuovi cattedratici. (Nota: per avere un valore della lira a quell'epoca, lo stipendio lordo annuo di Eduardo nel 1956 fu di 2.500.000 lire). In tali locali egli insedió la Cattedra di Fisica Teorica con autonomia amministrativa, avendo come aiuto Daniele Amati, assistente incaricato.

Nonostante la ricettivitá di tali locali fosse piuttosto scarsa, Eduardo cominció subito ad invitare fisici, soprattutto teorici, impegnati ai piú alti livelli scientifici. In particolare durante l'anno accademico 1956-57 visitarono Napoli molti teorici interessati al problema della rinormalizzabilitá delle teorie di campo, in cui Eduardo era personalmente impegnato. Questo problema era strettamente collegato al problema della regolarizzabilitá delle teorie, che mirava a dare delle regole per trattare le divergenze delle teorie relativistiche di campo e che Eduardo affrontava utilizzando il metodo di Hadamard.

Il finanaziamento per queste visite veniva in buona parte da grants americani. In quell'epoca, infatti, il governo americano aveva deciso di finanziare il rilancio della ricerca scientifica europea. Eduardo raccolse subito questa occasione e, in prossimitá del Natale 1956, si recó a Francoforte, con in mano un panettone, e illustró al colonnello americano incaricato i programmi scientifici che aveva in animo di portare avanti. Non si sa se fu la bontá del panettone o la forza dei suoi argomenti che consentirono a Eduardo di ricevere il grant n° 1 concesso dagli Stati Uniti nell'ambito del programma detto. Tale grant, dell' ammontare di

10.000 dollari, fu confermato fino al 1963 e certamente favorí la realizzazione delle iniziative scientifiche di cui si parlerá piu avanti.

Le condizioni difficili in cui la Cattedra di Fisica Teorica era costretta a vivere lo convinse che in quelle condizioni di ristrettezza non avrebbe potuto realizzare alcun programma di rilievo. I tentativi fatti presso il Rettore dell'Universitá di Napoli per avere spazi adeguati non ebbero successo, per cui si convinse a sondare strade alternative.

Il 1957 fu speso da Eduardo, oltre che per coltivare collaborazioni scientifiche all'epoca particolarmente intense, per portare avanti, insieme all'organizzazione didattica universitaria in senso moderno, sei diverse iniziative:

1) trasformazione della Cattedra in Istituto di Fisica Teorica;

2) reperimento di uno spazio abbastanza grande per ospitarlo;

3) creazione di una Scuola di Perfezionamento da affiancare all'Istituto;

4) avvio di una attivitá scientifica nel campo della cibernetica;

5) estensione dell'attivitá di ricerca nel campo della fisica nucleare;

6) reclutamento di un adeguato numero di ricercatori e tecnici.

1) Il primo problema fu in un certo senso il piú semplice da risolvere; le autoritá accademiche dettero il loro assenso alla trasformazione tra la fine del 1957 e i primi del 1958; Eduardo comunque giá da tempo avava provveduto a munirsi di carta intestata "Istituto di Fisica Teorica", come si evince dal rapporto a fianco riportato, sicuramente della fine del 1957, in cui si parla addirittura dell'attivitá svolta da tale Istituto nell'anno accademico 1956-57, anche se il Decreto Presidenziale istitutivo recherá la data del 18 giugno 1958 e sará pubblicato sulla Gazzetta Ufficiale del 20 settembre 1958. La dotazione ordinaria fu di 1.500.000 lire nel 1957-58 e fu elevata l'anno accademico successivo di 300.000 lire. Alla fine dell'anno accademico 1956-57, considerato che l'Universitá di Napoli non metteva a disposizione della Cattedra alcuna unitá di personale, Eduardo decise di "assumere" una segretaria utilizzando il grant americano; Anna Maria Mazzarella in Hilliard, che per moltissimi anni sará preziosa collaboratrice di Eduardo, prese servizio il 30 ottobre 1957. Per assicurare un minimo di agibilitá all'Istituto, Eduardo "as-

sunse" anche un bidello nella persona di Carmine Iscaro, che fu pagato per piú di un anno in modo precario sui fondi di ricerca; per la parte amministrativa, si avvalse della collaborazione del ricercatore Nello Onesto, finché non "assunse", agli inizi del 1959, un amministratore nella persona di Fulvio Calandrelli, che fu successivamente inquadrato in ambito I.N.F.N..

2) Il secondo problema, apparentemente insolubile, trovó una soluzione avventurosa insieme al terzo. Va premesso che da poco era stato istituito il C.N.R.N. (Comitato Nazionale per le Ricerche Nucleari), in uno con l'I.N.F.N. (Istituto Nazionale di Fisica Nucleare), che ne faceva parte. Il C.N.R., il cui presidente era Felice Ippolito, al fine di introdurre in Italia una cultura del nucleare, aveva deciso di appoggiare in modo deciso le attivitá scientifiche e didattiche in tale settore, favorendo fra l'altro la costituzione di Scuole di Perfezionamento in Fisica Teorica e Nucleare. Il C.N.R. era disponibile non solo al pagamento dei docenti delle scuole, ma anche, in caso di necessitá, al pagamento dell' affitto dei locali destinati ad ospitarle. Rimaneva il problema di trovare dei locali di cui l'Universitá non sembrava proprio disporre. Eduardo pensó allora di cercare una soluzione avvicinando l'ambiente industriale e manageriale napoletano. Per far questo si iscrisse ad un Lyons Club, dove ebbe modo di conoscere il presidente dell'Ente Mostra d'Oltremare (una fiera voluta da Mussolini per illustrare la presenza del lavoro italiano nel mondo), che gli offrí in fitto, per la modica cifra di 700.000 lire annue, il padiglione n° 19, progettualmente dedicato all' Estremo Oriente, incompleto e mai utilizzato. E' possibile che il presidente della Mostra d'Oltremare abbia fatto quell'offerta per togliersi di dosso un seccatore, cosí come é plausibile che nessun funzionario del C.N.R. abbia preso visione dei locali che si accingeva ad affittare. E' certo tuttavia che quando, a fine estate 1957, Eduardo accompagnó Arthur Wightman, che stava trascorrendo un periodo di ricerca a Napoli, sua moglie ed alcuni collaboratori (fra cui Nello Onesto, Giovanni Vingiani e Giuseppe Varcaccio) a visitare l'edificio destinato ad ospitare l'Istituto di Fisica Teorica, questi si trovarono di fronte ad un padiglione che aveva sí un tetto, ma non aveva un portone e, salvo che per un numero molto limitato di stanze, era senza pareti. Fu in

quell'edificio che a partire dal 4 novembre 1957, si trasferirono Eduardo ed i suoi collaboratori. All'inizio furono utilizzate solo le poche stanze agibili, ma agli inizi del 1959 gran parte del padiglione era stato ristrutturato, grazie ad un contributo straordinario del C.N.R. di 33 milioni.

3) Naturalmente anche la Scuola di Perfezionamento in Fisica Teorica e Nucleare, che sará formalmente istituita con lo stesso decreto dell'Istituto di Fisica Teorica, trovó ospitalitá nel padiglione 19. La prima lezione fu tenuta il 26 gennaio 1958; i docenti stabili del primo anno furono, oltre Eduardo, Daniele Amati, Francesco Mazzoleni, Nello Onesto, Luigi Salvatori, Roberto Stroffolini, Bruno Vitale; vi fu anche una notevole attivitá seminariale. Gli allievi erano Eugenio Galzenati, Francesco Lauria, Otello Mancini, Maria Marinaro , Pepé Speranza, Giuseppe Varcaccio, Francesco Verde e Giovanni Battista Vingiani. L'inaugurazione ufficiale della scuola avvenne il primo aprile dello stesso anno con una affollatissima lezione di Werner Heisemberg.

4) Per comprendere attraverso quali eventi Eduardo approdó alla cibernetica, bisogna fare un passo indietro. Nel 1954, sotto la spinta di Enrico Fermi che auspicava un maggiore interesse italiano nel campo dei calcolatori, fu organizzato presso l'Istituto di Fisica dell'Universitá di Roma un seminario sulle macchine calcolatrici, sulla teoria dell'informazione e sulla "cibernetica" di Norbert Wiener. Vi parteciparono, fra gli altri, Marcello Conversi, Bruno Touschek, Franzinetti, Gamba, Caracciolo, Eduardo Caianiello, allora assistente di quell'Istituto, e Valentino Braitenberg allora assistente volontario della clinica neuro-psichiatrica e "dilettante" di neuroatonomia. Alla fine del seminario, Eduardo e Valentino Braitenberg presero assieme la circolare esterna e, durante il tragitto, Eduardo manifestó il suo vivo interesse ad imparare qualcosa sul funzionamento del cervello.

Nella prima metá del 1955 Eduardo, a Copenaghen per un semestre, ebbe modo di imbattersi in un libro di Gray Walter, "The living brain", in cui si trattava il funzionamento del cervello con un linguaggio molto vicino alla fisica.

Eduardo, desideroso di approfondire le scarse conoscenze sul comportamento del cervello, prende sia l'iniziativa di invitare Braitenberg a trascorrere due o tre mesi a Napoli, sia quella di organizzare a Varenna una scuola estiva, che si terrá nell'estate 1958, cui parteciparono, fra gli altri, Wiener, Rosenblith, Bob Fano, Davenport, Bar-Hillel, Halle, Newman, Lee, Gabor, Rabin e Scott. Braitenberg arrivó a Napoli quando Eduardo era ancora a via Tari e, sebbene "stipendiato" per molti anni in modo precario, andrá via solo nel 1968. La collaborazione fra i due fu quanto mai fruttuosa; per limitarci al contributo di Eduardo citiamo un modello di rete nervosa in termini di oscillatori accoppiati, che precorre la moderna teoria sui neural networks, ed una nuova schematizzazione delle reti di neuroni digitali (reti di McCulloch e Pitts) in un linguaggio ("le equazioni del pazzo") proprio della fisica, che inauguró una tradizione nel campo. Inutile sottolineare il pionierismo di quegli anni, che troverá il suo coagulo alcuni anni dopo nel Laboratorio di Cibernetica del Consiglio Nazionale delle Ricerche.

5) Per poter avviare a Napoli una attivitá di Fisica Nucleare, Eduardo prese contatto con Edoardo Amaldi, che non mancó di aiutarlo nei momenti di difficoltá. Concretamente l'attivitá sperimentale fu avviata agli inizi dell' anno accademico 1958-59, prima con la venuta a Napoli di Renato Angelo Ricci, che avvió una attivitá sperimentale in Spettroscopia Nucleare, ed immediatamente dopo con la chiamata di Giulio Cortini che si occupava allora di fissione nucleare mediante l'uso di lastre nucleari. Per poter favorire il nascere sia di queste attivitá che di quelle connesse con le ricerche sperimentali nel settore della Cibernetica, Eduardo favorí l'organizzazione di una officina meccanica, utilizzando anche un contributo straordinario dell'Universitá di sette milioni. Inoltre ottenne una assegnazione una tantum dal Ministero della Pubblica Istruzione di circa 25 milioni, con cui fu tra l'altro acquistato un multicanale a 200 canali della Laben, con cui Ricci porterá avanti, insieme ai suoi collaboratori, una significativa ed apprezzata attivitá sperimentale. Ai primi del 1959 l'I.N.F.N. deliberava la creazione di una sua sottosezione a Napoli, che finanzió l'attivitá sulle lastre nucleari.

6) Giá alla fine dell'anno accademico 1958-59 Eduardo aveva la possibilitá di poter contare su un gruppo di ricercatori, neolaureati e laureandi che costituiranno il nucleo dell'attivitá scientifica della fisica napoletana.

ISTITUTO DI FISICA TEORICA
UNIVERSITÀ DI NAPOLI
VIA ANTONIO TARI, 3 - NAPOLI

ATTIVITA' DELL'ISTITUTO DI FISICA TEORICA DELL'UNIVERSITA' DI NAPOLI NELL'ANNO ACCADEMICO 1956-57

Didattica

Oltre ai corsi ufficiali di <u>Fisica Teorica</u> e <u>Spettroscopia</u> tenuti dal prof. E.R. Caianiello, sono stati tenuti i seguenti corsi:

1°) <u>Complementi di Fisica Teorica</u> (Dott. D. Amati)

2°) <u>Meccanica Analitica</u> (Dott. E. Tartaglione)

3°) <u>Meccanica Quantica</u> (Dr. W.I. Futtermann)

ed esercitazioni varie, destinati ad un gruppo abbastanza folto di laureati in anni precedenti e laureandi interni, ed intesi ad integrarne la preparazione. Inoltre si é organizzato un seminario interno trisettimanale sui <u>Metodi Matematici della Fisica</u>.

Questi corsi rappresentano un necessario primo passo verso la costituzione della Scuola di Perfezionamento in Fisica Teorica e Nucleare, il cui programma viene allegato apparte, e che si conta di iniziare regolarmente con il prossimo anno accademico.

Seminario di Fisica Teorica e Nucleare

Ha avuto luogo regolarmente durante l'anno con il duplice scopo di tenere gli studiosi locali aggiornati sui progressi piú recenti, e di porre in contatto visitatori di alto livello scientifico con gli elementi locali, per periodi di lunghezza determinata dalle particolari esigenze individuali. Ció si é rivelato fecondo di promettenti risultati, ed una attiva collaborazione ha avuto luogo, di conseguenza tra il nostro Istituto ed altri Istituti italiani e stranieri, con parecchi lavori pubblicati o in corso di preparazione d'intesa o in comune (si citano la Columbia University, Stanford, Princeton, Catania e Torino). Ogni sforzo sará in futuro rivolto a potenziare questa attivitá, che ha giá reso il nostro Istituto noto dovunque tali studi vengano compiuti, ed ha notevolissimi effetti morali, oltre che culturali, sulla massa degli studenti locali.

Per periodi varianti da qualche giorno a una o piú settimane, hanno partecipato al seminario:

Dott. T. Regge (Torino)

Prof. R. Gatto (Roma)

Dr. L. Spruch (U.C.L.A.)

Prof. G. Takeda (Tokio)

Dr. A. Garevich (General Electronic Lab.)

Dott. G. Luzzato (Genova)

Prof. D. Saxon (U.C.L.A.)

Dott. B. Vitale (Catania)

1956-1958

THE EROIC YEARS

by Bruno Preziosi

In 1957, on the occasion of a famous competitive examination presided by Enrico Fermi, the board of examiners asked the Minister to appoint Ettore Majorana professor of Theoretical Physics for his outstanding achievements "in a University of the Kingdom". The Minister complied with the request and gave Majorana a Chair at the University of Naples. The Chair was held by Majorana only for few months; later it went to Ezio Clementel and Luigi Radicati di Brozolo, both for less than a year. Eventually the Chair was assigned, after another national competition, in 1955.

The Board of examiners consisted of Bruno Ferretti, Nicoló Dallaporta and Giampiero Puppi. The Board completed its work in the autumn of 1955, and nominated the three winners in this order: Eduardo Renato Caianiello, Marcello Cini and Fausto Fumi. At the time when he was appointed professor of theoretical physics, Eduardo was in Princeton as Higgins visiting professor, giving lectures on his works in a course on "Advanced topics in quantum theory". Due to this commitment, he asked to postpone his nomination in Naples, where he started work at the beginning of February 1956. However he stayed in Naples only very shortly because he had to go back to Princeton to complete his scientific work. He actually started his academic work only in November 1956.

In order to understand the extraordinary impulse given by Eduardo to Neapolitan physics, one should bear in mind the situation at the time when Eduardo came back to Naples as professor. At the Physics Institute in via Tari there was only one professor, Antonio Carrelli, who was director of the Institute and professor of Experimental Physics. There were also three assistants, whose scientific research mainly concerned the macroscopic properties of fluids. This staff did all teaching and supervising. Moreover, due to the lack of a permanent theoretical physicist, Neapolitan students graduated in Physics without having followed a course in quantum mechanics. In order to allow Eduardo to start his teaching and scientific activity, Carrelli assigned him three rooms, previously used as store-rooms, at the last floor of the experimental building in via Tari. The rooms were furnished thanks to a one million Liras special fund granted at the time by the Naples University to new professors. To have an idea of the value of the Lira at the time, one should know that Eduardo's gross wage for the whole year 1956 was 2.500.000 Liras. In these three rooms he established the Chair of Theoretical Physics with administrative autonomy; Daniele Amati was his assistant.

Although he had not much room in his premises, Eduardo immediately started inviting physicists, mainly theorists, working at the highest scientific levels. In particular, during 1956-57, many theoretical physicists visited Naples, who were interested in the problem of the renormalizability of field theory, on which also Eduardo himself was working. This problem was tightly connected to the problem of regularization. Attempts were being made to find rules to handle divergences in relativistic field theories; Eduardo used for this purpose Hadamard's finite part integrals.

These visits were financed by American grants. At that time the American government had started financing the raise of the European scientific research; Eduardo immediately took this opportunity, and around Christmas 1956 went to Frankfurt to talk to the American colonel in charge of scientific programs. He was eloquent enough to obtain "Grant n° 1" of the program. This grant, amounting to 10.000 dollars, was continued until 1963 and greatly helped his scientific programs.

Life was hard for the Chair of Theoretical Physics; Eduardo soon realized that he could not pursue any really valuable program. He had in vain repeatedly asked the Rector of the University of Naples for larger premises. So, he decided to look for alternative solutions by himself.

Eduardo spent 1957 cultivating scientific relations and cooperation, which were particularly close for the time, organizing the teaching activity at the university in a modern manner, and fostering six different programs:

1) transforming the Chair in a Theoretical Physics Institute,

2) finding enough room to put up the Institute,

3) creating a specialization school together with the Institute,

4) starting scientific activity in the field of Cybernetics,

5) extending research to Nuclear Physics,

6) hiring enough researchers and technicians.

1) in a sense, the first problem was the easiest one to solve. The university authorities gave their consent to the transformation of the Chair into an Institute between the end of 1957 and the beginning of 1958. Long before that, however, Eduardo had supplied himself with "Institute of Theoretical Physics" headed paper, as one can see from the enclosed report, surely since the end of 1957. It is at the end of 1957 that the activity of the Institute during the 1957-58 is first mentioned. The founding Presidential Decree was issued on 18 June 1958 and published in the "Gazzetta Ufficiale" on 20 September 1958. The grant to the institute was of 1.500.000 Liras in the year 1957-58, and was increased of 300.000 Liras the following year. At the end of 1956-57, since the Chair had not yet been assigned any staff, Eduardo decided to hire a secretary, whom he would pay from the American grant. Anna Maria Mazzarella married Hilliard started working as secretary for Eduardo on 30 October 1957, and for many years she was a precious help for him. In order to make the Institute fit for work, Eduardo hired also a janitor, who was precariously paid from the research funds for over a year. As for administration, he was helped by the researcher Nello Onesto, until at the beginning of 1959 he could hire Fulvio Calandrelli as administrative director, who was later included in the permanent staff of I.N.F.N..

2) The second problem, which looked impossible to solve, found a fortunate solution together with the third one. It must be said beforehand that the C.N.R.N.

(National Committee on Nuclear Research) had shortly been established together with the I.N.F.N. (National Institute for Nuclear Physics). The C.N.R.N., whose president was Felice Ippolito, decisively supported scientific and teaching activities in its field. It also supported the creation of Specialization Schools in Theoretical and Nuclear Physics. Not only C.N.R.N. was willing to pay wages to the teachers in the schools, but also, where necessary, the rent of premises for to lodging these schools. Eduardo needed these badly, but the university did not seem to have any available; he therefore looked for a solution by tentatively approaching the industrial and managerial ambience in Naples, and thus entered the Lyons Club, where he met the president of the Ente Mostra d'Oltremare (an exhibition area created by Mussolini to illustrate the presence of Italian work in the world) who offered him to rent the incomplete and unused pavilion n° 19 for the reasonable price of 700.000 Liras a year. It is possible that the president of the Mostra d'Oltremare made Eduardo that offer to get rid of him; it is also likely that no official at the C.N.R.N. ever looked at the premises they were going to rent. It is sure, however, that, when in the summer 1957, Eduardo took Arthur Wightman, who was working on a research in Naples for a while, his wife and some of his cooperators (amongst whom Nello Onesto, Giovan Battista Vingiani and Giuseppe Varcaccio) to visit the building where the institute of theoretical physics would be lodged, they were faced with a pavilion which did not have a gate, and, apart from a very limited number of rooms, had very few walls indeed. Since 4 november 1957 Eduardo and his collaborators moved to that building. At they beginning the used only the few rooms available, but at the beginning of 1959 most of the pavilion had been restored thanks to a special 33 million grant-in-aid from C.N.R.N..

3) The Specialization School in Theoretical and Nuclear Physics was also formally established by the same Decree as the Institute of Theoretical Physics. Also the School was lodged in pavilion n° 19. The first lesson was held on 26 January 1958. The permanent teachers in the first year, besides Eduardo, were Daniele Amati, Francesco Mazzoleni, Nello Onesto, Luigi Salvadori, Roberto Stroffolini and Bruno Vitale. Also many seminars were held. The students were Eugenio

Galzenati, Francesco Lauria, Otello Mancini, Maria Marinaro, Pepé Speranza, Giuseppe Varcaccio, Francesco Verde and Giovanni Battista Vingiani. The official opening ceremony of the school was held on the first of April of that same year, with a very crowded audience owed by the lecturer Werner Heisemberg.

4) In order to understand the events through which Eduardo approached Cybernetics, we have to make a step backward. In 1954 Enrico Fermi, who wanted to arouse a greater interest in computers in Italy, supported the organization of a seminar in Rome on computers and on the Cybernetics of Norbert Wiener. In the conference participated also Marcello Conversi, Bruno Touschek, Franzinetti, Gamba, Caracciolo, Eduardo Caianiello, who was an assistant at that institute at the time, and Valentino Braitenberg, who was voluntary assistant at the neuropsychiatric clinic and "amateur" of neuroanatomy. At the end of the seminar Eduardo and Valentino Braitenberg took a train together, and Eduardo expressed his interest in learning something about the brain.

When he was in Copenhagen in the first half of 1955, Eduardo came across a book by Gray Walter, "The living brain" dealing with the functioning of the brain with a jargon very close to physics'. Eduardo, who very much desired to learn more about the brain, took the initiative to invite Braitenberg to spend two or three months in Naples. Moreover he organized a summer course in Varenna in the summer 1958. In the course participated also Wiener, Rosenblith, Bob Fano, Davenport, Bar-Hillel, Halle, Newman, Lee, Gabor, Rabin and Scott. Braitenberg arrived in Naples when Eduardo was still in via Tari. Although he was precariously paid for many years, he left only in 1968; the cooperation between the two of them was as fruitful as possible. Just to mention the contribution from Eduardo, he produced a model of nervous network consisting of coupled oscillators, anticipating the modern theory on neural networks, and an algebraic model of digital neuron networks (McCulloch and Pitts networks were only logical in a language fit for physics, he himself dubbed them as the "madman's equations", on the plausible tenet that only a mad brain could really work that way). These equation started a whole tradition in the field, and are known now in U.S.A. as McCulloch'equations (though McCulloch, a great estimator and friend of Eduardo, called them Caianiello' equations). Those pioneering years led to the creation of a Cybernetics Laboratory with the National Committee on Research.

5) In order to start activity in the field of nuclear physics, Eduardo contacted Edoardo Amaldi, who helped him through difficult times. The experimental activity practically started in 1958-59, when Renato Angelo Ricci first came to Naples. He started experiments in the field of nuclear spectroscopy, and immediately after him Giulio Cortini, dealing with nuclear fission through the use of nuclear plates, was called to Naples. In order to support the beginning of both these activities and of those related to experimental researches in the field of cybernetics, Eduardo obtain an unheard of thing, the creation of a mechanical workshop for his theoretical Institute aided also by a special grant of about 25 millions from the Ministry of Public Education. With that money Eduardo bought also a 200 channel device, which Ricci and his co-workers used for their significant and appreciated experimental activity. At the beginning of 1959 I.N.F.N. deliberated on creation of a subsection in Naples, which financially supported the activity on nuclear plates.

6) At the end of 1958-59, Eduardo could already count on a group of researchers, newlygraduated and last year students, who were to become the core of the scientific activity of all Neapolitan physics.

ISTITUTO DI FISICA TEORICA

 UNIVERSITÀ DI NAPOLI

VIA ANTONIO TARI, 3 - NAPOLI

ATTIVITA' DELL'ISTITUTO DI FISICA TEORICA DELL'UNIVERSITA' DI NAPOLI NELL'ANNO ACCADEMICO 1956-57

Didattica

Oltre ai corsi ufficiali di Fisica Teorica e Spettroscopia tenuti dal prof. E.R. Caianiello, sono stati tenuti i seguenti corsi:

1°) Complementi di Fisica Teorica (Dott. D. Amati)

2°) Meccanica Analitica (Dott. E. Tartaglione)

3°) Meccanica Quantica (Dr. W.I. Futtermann)

ed esercitazioni varie, destinati ad un gruppo abbastanza folto di laureati in anni precedenti e laureandi interni, ed intesi ad integrarne la preparazione. Inoltre si é organizzato un seminario interno trisettimanale sui Metodi Matematici della Fisica.

Questi corsi rappresentano un necessario primo passo verso la costituzione della Scuola di Perfezionamento in Fisica Teorica e Nucleare, il cui programma viene allegato apparte, e che si conta di iniziare regolarmente con il prossimo anno accademico.

Seminario di Fisica Teorica e Nucleare

Ha avuto luogo regolarmente durante l'anno con il duplice scopo di tenere gli studiosi locali aggiornati sui progressi piú recenti, e di porre in contatto visitatori di alto livello scientifico con gli elementi locali, per periodi di lunghezza determinata dalle particolari esigenze individuali. Ció si é rivelato fecondo di promettenti risultati, ed una attiva collaborazione ha avuto luogo, di conseguenza tra il nostro Istituto ed altri Istituti italiani e stranieri, con parecchi lavori pubblicati o in corso di preparazione d'intesa o in comune (si citano la Columbia University, Stanford, Princeton, Catania e Torino). Ogni sforzo sará in futuro rivolto a potenziare questa attivitá, che ha giá reso il nostro Istituto noto dovunque tali studi vengano compiuti, ed ha notevolissimi effetti morali, oltre che culturali, sulla massa degli studenti locali.

Per periodi varianti da qualche giorno a una o piú settimane, hanno partecipato al seminario:

Dott. T. Regge (Torino)

Prof. R. Gatto (Roma)

Dr. L. Spruch (U.C.L.A.)

Prof. G. Takeda (Tokio)

Dr. A. Garevich (General Electronic Lab.)

Dott. G. Luzzato (Genova)

Prof. D. Saxon (U.C.L.A.)

Dott. B. Vitale (Catania)

Higher Spin Symmetries and the Problem
of Unification of all Interactions

E. S. Fradkin

Lebedev Physical Institute
Moscow 117333, USSR

Preface. *It is a honour and a priviledge for me to have the opportunity of contributing to the celebration of the 70th birthday of Eduardo Caianiello, a great scientist with surprising breadth of interest and a wonderful friend.*

There are two problems of basic interest in modern theoretical physics. The first is to unify all the fundamental interactions <u>including gravity</u> at the quantum level. Nowadays it is a general point of view that the quantum gravity problem is in fact a problem of unification of all interactions.

The second is a problem of exact solvability in quantum field theory and quantum statistics. During the recent years there has been a considerable success in solving 2D conformal models, however the main problems in D>2 are still open. This is the <u>exact theory of phase transitions</u> in <u>D=3</u> and <u>a confinement problem in D=4</u>.

The question arising here is: What new methods should be developed to approach these problems? The development of theoretical physics during the whole century shows that practically all new fundamental results were obtained on the basis of discovering some novel symmetries. So it may be said: symmetry is a guide for physicists.

Then, what symmetries may be expected to appear to help us to solve the above problems? At least one thing can be said for sure: they will be infinite-dimensional symmetries.

Two types (classes) of infinite dimensional (super)symmetries were proposed by our group at Lebedev Physical Institute:

(1) Higher Spin Anti-de Sitter (*Fradkin & Vasiliev (1986), (1987c), (1988); Vasiliev (1988), (1989a,b), (1991)*) and Conformal (*Fradkin & Linetsky (1989a), (1990a); (1991a)*) supersymmetries.

(2) Virasoro-like (super)symmetries in dimension $D \geq 2$ or more generally a new class $AC(g)$ (and its central extension $\widetilde{AC}(g)$) of infinite-dimensional algebras, constructed by our method of infinite extension ("analytical continuation") of any semisimple finite Lie algebra (*Fradkin & Linetsky (1990b), (1991b,e,f)*).

From the mathematical point of view these infinite-dimensional algebras, containing the given semisimple finite-dimensional Lie algebras g as a maximal finite subalgebra, (for possibility of having a spontaneous broken phase with the lower-energy g-symmetry) are devided into two classes under the natural representation $[g, \cdot]$ of g. The first-class algebras (the higher spin or Kac-Moody algebras) are decomposed only into a direct sum of finite-dimensional irreducible g-modules. Second-class algebras (Virasoro like* or more generally $AC(g)$ and $\widetilde{AC}(g)$ algebras) involve also infinite-dimensional g-modules (irreducible or/and non-decomposable). We believe that these infinite-dim. algebras and their representation theory will play an important role in solving the problems of unification of all interactions and of exact solvability in QFT and statistics. In the present talk I shall try to give a brief account of our attempts to construct a local unified theory on the basis of the infinite-dimensional higher spin symmetries and the corresponding higher spin gauge theories proposed by us.

---

*The simplest example of second-class infinite-dimensional algebras is Virasoro-algebra $\left(\widetilde{AC}(sl_2)\right)$.

1. The Problem of Unification of all Fundamental Interactions Including Gravity (Basic Results).

During the whole twentieth century the selfconsistent unification of all the fundamental interactions including gravity has been a central problem of theoretical physics. This problem came into existence right after Einstein formulated General Relativity as a theory of gravity. At the beginning the problem consisted in unification of gravity and electrodynamics in the framework of classical field theory (it is sufficient to mention the Weyl conformally-invariant approach and the Kaluza-Klein approach). At the stage it was realized that gauge invariance principles could give keys to the unification of interactions, and all the attempts to construct a Unified Theory actually had been a search of some extension of the general coordinate transformation groups to incorporate also the gauge group of vector fields in a natural way.

The progress in physics has led to the establishment of the following main criteria that a Unified Theory must satisfy (for more details see *Fradkin (1989)*):

(1) it should be selfconsistent at the quantum level;

(2) it should give an adequate description of low-energy physics.

The anomaly cancellation conditions for all classical symmetries of the theory and the finiteness of the Unified Theory are understood as a selfconsistency. Simultaneously this criterium of anomaly cancellation imposes strong restrictions on the spectrum of elementary particles and the gauge group for vector fields. (*Fradkin E.S.* and *Tseytlin A.A. (1984)*, *Green M.B.* and *Schwarz J.H. (1984)*).

An adequate description of low-energy physics must include a solution of the cosmological constant problem, an answer to the question why the world we live in has just four observable dimensions, and it must predict the observed spectrum of elementary particles and fundamental forces, including the Clashow-Salam-Weinberg model, quantum chromodynamics and gravity, in final spontaneously broken phase.

Conventionally the evolution of theoretical physics on the way towards a Unified Theory can be divided into two stages. The first one is a lower spin stage, including only lower spin ($s \leq 2$) fields. A number of important discoveries has been obtained in this stage (the discovery of global non-Abelian symmetries among elementary particles led to their successful classification, the discovery of non-Abelian Yang-Mills gauge symmetries allowed one to unify the vector interactions and led to the Unification of electromagnetic, weak and strong interactions in the framework of Grand Unification Models; the discovery of global supersymmetry allowed one to unify Bose and Fermi particles together in unified supermultiplets and led to the Supersymmetric Grand Unification Models. At last, by gauging the global supersymmetry, supergravity was discovered).

The golden age of supergravity theories at the end of the seventies was a culmination of the lower spin stage on the way towards a Unified Theory. There had been discovered three types of supersymmetries and, correspondingly, three types of supergravity theories: Poincaré, anti-de Sitter and conformal ones. They describe systems of interacting gauge fields with spins $s = 2$ (graviton), $s = 3/2$ (gravitino), $s = 1$ (Yang-Mills vector fields), as well as matter fields with $s = 0, 1/2$ in extended models. (More details see in the review articles *Nieuwenhuizen (1981); Fradkin & Tseytlin (1985a).*)

However it did not become the end of the way. Eventually it turned out none of the known supergravity theories completely met the strong requirements for a Unified Theory. They were neither finite (except for the $N = 4$ conformal supergravity) nor contained the observable spectrum of elementary particles because there appeared a restriction $N \leq 8$ on the degree of N-extended supersymmetry* (or $N \leq 4$ in conformally invariant case).

In this way, realizing that it does not seem possible to construct a selfconsistent quantum theory including gravity and observed spectrum of elementary particles without introducing an infinite number of higher spin particles which are required both to make the theory finite and to get more extended supersymmetries, one comes to the next stage on the way towards a Unified Theory, the higher spin stage. According to the

---

*$N > 8$ supersymmetries involve higher spins $S > 2$ in the supermultiplets.

classification of supersymmetries and supergravities, there may exi. three kinds of theories involving infinite towers of higher spins along with the lower spin supergravity sector: Poincaré, anti-de Sitter and conformally-invariant theories.

The higher spin stage can be divided into two substages: a massive higher spin stage and massless gauge symmetry higher spin stage.

Superstring theory provides a first successful example which extends Einstein supergravity to the case of all massive higher spins. It describes gravity coupled with an infinite number of massive spin fields. In string theory there is a natural mass parameter inversely related with the square root of the string slope α'. In the zero-slope limit $\alpha' \to 0$ masses of all the higher spin excitations, as well as of the lower spin ones of higher levels, tend to infinity and only the massless (super)gravity and Yang-Mills sector remains as observed [in detail see: *Green, Schwarz* and *Witten*, String Theory (Cambridge (1987))].

However, as far as the higher energy domain is concerned, all the higher spin excitations of the string become equally essential and give their contributions to the interaction. This situation gives rise to several interesting higher energy phenomena. The interaction vertex in string theory contains increasingly higher derivatives (up to infinity). This in turn involves the negative power of the mass parameter to make the whole action dimensionless. As a result the interaction turns out to be non-analytical in the mass parameter* (which is proportional to $(\alpha')^{-1/2}$ in string theory) and this non-analyticity brings about the formal obstruction to the massless limit $\alpha' \to \infty$ on the flat background.

A striking example of the non-analyticity on the mass parameter $m \approx (\alpha')^{-1/2}$ of the interaction in string theory is the tree effectiv action for the electromagnetic field in string theory. As a consequence of the contribution of the interacting infinite tower of higher spin excitations in the effective action of strings (*Fradkin & Tseytlin (1985c-e)*), the effective Lagrangian of the electromagnetic field has a Born-Infeld form (see:

---

*However, it is natural to wait that a nontrivial phase transition with a drastical change in the structure of the flat vacuum in string theory makes possible the continuous massless limit for the higher spins.

Fradkin & Tseytlin (1985f))

$$L_{eff} = \frac{1}{2\pi\alpha'} \sqrt{det(\delta_{\mu\nu} + 2\pi\alpha' \, F_{\mu\nu})} \, . \tag{1.1}$$

An other important and closely related consequence of the infinite interacting tower of spins is the upper bound[*] on the stress tensor $F_{\mu\nu}^{max}$ in (1.1). Moreover, when the electric field E tends to its critical value $E^{max} = (2\pi\alpha')^{-1}$ the mass of the higher spins tends to zero. This illustrates the possibility of a phase transition from the massively higher spin stage to the "non-broken" phase – the massless (gauge) higher spin stage. (Moreover it is natural to suppose that the masses of the higher spins in the massively higher spin stage are resulted by spontaneous breakdown of the higher spin gauge symmetry.)

The massless higher spin gauge stage of Unification of all interactions was initiated and developed in our group at Lebedev Physical Institute. Two types of such gauge theories describing infinite towers of massless higher spin gauge fields coupled to supergravity: Anti-de Sitter Higher Spin and Conformal Higher Sping Gauge Theories have been obtained.

Anti-de Sitter higher spin gauge theories in four space-time dimensions have been developed by *Fradkin* and *Vasiliev (1986)*, *(1987a-c)*, *(1988)* and *Vasiliev (1988)*, *(1989a,b)*, *(1990)*, *(1991)*. They are based on a new infinite-dimensional higher spin global supersymmetry generalizing the ordinary anti-de Sitter supersymmetry to all higher spins which was discovered by *Fradkin* and *Vasiliev (1986)*, *(1987c)* and *Vasiliev (1988)*. From the mathematical point of view this is an infinite-dimensional Lie superalgebra containing $osp(N \mid 4)$ as its maximal finite subalgebra. Its representation theory was studied by *Vasiliev (1987)* and *Konstein* and *Vasiliev (1989)*, *(1990)*. These global higher spin symmetries may be gauged similarly to the AdS supergravity. The corresponding gauge theory was constructed by *Fradkin* and *Vasiliev (1987a,b)* in the cubic approximation in the framework of Lagrangian field theory, and then extended to

---

[*]The critical bound of the physical quantities (electromagnetic stress tensor, gravity curvature, temperature, etc.), and the essential non-linearity of the tree effective actions for the fundamental fields is typical for a finite quantum theory and this makes finite also the classical limit of such theories.

all orders in the interaction by *Vasiliev (1990), (1991)* in the framework of the invariant equations of motion. Two most interesting qualitative physical properties of the theory are the presence of higher derivatives in the interaction vertices, as in string theory, and a peculiar non-analyticity of the interaction in the anti-de Sitter cosmological constant Λ. This non-analyticity does not allow one to pass to the naive flat limit $\Lambda \to 0$. The situation here is quite similar to one in the closed string theory where one is not allowed to pass to the massless limit $\alpha' \to \infty$ ($T \to 0$) due to the non-analyticity of string vertices in the string tension.

It should be mentioned also that our results disprove the common belief that consistent gauge-invariant interaction among massless higher spins and gravity does not exist (the so-called "no-go" theorem*). A key point here is just the above mentioned non-analyticity in the cosmological constant and the presence of higher derivatives interaction vertices in our higher spin gauge theories.

The conformal higher spin gauge theory has been developed by *Fradkin* and *Linetsky (1989a), (1990a), (1991a)*. It generalizes the conformal supergravity (supersymmetric extension of the C^2 conformal gavity discovered by Hermann Weyl) and describes a conformally invariant interaction among higher spin fields in four space-time dimensions. This gauge theory is based on the global infinite-dimensional higher-spin conformal supersymmetry discovered by *Fradkin* and *Linetsky (1990a)*. From the mathematical point of view it is an infinite-dimensional Lie superalgebra containing $SU(2, 2 \mid N)$ as its miaximal finite subalgebra. The invariant Lagrangian generalizing the Weyl Lagrangian $C^2_{\mu\nu\rho\sigma}$ (C-Weyl tensor) to the interaction of all higher spins was constructed by *Fradkin*

---

*It was demonstrated that the <u>minimal</u> gravitational interaction introduced by means of covariantizing the derivatives violated the higher spin gauge symmetry under which the free higher spin massless action was invariant. Precisely, in the higher spin gauge variation of the higher spin action there appear some terms including gravitational Weyl tensor (already in the cubic order). Such terms cannot be compensated not to violate the gauge symmetry of the higher spin action via any modifications of the transformational law of the gravitational field (metric) because the variation of the Einstein action is proportional only to the Ricci tensor, no longer the whole Riemann one. (More details see: *C. Aragone* and *S. Deser*, Phys. Lett. <u>B86</u> (1979) 161; *S. Christensen* and *M. Dutt*, Nucl. Phys. <u>B154</u> (1979) 301; *F.A. Berends et al.*, J. Phys. <u>A13</u> (1980) 1643; *B. de Wit* and *D.Z. Freedman*, Phys. Rev. <u>D21</u> (1980).) This no-go theorem had served for a long time as an obstruction to all the attempts to build up a massless higher spin gauge theory coupled with gravity.

and *Linetsky (1989a)*, *(1991a)* in the cubic order approximation. Most important qualitative feature of this theory is the absence of any scale parameters. The only arbitrary parameter in the theory is a dimensionless Weyl gravity coupling constant. Hence such a theory might be regarded as an asymptotic theory of gravity and higher spins in the ultra-high energy domain where any mass parameters become insufficient.

In particular, all the three theories, string theory, anti-de Sitter higher spin gauge theory and conformal higher spin theory, though they are essentially different, might turn out eventually to be actually three different phases of one and the same Unified field theory (see *Fradkin* and *Linetsky (1991g)*) with new forces conditioned by higher spin gauge fields.

2. Anti-de Sitter Higher Spin Gauge Theory

Below I will draw more details upon the ideas, methods and the structure of the Higher Spin Gauge Theories constructed by us. The cornerstone of our approach is a new class of symmetries, so-called higher spin symmetries and their localization.

The mathematical basis to describe AdS higher spin symmetries, generalizing the usual finite space-time (super)symmetries is an infinite-dimensional Lie superalgebra which contains $osp(N \mid 4)$ as its maximal finite-dimensional subalgebra.

2.1. The superalgebras for higher spins in AdS$_4$.

There is more than one way to construct the infinite-dimensional global superalgebras. One way is to postulate the superalgebra spectrum with respect to a finite-dimensional subalgebra g and, by imposing some additional restrictions, solve the Jacobi identities for the structure constants of the infinite-dimensional superalgebra. In this way we (*Fradkin & Vasiliev (1986)*, *(1987c)*) first obtained the simplest superalgebra of higher spins in AdS$_4$, denoted by $shs(1 \mid 4)$ (*shs* means super higher spin). The anti-de Sitter superalgebra $osp(1 \mid 4)$ is its maximum finite-dimensional subalgebra. The gauge fields of this superalgebra have the labelling $\omega_{\mu,\alpha(n),\dot\beta(m)}$ with arbitrary $n, m \geq 0$. The fields ω with an even (odd) number of spinor indices are commuting (anticommuting). They also satisfy the hermiticity relation[*]

$$\omega^{+}{}_{\mu,\alpha(n),\dot\beta(m)} = \omega_{\mu,\beta(m),\dot\alpha(n)} \tag{2.1}$$

The curvatures of $shs(1\,|\,4)$ are of the form (*Fradkin & Vasiliev (1987c)*)

$$
\begin{aligned}
R_{\mu\nu,\alpha(n),\dot\beta(m)} &= \partial_{[\mu}\omega_{\nu],\alpha(n),\dot\beta(m)} + \\[2mm]
&+ \sum_{p,q,s,k,l,t=0}^{\infty} i^{s+t-1}\,\frac{n!m!}{p!q!s!k!l!t!}\,\times \\[2mm]
&\times\, \lambda^{1+\frac{1}{2}\left(|n-m|-|p+s-k-t|-|q+s-l-t|\right)}\,\times \\[2mm]
&\times\, \delta\left(|\,(p+k)(q+l)+(p+k)(s+t)+(q+l)(s+t)+1\,|_2\right)\,\times \\[2mm]
&\times\, \delta(n-p-q)\,\delta(m-k-l)\,\omega_{\mu,\alpha(p)\gamma(s),\dot\beta(k)\dot\delta(t)}\,\times \\[2mm]
&\times\, \omega_{\nu,\alpha(q)}{}^{\gamma(s)}{}_{,\dot\beta(l)}{}^{\dot\delta(t)}\,.
\end{aligned}
\tag{2.2}
$$

Here $\delta(n)$ equals 1 or 0 for $n=0$ and $n\neq 0$ respectively, and $\delta(|\,n\,|_2)$ equals 1 or 0 for n even and n odd respectively. The real parameter λ coincides with the inverse radius of the background anti-de Sitter space. The cosmological constant Λ is proportional to $-\lambda^2$.

The curvatures (2.2) contain both positive and negative powers of λ. This is the reason for the non-analyticity of the higher-spin interaction in the cosmological constant. Note that the linearized curvatures, unlike the full ones, are analytic in λ and admit the flat limit $\lambda \to 0$.

---

[*]Spins s of the field $\omega_{\nu,\alpha(n),\dot\beta(m)}$ are connected with n and m by the relation $n+m = 2(s-1)$. Each set of spinor indices $\alpha(n)$ and $\beta(m)$ takes the values 1,2 and is symmetrized separately. The indices are raised and lowered by the respective antisymmetric symbols $\varepsilon_{\alpha\beta}$; $\varepsilon_{\dot\alpha,\dot\beta}$. The dynamical fields are components of $\omega_{\nu,\alpha(n),\dot\beta(m)}$ with $|\,n-m\,|\leq 1$. The components with $n-m=2$ are auxiliary fields. The components with $|\,n-m\,|>2$ are called extra fields. The extra and auxiliary fields can be expressed through the dynamical fields and are needed for the construction of the curvatures. For example, in the case of gravitational fields, $\omega_{\nu,\alpha\dot\beta}$ is the vierbein, and the auxiliary fields $\omega_{\nu\alpha(2)}$, $\omega_{\nu\dot\beta(2)}$ are components of the Lorentz connection. For more details of other notations see appendix.

This method, connected with the explicit solving of the Jacobi identities, is, however, very cumbersome and, therefore, practically inapplicable to more complicated cases such as the conformal theory. Another method, based on the operator realization of the algebra, is much simpler and more illuminating. This method for the construction of the infinite-dimensional superalgebra G containing a given finite-dimensional subalgebra g consists of the following (*Vasiliev (1988)*). First, one chooses an appropriate operator realization of the subalgebra g by quadratic and linear combinations (generally, with additional constraints) of the generators $\hat{Z}_A$ of the Heisenberg-Clifford superalgebra. (The supercommutator of $\hat{Z}_A$ is $[\hat{Z}_A, \hat{Z}_B] = \mathbb{C}_{AB}$ where $\mathbb{C}_{AB}$ is a constant orthosymplectic matrix.) Next, one considers an associative algebra of polinoms of all powers in the operators $\hat{Z}_A$. In this algebra one introduces the Grassman parity $\varepsilon(A)$ and the supercommutator

$$[\hat{A}, \hat{B}] = \hat{A}\hat{B} - (-1)^{\varepsilon_A \varepsilon_B} \hat{B}\hat{A} . \tag{2.3}$$

Technically, the problem is to find the structure constants of this Lie superalgebra (G). Its generators T_A are expressed through the powers of the operators $\hat{Z}_A$ in the Weyl (symmetric) normal form, and one must compute $[T_A, T_B] = F^C_{AB} T_C$. Such problems are most conveniently solved, by the method of symbols (*Beresin (1966)*). Let the Weyl symbol of the gauge field $\omega_\mu(\hat{Z})$ corresponding to the superalgebra G be $\omega_\mu(Z)$. Then for the operator curvature one has

$$R_{\mu\nu}(\hat{Z}) = \partial_{[\mu}\omega_{\nu]}(\hat{Z}) + [\omega_\mu(\hat{Z}), \omega_\nu(\hat{Z})] \tag{2.4}$$

and the corresponding symbol is of the form

$$R_{\mu\nu}(Z) = \partial_{[\mu}\omega_{\nu]}(Z) + [\omega_\mu(Z), \omega_\nu(Z)\}^*, \tag{2.5}$$

$$[A, B\}^* = A * B - (-1)^{\varepsilon(A)\varepsilon(B)} B * A \tag{2.6}$$

where $*$ denotes the associative multiplication of Weyl symbols:

$$A * B = A(Z) \exp(\overset{\leftrightarrow}{\triangle}) B(Z), \tag{2.7}$$

$$\overleftrightarrow{\triangle} \;=\; \frac{1}{2}\,\frac{\overleftarrow{\partial}}{\partial Z_A}\,C_{A,B}\,\frac{\overrightarrow{\partial}}{\partial Z_B}\,, \quad [Z_A, Z_B\}^* \;=\; C_{A,B}\,. \tag{2.8}$$

Now let us consider the higher spin extension of AdS supersymmetry (where $g = osp(N \mid 4)$). First we choose an appropriate operator realization of $osp(N \mid 4)$.

Let $q_\alpha, r_{\dot\beta}$ and ψ_i be Heisenberg-Clifford generating elements with non-zero supercommutators

$$[q_\alpha, q_\beta] \;=\; 2i\varepsilon_{\alpha\beta}, \quad [r_{\dot\alpha}, r_{\dot\beta}] \;=\; 2i\varepsilon_{\dot\alpha\dot\beta}, \tag{2.9a}$$

$$\{\psi_i, \psi_j\} \;=\; 2\delta_{ij}, \tag{2.9b}$$

where $\dot\alpha, \dot\beta = 1, 2$ and $\alpha, \beta = 1, 2$ are dotted and undotted two-component $sl(2; C)$ multispinorial indices, $\varepsilon_{\alpha\beta} = -\varepsilon_{\beta\alpha}, \varepsilon_{\dot\alpha\dot\beta} = -\varepsilon_{\dot\beta\dot\alpha}$ are the symplectic metrics; the Hermitian conjugation is defined as follows

$$(q_\alpha)^+ \;=\; r_{\dot\alpha}, \quad (r_{\dot\alpha})^+ \;=\; q_\alpha, \quad (\psi_i)^+ \;=\; \psi_i, \tag{2.9c}$$

and $i = 1, \cdots, N$ are SO(N) internal indices. Then the second-order polynomials on the generating elements q, r and ψ with the basis

$$M_{\alpha\beta} \;=\; \frac{1}{2}\,q_{(\alpha}q_{\beta)}, \quad \bar{M}_{\dot\alpha\dot\beta} \;=\; \frac{1}{2}\,r_{(\dot\alpha}r_{\dot\beta)}, \quad P_{\alpha\dot\beta} \;=\; q_\alpha r_{\dot\beta}$$

$$T_{ij} \;=\; \frac{1}{2}\,\psi_{[i}\psi_{j]}, \quad Q_{i\alpha} \;=\; q_\alpha\psi_i, \quad \bar{Q}_{i\dot\alpha} \;=\; r_{\dot\alpha}\psi_i \tag{2.10a}$$

form a Lie superalgebra isomorphic to $osp(N \mid 4)$ ($M_{\alpha\beta} = M_{\beta\alpha}$, $\bar{M}_{\dot\alpha\dot\beta} = \bar{M}_{\dot\beta\dot\alpha}$ are the $sl(2; C)$ Lorentz generators, $P_{\alpha\dot\beta}$ - AdS$_4$ translations, $T_{ij} = -T_{ji} - so(N)$-generators, and $Q, \bar{Q}$-supersymmetry generators).

An idea to generalize this construction to the case of all higher spins consists in introducing into the game polynomials of all even orders on the generating elements q, r and ψ (the restriction to only even polynomials here is for convenience; it does not lead to any loss of the information, see *(Vasiliev (1988))*.) According to this idea, the gauge fields of all higher spins are generated from the single object $\omega_\mu(q, r, \psi \mid x)$ as follows:

$$\omega_\mu(q, r, \psi \mid x) = \sum_{n,m=0}^{\infty} \sum_{k=0}^{N} \frac{1}{n!m!k!}\, \omega_{\mu,\alpha_1\cdots\alpha_n,\dot\beta_1\cdots\dot\beta_m,i_1\cdots i_k}$$

$$\times\ q^{(\alpha_1}\ldots q^{\alpha_n)}\, r^{(\beta_1}\ldots r^{\beta_m)}\, \psi^{[i_1}\ldots\psi^{i_k]},\quad n+m+k \text{ is even}$$

$$(2.10b)$$

(The round (square) brackets mean complete symmetrization (anti-symmetrization), that corresponds to the Weyl ordering of the bosonic $q_\alpha, r_{\dot\beta}$ and fermionic ψ_i operators.) The terms with $n + m + k = 2$ in the above expansion corresponds to the N-extended AdS$_4$ supergravity gauge fields, while the collection of fields with $n+m+k=2(s_m\text{-}1)$, form $osp(N \mid 4)$-irreducible higher spin supermultiplets with the maximal spin s_m (take into account additional unity due to the vector index μ of ω_μ). Such an infinite-dimensional Lie superalgebra formed by the generators

$$T_{\alpha(n),\dot\beta(m),i(k)} = \frac{1}{n!m!k!}\, q_{(\alpha_1}\ldots q_{\alpha_n)}\, r_{(\dot\beta_1}\ldots r_{\dot\beta_m)}\, \psi_{[i_1}\ldots\psi_{i_k]} \qquad (2.11a)$$

was denoted *(Vasiliev (1988))* as $shs^E(N \mid 4)$ (*shs* means super higher spin). The curvatures of the AdS$_4$ higher spin gauge theory based on *shs*-algebras are defined as usual

$$R_{\mu\nu} = \partial_\mu\omega_\nu - \partial_\nu\omega_\mu + [\omega_\mu, \omega_\nu\} . \qquad (2.11b)$$

To calculate the supercommutator of two gauge fields $\omega_\mu(q, r, \psi)$ and $\omega_\nu(q, r, \psi)$ it is helpful to employ the very convenient formalism of the Weyl symbols corresponding to quantum operators. As a result one arrives at the final expression for the curvature

(Vasiliev (1988)).

$$
\begin{aligned}
R_{\mu\nu,i(f),\alpha(n),\dot\beta(m)} =\ & \partial_\nu \omega_{\mu,i(f),\alpha(n),\dot\beta(m)} - \partial_\mu \omega_{\nu,i(f),\alpha(n),\dot\beta(m)} \\
& + \sum_{p,q,s,k,l,t,u,v,r=0}^{\infty} i^{s+t-1}(-1)^{r(r-1)/2}\,\frac{f!\,n!\,m!}{p!\,q!\,s!\,k!\,l!\,t!\,u!\,v!\,r!} \\
& \times (\hbar)^{r+s+t-1}\,\lambda^{1+(|n-m|-|p+s-k-t|-|q+s-l-t|)/2} \\
& \times \delta(f-u-v)\,\delta(n-p-q)\,\delta(m-k-l) \\
& \times \delta\big(|\,(p+k)(q+l) + (p+k)(s+t) \\
& \quad + (q+l)(s+t) + u\cdot v + u\cdot r + v\cdot r + 1\,|_2\big) \\
& \times \omega_{\nu,i(u)j(r),\alpha(p)\gamma(s),\dot\beta(k)\dot\delta(t)}\ \omega_{\mu,}{}^{j(r)}{}_{i(v),\alpha(q)}{}^{\gamma(s)}{}_{,\dot\beta(l)}{}^{\dot\delta(t)}
\end{aligned}
\tag{2.12}
$$

In the case when $N = 1$, (2.12) coincides with (2.2).

In the structure constants of the AdS_4 higher spin superalgebra and, consequently, in the expressions for the curvatures there is a dimensional parameter λ. All the dependence on λ is uniquely fixed by the AdS_4 commutation relations $[P_\mu, P_\nu] = -\lambda^2 M_{\mu\nu}$ (that is $-\lambda^2 \sim \Lambda$ is the AdS cosmological constant) and the requirement on the gauge higher spin fields to have correct physical dimensions to describe free dynamics. The crucial point here is a non-analyticity of the structure constants and curvatures in the AdS cosmological constant $\Lambda \sim -\lambda^2$. Indeed it is easy to see that in the above expression for the curvatures both positive and negative powers of λ are involved. This non-analyticity results in the non-analyticity of the higher-spin-gravitational interaction and does not permit one to pass to the naive flat background $\lambda \to 0$.

It should be mentioned that we have presented here only the simplest versions of the AdS_4 higher-spin superalgebras to illustrate the very idea of construction. In fact, in practice some more complications are needed to obtain also the necessary sets of auxiliary fields and to meet the requirement of unitarity.

It is known that in supergravity, without auxiliary fields, the numbers of boson and fermion off-shell degrees of freedom are not equal to each other, and the local gauge algebra is open. The introduction of auxiliary fields closes the algebra off shell. This essentially facilitates both the construction of the full Lagrangian and the quantization procedure. Such a superalgebra is obtained *(Fradkin & Vasiliev (1988))* which, in addition to the higher-spin gauge fields, gives rise to auxiliary fields generalizing the "new minimal" formulation of $N = 1$ supergravity *(Sohnius & West (1981))*. This superalgebra $shs\alpha(1)$ ("α" means auxiliary) admits a natural operator realization generalizing the previous construction. Namely, in addition to the spinor operators (2.11a) one introduces the operators $\hat{Q}$ and $\hat{R}$ such that

$$\hat{Q}^2 = 1, \quad \hat{R}^2 = 1, \quad [\hat{Q}, \hat{R}] = 0, \quad \{\hat{Q}, \hat{q}_\alpha\} = 0,$$

$$[\hat{Q}, \hat{\tau}_{\dot{\beta}}] = 0, \quad \{\hat{R}, \hat{\tau}_{\dot{\beta}}\} = 0, \quad [\hat{R}, \hat{q}_\alpha] = 0,$$

$$\hat{R}^+ = \hat{Q}, \quad \hat{Q}^+ = \hat{R}. \tag{2.13}$$

The gauge fields of $shs\alpha(1)$ are of the form

$$\omega_\mu = \sum_{A,B=0,1} (\hat{Q})^A (\hat{R})^B \, \omega_\mu^{AB}(q,\tau), \tag{2.14}$$

and for the curvatures one finds *(Fradkin & Vasiliev (1988))*

$$
\begin{aligned}
R^{FG}_{\nu\mu,\alpha(n),\dot{\beta}(m)} =\ & \partial_\nu \omega^{FG}_{\mu,\alpha(n),\dot{\beta}(m)} - \partial_\mu \omega^{FG}_{\nu,\alpha(n),\dot{\beta}(m)} \\
& + \sum_{p,q,s,k,l,t;A,B,C,D} \delta(|\,F + A + C\,|_2)\,\delta(|\,G + B + D\,|_2)\,\delta(n - p - q) \\
& \times \delta(m - k - l)\,i^{s+t-1}\,(-1)^{C(p+s)+D(t+k)} \times \frac{n!\,m!}{p!\,q!\,s!\,k!\,l!\,t!} \\
& \times \delta(|\,C(p + s) + D(k + t) + A(q + s) + B(l + t) \\
& \quad + s + t + (p + k + s + t)(q + l + s + t) + 1\,|_2) \\
& \times \omega^{AB}_{\nu,\alpha(p)\gamma(s),\dot{\beta}(k)\dot{\delta}(t)}\,\omega^{CD}_{\mu,\alpha(q)}{}^{\gamma(s)}{}_{,\dot{\beta}(l)}{}^{\dot{\delta}(t)}.
\end{aligned} \tag{2.15}
$$

The fields ω^{AB} with $A = B = 0, 1$ describe higher spins and ω^{AB} with $A = 0$, $B = 1$ or $A = 1$, $B = 0$ are auxiliary fields as one can see after the localization of this algebra.

In the superalgebra $shs\alpha(1)$ all spins are contained twice. That such a doubling should indeed take place is seen from the following considerations. A necessary condition for the existence of a <u>full</u> theory of massless higher-spin fields is the existence of massless unitary particle-like representations of the global superalgebra describing the higher-spin symmetry, which have the same spectrum as gauge fields have. The existence of such representations requires the doubling of the fields of higher spins *(Konstein & Vasiliev (1989), (1990))*. This fact can be illustrated as follows. The anti-de Sitter group $0(3,2)$ possesses two elementary representations (singletons) discovered by Dirac*. They are denoted by Di and Rac.

The supersingleton $S = Di \oplus Rac$ forms a representation of $osp(1 \mid 4)$ which can be continued to a representation of the higher-spin superalgebra. The square of the supersingleton equals

$$S \otimes S = (Di \otimes Di) \oplus (Rac \otimes Rac) \oplus$$
$$\oplus (Di \otimes Rac) \oplus (Rac \otimes Di) \tag{2.16}$$

and the spin content of these representations is as follows

$$(Rac \otimes Rac) = (0^1, 1, 2, \cdots),$$
$$(Di \otimes Di) = (0^2, 1, 2, \cdots),$$
$$(Di \otimes Rac) = \left(\frac{1}{2}, \frac{3}{2}, \frac{5}{2}, \cdots\right) \tag{2.17}$$

where 0^1 and 0^2 are masless representations with zero spin: $(1,0)$ and $(2,0)$ respectively. (It is important that the unitary representations contain all lower-order spins including spin 0.) The square $S \otimes S$ contains all massless unitary representations of $0(3,2)$ twice and can be continued to a unitary representation of the full superalgebra $shs^f(1)$ which contains all higher spins twice and is a subalgebra of $shs\alpha(1)$. On the other hand, the superalgebra $shs(1 \mid 4)$ which contains all spins only once has no unitary representations with the required properties.

The fact of the doubling of all spins raises the question of their interpretation, particularly of spin 2. One possibility is that only some specific combinations of the spin-2

---

*See also: *Flato M.*, and *Fronsdal C.*, Phys. Lett. **97B** (1980) 236; *Fronsdal C.*, Phys. Rev. **D26** (1982) 1988.

fields will remain massless in a physically relevant phase with spontaneously broken higher spin symmetries, and this combination will be the only candidate for the role of the gravitational field.

To summarize, higher spin symmetries constitute a novel class of fundamental symmetries generalizing the space-time symmetries, somewhat like supersymmetries. The discovery of supersymmetry introduced in physics spin $1/2$ supersymmetry generators and spin $3/2$ gauge fields. Similar to this, higher spin symmetries introduce into the game an infinite tower of higher spin generators and gauge fields, which are also obligatory for finiteness of the local theory.

2.2. <u>AdS$_4$ Higher Spin Gauge Theory.</u>

Having the infinite-dimensional higher-spin superalgebra, corresponding gauge fields and curvatures at one's disposal and basing on the supergravity experience, one can try to construct an interacting gauge theory for higher spins. One of the elegant approaches to build up an action starting from the known curvatures is the higher spin extension of the $R_A \wedge R_B Q^{AB}$ approach*. This $R \wedge R$ approach was applied first to the AdS$_4$ higher-spin gauge theory based on the higher-spin superalgebras $shs(N \mid 4)$ *(Fradkin & Vasiliev (1987a))*. It turned out that there exists a unique real, P-invariant action which, when supplemented with some suitable curvature constraints, currently describes the free fields of higher spins in AdS$_4$ as well as their cubic interaction. This action is of the form

$$
S = \frac{\beta}{2} \sum_{n,m=0}^{\infty} \frac{i^{n+m+1}}{n!m!} \lambda^{-|n-m|} \varepsilon(n-m) \times
$$
$$
\times \int d^4x\, \varepsilon^{\mu\nu\rho\sigma}\, R_{\mu\nu,\alpha(n),\dot\beta(m)} \, R_{\rho\sigma,}{}^{\alpha(n),\,\dot\beta(m)}
$$

$$(2.18)$$

and the additional constraints read

$$
\varepsilon^{\nu\mu\rho\sigma}\, R_{\mu\nu,\alpha(n),\dot\beta(m-1)\dot\delta}\, h_{\rho,\alpha}{}^{\dot\delta} = 0 \text{ for } n \geq m > 0,
$$
$$(2.19)$$

---

*This approach was introduced in gravity by *MacDowell S.* and *Mansouri F.*, (1977).

$$\varepsilon^{\nu\mu\rho\sigma}\, R_{\nu\mu,\alpha(n-1)\gamma,\dot\beta(m)}\, h^{\gamma}_{\rho,\dot\beta} = 0 \text{ for } m \geq n > 0. \tag{2.20}$$

Here $\varepsilon(n-m)$ equals 1 for $n > 0$, 0 for $n = m$ and -1 for $n < m$. The action S contains two independent parameters – the cosmological constant λ and the overall factor β. To arrive at the correctly normalized Einstein (spin-2) action – $(1/4\kappa^2)\int d^4x\sqrt{-g}\, R$ the constant β should be $\beta = -1/16\kappa^2$.

To analyze the action S one may expand it to the point where all fields, except the gravitational field, vanish and

$$h_{\mu\alpha\dot\beta} = \overset{\circ}{\omega}_{\mu,\alpha,\dot\beta}, \quad W_{\mu,\alpha(2)} = \overset{\circ}{\omega}_{\mu,\alpha(2)}, \quad \bar W_{\mu,\dot\beta(2)} = \overset{\circ}{\omega}_{\mu,\dot\beta(2)} \tag{2.21}$$

where h and $W, \bar W$ are respectively the vierbein and Lorentz connection of the background AdS$_4$ space. It can be shown that the fields $\omega(n,m)$ with $\mid n - m \mid > 2$ (the so-called "extra" fields) fall out of the action (more precisely, they enter the total-derivative terms). However, these fields enter the full curvatures of $shs(1 \mid 4)$. This is the reason why the action should be supplemented with the constraints (2.19), (2.20). The constraints and equations of motion make it possible to express the curvature as follows *(Vasiliev (1987))*.

$$R^l_{\mu\nu,\alpha(n),\dot\beta(m)} = \delta(n)\, h_{\mu,\gamma}\,{}^{\dot\delta} h_{\nu,}{}^{\gamma\dot\delta}\, \bar C_{\dot\beta(m)\dot\delta(2)} +$$
$$+ \delta(m)\, h^{\gamma}_{\mu\dot\delta}\, h_{\nu}^{\gamma\dot\delta}\, C_{\alpha(n)\gamma(2)} \tag{2.22}$$

where C and $\bar C$ generalize the gravitational Weyl tensor ($C_{\alpha(4)}$ and $\bar C_{\dot\beta(4)}$) to the case of higher spins. (2.22) generalizes the linearized Einstein equations and expresses all curvatures through the higher-spin Weyl tensors. The Weyl tensors of the AdS$_4$-space satisfy the equations

$$h^{\rho\gamma}_{\dot\beta}\, \mathcal{D}^L_\rho\, C_{\alpha(n-1)\gamma} = 0, \tag{2.23}$$

$$h^{\rho\dot\delta}_\alpha\, \mathcal{D}^L_\rho\, \bar C_{\dot\beta(m-1)\dot\delta} = 0 \tag{2.24}$$

which follow from (2.22) and the linarized Bianchi identities for the curvatures of $shs(1 \mid 4)$. In (2.23) and (2.24), $\mathcal{D}^L$ is the Lorentz-covariant derivative with the connection $W, \bar W$.

The gauge invariance of the action in the trilinear approximation can be verified as follows. The variation of the action in this approximation has the structure $\int R^l R^l \varepsilon$. By virtue of the linearized equation (2.22), the expression contains only terms of the type CC and $\bar{C}\bar{C}$ (the terms $C\bar{C}$ do not enter δS owing to the identity $\varepsilon^{\mu\nu\rho\sigma} h_{\mu\alpha\dot{\delta}} h_{\nu\alpha}{}^{\dot{\delta}} h_{\rho\gamma\dot{\beta}} h^{\gamma}{}_{\sigma\dot{\beta}} \equiv 0$). The condition for the cancellation of these terms leads to an equation for the bilinear form Q^{AB} in the action $\int R_A \wedge R_B Q^{AB}$, which has the unique solution consistent with the free theory (see *Vasiliev (1987)*) of higher spins in AdS$_4$. This unique solution is given by (2.18). So the above action and constraints provide a consistent description of the interacting collection of massless spins $3/2 \leq s \leq \infty$, in the cubic approximation. Taking into account that the no-go theorem also dealt with just the cubic approximation, one comes to the conclusion that the no-go theorem does not hold well on the non-trivial AdS background. The key point here is the celebrated non-analyticity in Λ, and the existence of non-minimal higher-derivative interaction vertices.

Actually this is a quite non-perturbative result with respect to the flat background, because in this case cosmological terms cannot be treated as perturbations over the trivial flat vacuum. In fact, the anti-de Sitter vacuum and the flat vacuum may be related only by some non-trivial phase transition, rather than a naive flat limit $\Lambda \to 0$, when massless higher spins are presented in the AdS phase.

It is worthwhile to emphasize now that in the AdS higher spin gauge theory there are only two independent coupling constants. These are gravitational κ and cosmological $\Lambda \sim -\lambda^2$ constants. In the N-extended theories in the action (or cubic order) there is also an additional Yang-Mills term integration with a constant e but it is in fact related to the first two $e^2 = 4\lambda^2\kappa^2$ *(Fradkin & Vasiliev (1987b))*.

The same situation with coupling constants takes place also in the AdS$_4$ extended supergravity and, therefore, the introducing of an infinite tower of higher spin fields does not bring any new essential independent physical parameters in the theory. All higher spin interaction coupling constants turn out to be related with the only fundamental ones Λ, κ and e by the very powerful infinite-dimensional higher spin symmetry.

This situation is quite similar to string theory with the string tension $T = (2\pi\alpha')^{-1}$ on the place of the cosmological constant Λ. Moreover, the connection of e^2 and κ^2 is quite similar to the heterotic superstring case with the accuracy of the interchange $\Lambda \leftrightarrow T$. So to say, we come back again to our parallel between the AdS radius and the string "length".

Above we have considered the higher spin dynamics in the cubic approximation. The way beyond this approximation requires knowledge of the complete constraints up to all orders, along with the very Lagrangian. It produces some serious difficulties which have not been overcome in this approach up to now.

However, a very important progress has been recently achieved in quite another approach *(Vasiliev (1990), (1991))* where on the basis of superalgebra $shs\alpha(1)$ (2.13) - (2.15) and its conformal extensions consistent equations of motion desribing interacting massless fields of all spins, from zero to infinity, have been constructed in the framework of the so-called free differential algebra.

This system of equations describes higher spin interactions up to all orders and it is gauge invariant under an infinite-dimensional algebra. They have a rather specific form, like some zero-curvature conditions for an enlarged infinite algebra, supplemented with some additional constraints. The problem of corresponding complete Lagrangian is now under investigation.

3. Conformal Higher Spin Theory

Let us turn to the second possibility for a higher spin gauge theory, which is a higher spin $(s > 2)$ generalization of usual conformal supergravity.

In a series of papers *(Fradkin & Linetsky (1989a,b) (1990a) (1991a))* a superconformal higher spin gauge theory has been proposed in dimensions $D \leq 4$ (but only for lower dimensions $(D < 4)$ the action is constructed exactly).

Below I adduce more details of the main results of our attempt to construct Super-conformal Higher Spin Theory in three and four dimensions.

3.1. Conformal Higher Spin Gauge Theory in $D = 2 + 1$.

For the construction of a conformal higher spin theory in $D = 2 + 1$ first of all we choose a suitable operatorial realization of the conformal superalgebra in $D = 2 + 1$, $osp(N \mid 4)$. Let $\hat{a}_\alpha, \hat{b}_\alpha$ and $\hat{\psi}_i$ be the generating elements with non-zero commutators

$$[\hat{a}_\alpha, \hat{b}_\beta] = 2i\varepsilon_{\alpha\beta} \quad , \quad \hat{a}_\alpha^+ = \hat{a}_\alpha \quad , \quad \hat{b}_\alpha^+ = \hat{b}_\alpha , \tag{3.1a}$$

$$\{\hat{\psi}_i, \hat{\psi}_j\} = 2\delta_{ij} \quad , \quad \hat{\psi}_i^+ = \hat{\psi}_i \tag{3.1b}$$

where the Greek letters $\alpha, \beta, \cdots = 1, 2$ are the $so(2, 1)$-spinorial indices and $i, j = 1, \cdots, N$ are the internal indices. Then the $osp(N \mid 4)$ generators (translation, $SO(2, 1)$, conformal boost, dilatation, $SO(N)$, supersymmetry and special conformal supersymmetry generators) are $P_{\alpha(2)} = \frac{1}{4i}\hat{a}_\alpha \hat{a}_\alpha$, $M_{\alpha(2)} = \frac{1}{4i}(\hat{a}_\alpha \hat{b}_\alpha + \hat{b}_\alpha \hat{a}_\alpha)$, $K_{\alpha(2)} = \frac{1}{4i}\hat{b}_\alpha \hat{b}_\alpha$, $D = \frac{1}{8i}(\hat{a}_\alpha \hat{b}^\alpha + \hat{b}^\alpha \hat{a}_\alpha)$, $T_{ij} = \frac{1}{4}\hat{\psi}_i \hat{\psi}_j$, $Q_{i\alpha} = \frac{1}{2}\hat{a}_\alpha \hat{\psi}_i$ and $S_{i\alpha} = \frac{1}{2}\hat{b}_\alpha \hat{\psi}_i$, respectively. Generators of the infinite-dimensional generalization of the conformal superalgebra $osp(N \mid 4)$ can be chosen in the form of all order polynomials that commute with "particle number" operator $N = a^\alpha b_\alpha$

$$T^{(s,c)}_{i(k),\alpha(2l)} = \left[\frac{(2l+1)!}{(s-l)!(l+c)!(l-c)!(s+l+1)!}\right]^{1/2}$$
$$\times \psi_{i_1} \cdots \psi_{i_k} \underbrace{a_\alpha \cdots a_\alpha}_{l-c} \underbrace{b_\alpha \cdots b_\alpha}_{l+c} (a^\alpha b_\alpha)^{s-l} , \tag{3.2}$$

where (and below) we have passed to the Weyl symbols of the operators and a, b, ψ are the symbols of the generating elements in (3.1). In (3.2) the indices have the following meaning: s determines the $SO(3, 2)$-irreducible representation space with the dimension $d(s) = (2s+3)(2s+1)(s+1)/3$; c is a conformal weight of the generators ($[D, T^c] = cT^c$); l is the $SO(2, 1)$-signature; k is a number of the internal indices and

$$s = 0, \frac{1}{2}, 1, \cdots ; \quad c = -s, \cdots, s; \quad l = \mid c \mid, \cdots, s; \tag{3.3}$$

$k = 0, 1, \cdots N$ and $k + 2s$ is even.

The last restriction has been introduced for convenience. The generators (3.2) form a conformal basis (D is diagonal) in a conformal higher spin superalgebra which we call $shsc(N \mid 3)$. The structure constants of $shsc(N \mid 3)$ in the conformal basis associated with the reduction to a subalgebra $osp(N \mid 4) \rightarrow SO(3,2) \oplus SO(N) \rightarrow SO(2.1) \oplus SO(1,1) \oplus SO(N)$. Gauge fields and curvatures corresponding to $shsc(N \mid 3)$ have the form *(Fradkin & Linetsky (1989b), (1990a))*

$$\omega_\mu = \frac{1}{2i} \sum i^{-\varepsilon_s} \, \omega_\mu^{(s,c)i(k),\alpha(2l)} \, T_{i(k),\alpha(2l)}^{(s,c)} , \tag{3.4}$$

$$
\begin{aligned}
R_{\mu\nu,i(k),\alpha(2l)}^{(s,c)} &= \partial_\mu \, \omega_{\nu,i(k),\alpha(2l)}^{(s,c)} - (\mu \leftrightarrow \nu) \\
&+ \sum i^{s'+s''-s+r-|r|_2-1} \frac{k!}{u!v!r!} \, \delta(k - u - v) \\
&\times \, \delta(p - l' - l'' + l) \, \delta(q - l' + l'' - l) \, \delta(t - l'' + l' - l) \\
&\times \, \pi\left(4s's'' + s' + s'' - s + uv + r(u + v) + 1\right) \\
&\times \begin{pmatrix} s' & s'' & s \\ c' & c'' & c \\ l' & l'' & l \end{pmatrix} \, \omega_{\mu,i(u)j(r),\alpha(q)\gamma(p)}^{(s',c')} \\
&\times \, \omega_{\nu,i(v)}^{(s'',c'')j(r) \quad \gamma(\mu)} \,_{,\,\alpha(t)} , \tag{3.5}
\end{aligned}
$$

where $\begin{pmatrix} s & s' & s'' \\ c & c' & c'' \\ l & l' & l'' \end{pmatrix} = \delta(c + c' - c'') \varepsilon(s, s', s'') \varepsilon(l, l', l'')$

$$\times \left[\frac{(2l + 1)!(2l' + 1)!(2l'' + 1)!}{(l + l' - l'')!(l - l' + l'')!(l' - l + l'')!(l + l' + l'' + 1)!} \right]^{1/2}$$

$$\times \sum_{k,k',k''} (-1)^{1/2(s+s'-s''-k-k'+k'')} \left\{ \begin{array}{ccc} \dfrac{s+k}{2}, & \dfrac{s-k}{2}, & l \\[2mm] \dfrac{s'+k'}{2}, & \dfrac{s'-k'}{2}, & l' \\[2mm] \dfrac{s''+k''}{2}, & \dfrac{s''-k''}{2}, & l'' \end{array} \right\}$$

$$\times \ \frac{d^{l}_{c,k}\left(\frac{\pi}{2}\right) d^{l'}_{c',k'}\left(\frac{\pi}{2}\right) d^{l''}_{k'',c''}\left(-\frac{\pi}{2}\right)}{\left[\triangle\left(\frac{s+k}{2},\frac{s'+k'}{2},\frac{s''+k''}{2}\right)\triangle\left(\frac{s-k}{2},\frac{s'-k'}{2},\frac{s''-k''}{2}\right)\right]^{1/2}},$$

$$\varepsilon(a,b,c) \ = \ 1(0) \quad \text{for} \quad c \in \{|a-b|,\cdots a+b\}\,(c \notin \{|a-b|,\cdots,a+b\}),$$

$$\triangle(a,b,c) \ = \ \frac{(a+b-c)!(a-b+c)!(b-a+c)!}{(a+b+b+1)!},$$

$$k \ = \ -l,\cdots,l, \quad k' \ = \ -l',\cdots,l', \quad k'' \ = \ -l'',-l''+1,\cdots,l''$$

$$\pi(n) \ = \ 0(1) \quad \text{for} \quad n \ \text{even (odd)}. \tag{3.5a}$$

We have expressed the structure coefficients of the $*$-product in the conformal basis in terms of the 9 j-symbols and particularly values of the Wigner d-functions $d\left(+\frac{\pi}{2}\right)$.

The gauge field ω_μ is the element of the second-class Grassmann shell of $shsc(N\mid 3)$ $(\omega_\mu^s T^{s'} = (-1)^{4ss'} T^{s'} \omega_\mu^s$, Grassmann parity is $\varepsilon(T^s) = \varepsilon(\omega^s) = \varepsilon_s = 0(1)$ for s integer (half)) and the hermiticity condition is read

$$\omega_\mu^+ \ = \ -\omega_\mu, \quad \left(\omega^{(s,c)}_{\mu,i(k),\alpha(2l)}\right)^+ \ = \ (-1)^{\frac{k(k-1)}{2}}\, \omega^{(s,c)}_{\mu,i(k),\alpha(2l)}. \tag{3.6}$$

In three dimensions, the action invariant under the gauge transformations of the algebra $shsc(N\mid 3)$ has the Chern-Simons form *(Fradkin & Linetsky (1990a))*

$$S \ = \ \int tr\left(\omega \wedge d\omega + \frac{2}{3}\,\omega \wedge \omega \wedge \omega\right), \tag{3.7}$$

where $\omega \wedge \omega = \omega_\mu * \omega_\nu dx^\nu \wedge dx^\nu$ and the $*$-multiplication of Weyl symbols. This action generalizes the three-dimensional conformal supergravity action. The conformal supergravity fields in our notation are

$$\left(e_{\mu\alpha(2)},\ f_{\mu\alpha(2)},\ \omega_{\mu\alpha(2)},\ b_\mu,\ A_{\mu ij},\ \phi_{\mu i\alpha},\ \psi_{\mu i\alpha}\right)$$
$$\sim \left(\omega^{(1,-1)}_{\mu\alpha(2)},\ \omega^{(1,1)}_{\mu\alpha(2)},\ \omega^{(1,0)}_{\mu\alpha(2)},\ \omega^{(1,0)}_\mu,\ \omega^{(1,0)}_{\mu ij},\ \omega^{(1/2,1/2)}_{\mu i\alpha},\ \omega^{(1/2,-1/2)}_{\mu i\alpha}\right). \tag{3.8}$$

The equations of motion of the conformal higher spin fields in $D = 2 + 1$ have the form

$$R^{(s,c)}_{\mu\nu,i(k),\alpha(2l)} = 0 \,.\tag{3.9}$$

These equations generalize the $D = 2 + 1$ conformal supergravity equations.

For the quantization of Chern-Simons theory (the rang-one theory) we introduce a ghost and antighost fields $C = \sum_A C_A T^A, \bar{C} = \sum_A \bar{C}^A T_A, \varepsilon(C_A) = \varepsilon(\omega_{\mu,A}) + 1$ and a gauge fermion $\Psi = \sum_A \int d^3 x \bar{C}^A \Psi_A$ with gauge conditions Ψ_A (A is the collective index in the algebra). Then the exactly generating functional for the three-dimensional superconformal higher spin quantum theory reads as follows

$$\mathcal{Z} = \int \mathcal{D}\omega \mathcal{D}C \mathcal{D}\bar{C} \mathcal{D}\pi \, \exp\{iS_{eff}\} \,,\tag{3.10}$$

$$\begin{aligned}
S_{eff} = S + \int d^3 x tr \Bigg(\frac{\delta r \Psi}{\delta \omega_\mu} * \mathcal{D}_\mu C \\
+ \frac{\delta r \Psi}{\delta C} * C * C + \frac{\delta r \Psi}{\delta \bar{C}} * \pi + \mathcal{J}^\mu * \omega_\mu \Bigg) \,,
\end{aligned}\tag{3.11}$$

where S is the origin Chern-Simons action, π is the Lagrangian multiplier for the gauge conditions, $\frac{\delta \Psi}{\delta \mathcal{Z}} = \sum_A \frac{\delta \Psi}{\delta \mathcal{Z}_A} T_A$ for $\mathcal{Z} = (\omega, C, \bar{C}, \pi)$ and $\mathcal{D}_\mu$ is the covariant derivative.

To conclude this section let us consider the compactification of Chern-Simons theory. Introducing a notation $\phi = \omega_2$ and expanding the gauge fields by coordinate $x^2, \omega_\mu = \sum_{n=-\infty}^{\infty} \omega_\mu(n) e^{in \, r^2/r_c}$, where r_c is a compactification radius, we have an action in $D = 1 + 1$

$$S = \sum_{n=-\infty}^{\infty} \int tr \left(\phi(n) * R(-n) + \frac{in}{r_c} \omega(-n) \wedge \omega(n) \right) \,.\tag{3.12}$$

In addition to usual action in $D = 1 + 1$, the action (3.12) also contains a "topological mass term" $tr \sum_n \frac{in}{r_c} \omega(-n) \wedge \omega(n)$. The corresponding ghost Lagrangian has the form

$$\begin{aligned}
\mathcal{L}_{gh} = \sum_{n=-\infty}^{\infty} tr \Bigg(\frac{\delta r \Psi}{\delta \omega_\mu(n)} * (\mathcal{D}_\mu C)(-n) \\
+ \frac{in}{r_c} \frac{\delta r \Psi}{\delta \phi(-n)} * C(n) + \frac{\delta r \Psi}{\delta C(n)} * \sum_{m+k=-n} C(m) * C(k) \\
+ \frac{\delta r \Psi}{\delta \phi(-n)} * [\phi, C](n) \Bigg) \,.
\end{aligned}\tag{3.13}$$

3.2. <u>Conformal Higher Spin Theory in $D = 3 + 1$.</u>

Now let us turn to the second possibility for a higher spin gauge theory in four dimensions as a higher spin extension of conformal supergravity. We start with the conformal supergravity which is a supersymmetry generalization of the Weyl gravity with the quadratical Weyl tensor action.

Similar to the AdS_4 supergravity, conformal supergravity is constructed as a gauge theory for a conformal superalgebra $SU(2, 2 \mid N)$ [see rev. of conformal supergravity *(Fradkin & Tzeytlin (1985a))*].

Since we are going to construct higher spin generalizations of the conformal supergravity, a first thing to do is to find a higher spin generalization of the four-dimensional conformal supersymmetry. It was done *(Fradkin & Linetsky (1989a), (1990a))* by the method of the operator realization.

A conformal higher spin superalgebra can be simply constructed by employing <u>quantum (super)twistors</u>: ($\hat{\alpha}$ and $\hat{\alpha}^+$ are Fermi-operators)

$$Z_A = \left(a^\alpha, a_{\dot{\beta}}, \hat{\alpha} \right), \quad \bar{Z}^B = \left(\bar{a}_\alpha, \bar{a}^{\dot{\beta}}, \hat{\alpha}^+ \right); \quad [Z_A, \bar{Z}^B\} = 2\delta_A^B . \tag{3.14}$$

Then the conformal superalgebra $SU(2, 2 \mid N)$ is realized as an algebera of second-order polynomials that commute with the "particle number" operator $T = \bar{Z}^A Z_A$, i.e. the $SU(2, 2 \mid N)$ generators are:

$$T_A^B = \frac{1}{2} \left(\bar{Z}^A Z_B + (-1)^{P_A P_B} Z_B \bar{Z}^A \right) - \frac{1}{4 - T} \delta_A^B T . \tag{3.15}$$

Then an infinite-dimensional conformal higher spin superalgebra $shc^\infty(4 \mid 1)$ is defined as an algebra at all order polynomials that commute with T, excluding powers of T itself.

To construct a gauge theory it is necessary to introduce the basis on $so(4, 2)$-irrepses connected with the decomposition $so(4, 2) \oplus so(3, 1)$, where $so(3, 1)$ is the Lorentz subalgebra and $so(1, 1)$ generated by dilatation generator D. In this basis all generators and the gauge fields will have the defined Weyl weight and the manifest four-dimensional index Lorentz structure. Such a superconformal basis was constructed by standard group theory methods and the structure constants of $shsc^\infty(4\mid 1)$ in this basis was obtained (see *Fradkin & Linetsky (1990a), (1991a)*).

As a result, the gauge field corresponding to $shsc^\infty(4\mid 1)$ is a function of supertwistors $(\omega_\mu(z,\bar z\mid x);\ [T,\omega_\mu(z,\bar z\mid x)]=0)$ and has the form*

$$
\omega_\mu = \sum_{N=1}^{\infty}\sum_{s=1}^{N}\sum_{(\mathfrak{s},c,u,l,j)} i^{-\varepsilon_{\mathfrak{s}}}\, \omega^{(N,s,\mathfrak{s},c,u)}_{\mu,\alpha(2l),\dot\beta(2j)}
$$
$$
\times\ T^{(N,s,\mathfrak{s},c,u)\,\alpha(2l),\dot\beta(2j)}
\tag{3.16}
$$

$\varepsilon_{\mathfrak{s}} = 0(1)$ for $\mathfrak{s}$ integer (half-integer).

Here indices of the fields and generators have the following meaning. The index $N = 1,2,\cdots$ is a number of the level $L(N)$. The index $s = 1,2,\cdots,N$ defines $su(2,2\mid 1)$-irreps $V(s)$ (the gauge fields with the fixes s form the $su(2,2\mid 1)$ -super-multiplet with maximal spin $s+1$; 1 due to the additional vector index μ). Thus on the N-th level there are n conformal supermultiplets with the maximal spins from $N+1$ to 2. The index $\mathfrak{s} = s-1,\ s-1/2,\ s$ defines the conformal multiplet (the gauge fields describe the spin $\mathfrak{s}+1$). The index u is a chiral weight and $u=0$ for Bosons ($\mathfrak{s}=$ integer) and $u=\pm 1/2$ for Fermions ($\mathfrak{s}=$ halfinteger). The index $c = -\mathfrak{s},\ -\mathfrak{s}+1,\cdots,\mathfrak{s}$ is a conformal weight of the generators and the indices $l,j = 0,1/2,1,\cdots$ define a Lorentz

---

*Under the $su(2,2\mid 1)$ representation each level decomposes into the direct sum of $su(2,2\mid 1)$ irreducible representation spaces (irrepses) $L(N) = \bigoplus_{s=1}^{N} V(s)$. In its turn the irreps $V(s)$ decomposes into the sum of $so(4,2)$ irrepses, $V(s) = D(s,s,0)\oplus D\left(s-\frac{1}{2},s-\frac{1}{2},\frac{1}{2}\right)\oplus D\left(s-\frac{1}{2},s-\frac{1}{2},-\frac{1}{2}\right)\oplus D\left(s-1,s-1,0\right)$, where $D(n_1,n_2,n_3)$ is the $so(4,2)$ irreps with the highest weight (n_1,n_2,n_3) under the Cartan subalgebra of $so(4,2)$. Here n_1 is the maximal conformal weight in the representation and $(n_2-n_3)/2$ and $(n_2+n_3)/2$ define a Lorentz signature $((n_2-n_3)$ and (n_2+n_3) are the numbers of dotted and undotted indices, respectively) of the vector with highest conformal weight. Note that the representations $\left(\mathfrak{s}-\frac{1}{2},\mathfrak{s}-\frac{1}{2},\pm\frac{1}{2}\right)$ are mutually conjugated under the Weyl reflection (these are usually called chirally or complex conjugated representations). The dimension of the $so(4,2)$ irreps $D(\mathfrak{s},\mathfrak{s},u)$ is equal to $d(\mathfrak{s},\mathfrak{s},u) = \frac{1}{12}(2\mathfrak{s}+3)(\mathfrak{s}+u+1)(\mathfrak{s}+u+2)(\mathfrak{s}-u+1)(\mathfrak{s}-u+2)$. The first level consists of only one $su(2,2\mid 1)$ irreps $V(1)$ which is the adjoint representation, and under $so(4,2)$ we have $V(1) = D(1,1,0)\oplus D\left(\frac{1}{2},\frac{1}{2},\frac{1}{2}\right)\oplus D\left(\frac{1}{2},\frac{1}{2},-\frac{1}{2}\right)\oplus D(0,0,0)$. The basis in these irrepses can be chosen as follows: $D(1,1,0):\{P,K,M,D\}$, $D\left(\frac{1}{2},\frac{1}{2},\frac{1}{2}\right):\{S_\alpha,Q_\beta\}$, $D\left(\frac{1}{2},\frac{1}{2},-\frac{1}{2}\right):\{S_\beta,Q_\alpha\}$, $D(0,0,0):\{U\}$.

signature ($2l$ and $2j$ are the numbers of dotted and undotted indices) and the following restrictions take place

$$l + j \leq s, \quad l \geq \left| \frac{c-u}{2} \right|, \quad j \geq \left| \frac{c+u}{2} \right|, \tag{3.17}$$

with $l + \dfrac{c-u}{2}$ and $j + \dfrac{c+u}{2}$ integers.

It should be mentioned that the structure of the gauge field (2.13) is analogous to the structure of the string field $\Phi[X]$ in the string field theory. There is an infinite tower of levels and there are fields with all spins from maximal ($s = N + 1$) to minimal ($s = 1$) on each N-th level. For effective work with the infinite tower of levels one can use the following approximation procedure. The number N should be limited by N_{max} and then $N_{max} \to \infty$.

Note that the gauge fields of usual conformal supergravity ($N = s = 1$) are, in our notations*,

$$\left(e_{\mu\alpha\dot\beta}, \omega_{\mu\alpha(2)}, \omega_{\mu\dot\beta(2)}, b_\mu, f_{\mu\alpha\dot\beta}, A_\mu, \psi_{\mu\alpha}, \psi_{\mu\dot\beta}, \psi_{\mu\alpha}, \psi_{\mu\dot\beta} \right)$$
$$\sim \left[\omega^{(1,1,1,-1,0)}_{\mu,\alpha,\dot\beta}, \omega^{(1,1,1,0,0)}_{\mu,\alpha(2)}, \omega^{(1,1,1,0,0)}_{\mu,\dot\beta(2)}, \omega^{(1,1,1,0,0)}_{\mu}, \omega^{(1,1,1,1,0)}_{\mu,\alpha,\dot\beta}, \omega^{(1,1,0,0,0)}_{\mu}, \right.$$
$$\left. \omega^{(1,1,1/2,-1/2,1/2)}_{\mu,\alpha}, \omega^{(1,1,1/2,-1/2,-1/2)}_{\mu,\dot\beta}, \omega^{(1,1,1/2,1/2,-1/2)}_{\mu,\alpha}, \omega^{(1,1,1/2,1/2,1/2)}_{\mu,\dot\beta} \right] . \tag{3.18}$$

Now we can give explicit formulae for the curvatures of $hsc^\infty(4 \mid 1)$ *(Fradkin & Linetsky (1990a), (1991a)).*

$$R_{\mu\nu} = \partial_\mu \omega_\nu - \partial_\nu \omega_\mu + [\omega_\mu, \omega_\nu \},$$
$$R^A_{\mu\nu} = \partial_{[\mu} \omega^A_{\nu]} + \int^A_{BC} \omega^B_\mu \, \omega^C_\nu , \tag{3.19}$$

---

*The physical spin-$(s+1)$-fields in $\omega^{(Nscu)}_{\mu,\alpha,\dot\beta}$ are: $\omega^{(N,s,c,-c0)}_{\mu,\alpha(c),\dot\beta(c)}$ (for integer s); or $\omega^{(N,s,c,-c,1/2)}_{\mu,\alpha(c+1/2),\dot\beta(s-1/2)}$ and $\omega^{(N,s,c,-c,-1/2)}_{\mu,\alpha(c-1/2),\dot\beta(s+1/2)}$ (for half-integers s). These fields generalize the conformal supergravity ones $e_{\mu\alpha\dot\beta}$, $\psi_{\mu\alpha}$, $\psi_{\mu\dot\beta}$, A_μ. The constraints allow to express all other (auxiliary) fields through the physical ones up to a pure gauge part.

$$R^{(N,s,\dot{s},c,u)}_{\mu\nu,\alpha(2l),\dot{\beta}(2j)} = \partial_\mu \omega^{(N,s,\dot{s},c,u)}_{\nu,\alpha(2l),\dot{\beta}(2l)} - (\mu \leftrightarrow \nu)$$

$$+ \sum \delta(c' + c'' - c)\, \delta(u' + u'' - u)\, \delta(m - l' - l'' + l)\, \delta(r - l'' + l' - l)$$

$$\times\ \delta(t - l' + l'' - l)\, \delta(p - j' - j'' + j)\, \delta(q - j'' - j + j')$$

$$\times\ \delta(k - j - j' + j'')\, \pi(N + N' + N'')\, i^{j' + j'' - j - l}$$

$$\times\ \left\{ \begin{matrix} N' & s' & \dot{s}' & c' & u' & l' & j' \\ N'' & s'' & \dot{s}'' & c'' & u'' & l'' & j'' \\ N & s & \dot{s} & c & u & l & j \end{matrix} \right\}$$

$$\times\ \omega^{(N',s',\dot{s}',c',u')}_{\mu,\alpha(l)\gamma(m),\dot{\beta}(k)\dot{\delta}(p)}\ \omega^{(N'',s'',\dot{s}'',c'',u'')\gamma(m)\ \ \ \dot{\delta}(p)}_{\nu,\alpha(r)\ \ \ \ \ \ \ \ \ ,\dot{\beta}(q)} . \tag{3.20}$$

Here the parity function $\pi(n) = 0(1)$ for n even (odd). They are expressed through the group-theoretical factors which are well known from angular momentum theory (Clebsch-Gordan coefficients, $9j$-symbols etc.). Some simple symmetry properties of the structure coefficients provide the existence of invariant bilinear form on $shc^\infty(4 \mid 1)$ $(A, B) = tr(A * B)$, where the trace is defined by $tr(A(z, \bar{z})) = A(0,0)$ and $A * B$ is the Weyl product of the symbols A and B (see in detail: *Fradkin & Linetsky (1991a)*).

3.3. <u>The Cubic Invariant Action in Superconformal Higher-Spin Theory.</u>

The action based on the conformal higher-spin superalgebra $shsc^\infty(4 \mid 1)$ of the type (3.20) can be written in the form *(Fradkin & Linetsky (1991a))*

$$A_0 = \frac{-1}{8\alpha^2} \sum_{N=1}^{\infty} \sum_{s=1}^{N} \sum_{\dot{s},c,u,n,m} (-1)^{N-s}\, j^{n+m+1}\, \mathcal{E}(n - m)$$

$$\times \int R^{(N,s,\dot{s},c,u)}_{\alpha(n),\dot{\beta}(m)} \wedge R^{(N,s,\dot{s},-c,-u)\alpha(n),\dot{\beta}(m)} \tag{3.21}$$

and the Yang-Mills term for vector fields $\omega^{(N,1,0,0,0)}_\mu$ reads

$$A_{YM} = \frac{-1}{4\alpha^2} \sum_{N=1}^{\infty} (-1)^{N-1} \int d^4x \sqrt{-g}\ g^{\mu\rho}\, g^{\nu\sigma}\, R^{(N,1,0,0,0)}_{\mu\nu}\, R^{(N,1,0,0,0)}_{\rho\sigma}. \tag{3.22}$$

and the following constraints*

$$KR = 0 \,. \tag{3.22a}$$

The action $A = A_0 + A_{YM}$, does not contain any dimensionful parameters. The metric in (3.22) is defined as

$$g_{\mu\nu} = \frac{1}{2}\,\omega^{(1,1,1,-1,0)}_{\mu,\alpha,\dot\beta}\,\omega^{(1,1,1,-1,0)\alpha,\dot\beta}_{\nu}\,,$$

$$g^{\mu\nu} = \left(g_{\mu\nu}\right)^{-1}, \quad g = \det\left(g_{\mu\nu}\right)\,. \tag{3.23}$$

First it was demonstrated *(Fradkin & Linetsky (1991a))* that (3.21)-(3.22a) determined the linearized theory completely. The linearized curvatures $R^l = d\omega + \mathcal{P}\omega$ are constructed with the help of the nilpotent $\hat{P}$ operator, $\mathcal{P}\omega = [P^{\alpha\beta}, \sigma_{\alpha\beta} \wedge \omega]$, acting on the differential forms taking their values in $shs^\infty(4 \mid 1)$. With the help of the generalization $\circledast = \star \circ \mathcal{R}$ of the Hodge star $\star$ including Weyl reflection $\mathcal{R}$ (in the $so(4,2)$ representations it changes the sings of c and u), the nilpotent operator $\mathcal{K} = \circledast \mathcal{P} \circledast$ conjugated with $\mathcal{P}$ under some natural scalar product $\int tr(A \wedge \circledast B)$ was introduced. The operators $\mathcal{P}$ and $\mathcal{K}$ converted the sequence of linear spaces of q-forms into the conformal cohomological complex which is analogous to the de Rham complex on the Riemann manifold. The linearized constraints $KR^l = 0$ allow us to express all auxiliary fields through the physical ones up to a pure gauge part. The general solution of these constraints in terms of the curvatures was obtained. It turned out that all curvatures can be expressed through (derivatives on) the Weyl multispinors representing the Weyl tensors and spinor-tensors. The Weyl tensors in this context are non-trivial cohomological classes (harmonic forms) for the conformal cohomological complex. The linearized actions quadratic on the curvatures $R^l \wedge R^l$ both for integer and half-integer spins were brought to the free C^2-form as in the tensor formalism. In this way the equivalence of our geometrical formulation and the usual formulation for free higher spins in the symmetric tensor formalism was established (see *Fradkin & Linetsky (1991a)*).

---

*Two nilpotent operators P and K, which are very useful on conformal higher spin theory, was introduced by us. The operator P increases the rank and decreases the conformal weight of an arbitrary differential form at unity and anticommutes with the usual exterior differential. The operator K decreases the rank and increases the conformal weight of a differential form at unity.

The (3.21) is a special form of our general extension of MacDowell-Mansouri actions $i \int [R^A \wedge R^B G_{AB} - R^{\bar{A}} \wedge R^{\bar{B}} G_{\bar{A}\bar{B}}$ for higher spins, where $R^A(R^{\bar{A}})$ is the set of curvatures $R^{\Omega}_{\alpha(n),\dot\beta(m)}$ with $n > m(n < m)$ (the curvatures R^a with $n = m$ do not enter the action), and $G_{AB}, G_{\bar{A}\bar{B}}$ are the blocks of the invariant bilinear form in the algebra. The cubic invariant action for conformal higher-spin theory (Bose case) then is chosen in the above form. The proof of cubic gauge invariance has the following steps. The general structure of the gauge variation in cubic order is $R^l \wedge R^l \mathcal{E}$. Firstly, due to the symmetry properites of G_{AB} and $G_{\bar{A}\bar{B}}$, the terms $R^A \wedge R^B \mathcal{E}$ and the complex conjugates $R^{\bar{A}} \wedge R^{\bar{B}} \mathcal{E}$ are cancelled separately. Secondly, taking into account the self-duality of $R^{l\bar{A}}$ and the antiself-duality of R^{lA} as follows from the linearized conventional constraints, the terms $R^{lA} \wedge R^{l\bar{B}} \mathcal{E}$ vanish identically. Finally, among the remaining terms $R^{la} \wedge R^A \mathcal{E}$ and $R^{la} \wedge R^{\bar{A}} \mathcal{E}$ only the terms with $R^{la} = R^{l(N,s,s)}_{\alpha(s),\dot\beta(s)}$ are non-zero due to the constraints. They must be compensated for by some deformations $\triangle \omega$ in the gauge transformation law for auxiliary fields. Obtaining $\triangle \omega$ from the requirement $\delta_g A + \triangle \omega = 0$, one should verify that these deformations are compatible with the conventional constraints. In this way we find the second-order constraints for the part of auxiliary fields (only for those which get the deformations $\triangle \omega$). The action found in such a way is unique.

In the superconformal case the Yang-Mills term for the vector fields must be added along with the above-considered action. The proof of invariance here is analogous to the purely bosonic case with some technical complications stipulated by the increased number of terms (more details see: *Fradkin & Linetsky (1991a)*).

Let us briefly sum up the main results and point out a number of problems that need further study in conformal higher spin theory.

We have shown that there exists a gauge invariant cubic interaction among bosonic and fermionic conformal higher-spin fields incorporating conformal supergravity. This result opens up the possibility of constructing a self-consistent interacting conformally invariant higher-spin theory. Together with previous results about higher-spin interaction in AdS_4, it gives hope to solve the longstanding higher-spin problem which would be a considerable step towards a unified theory. It seems natural that in our construction there is an infinite number of fields of each spin. It is completely analogous to string field theory. Each level contains all spins from maximal to minimal (spin 1). Such a structure of levels also looks natural from the point of view of spontaneous symmetry breaking. Only the first level (spins ≤ 2) might remain massless; the other higher levels should become massive.

However regarding the infinite multiplicity of spins in the gauge invariant conformal higher-spin interacting theory one should keep the following circumstances in mind. Right from the start we have dealt with the superalgebra $shs^\infty(4 \mid 1)$. In principle it is not impossible for the invariant interaction to be based on another superalgebra containing each spin with a finite multiplicity. We constructed *(Fradkin & Linetsky (1990a))* a whole family of such superalgebras $shsc_\rho^{(n)}(4 \mid N)$, where $n = 1, 2, \cdots$ is the multiplicity of $SU(2, 2 \mid N)$-supermultiplet with the fixed maximal spin in the algebra and $\rho \in \mathbb{R}$ some numerical parameter. They are factor-algebras of the original superalgebra $shsc^\infty(4 \mid 1)$. However all those superalgebras seemingly may not be localized. To build a cubic gauge invariant interaction it is necessary that some invariant bilinear form exists on the algebra. But $shsc_\rho^{(n)}(4 \mid 1)$ apparently does not possess any invariant bilinear form (the structure constants of factor-algebras have no simple symmetry properties which differ from $shsc^\infty(4 \mid 1)$). Meanwhile it cannot be excluded that any such superalgebras [e.g. certain factor-algebras of the universal enveloping algebra $U(so(4, 2))$] with a finite multiplicity of spins exist.

There are a number of problems that require a further study. The first is to expand the construction presented here to all orders in the interaction; in particular, to find a non-linear version of the standard constraints. Another problem is to construct N-extended theories. These theories may be based on the N-extended conformal higher-spin superalgebras $shsc^\infty(4 \mid N)$ constructed by us *(Fradkin & Linetsky (1990a))*. To transfer the theory presented here to higher dimensions and the extensions with non-trivial (in general an operator) central terms as well as certain non-linear terms in the right side of this supercommutator in the higher spin symmetry are also important tasks.

4. Conclusion

Now it is time to summarize everything. Thus, there are just three principal possibilities.

- Supergravity (without cosmological term) and an infinite tower of massivly higher spins (superstrings, or string-like models);
- AdS supergravity and an infinite tower of massless higher spin fields with an infinite-dimensional higher spin gauge symmetry (AdS higher spin theory);
- Conformal supergravity and an infinite tower of conformal higher spin fields with infinite-dimensional conformal higher spin symmetry (conformal higher spin theory).

These three seem drastically different. However, a synthesis could be conjectured, when with the increase of temperature there is a phase transition in strings. Then a higher spin symmetry is restored, higher spin excitations become massless and a cosmological constant of the Planck order is induced.

However, there is still a dimensionful parameter, a cosmological constant, in the theory. Meanwhile any quantum theory in the ultra high energies domain may be described by some scale-free conformally invariant theory. Conformal higher spin theory might play a role of such an asymptotical theory.

Summing-up the above expounded arguments, the following scenario may be suggested. In the ultra high-energy domain a unified theory is effectively described as a conformal higher-spin theory generalizing Weyl gravity. The spontaneous conformal symmetry breaking leads to the massless higher-spin theory in the anti-de Sitter universe generalizing the AdS supergravity. Further the AdS higher-spin symmetry breaking leads to the string-like phases in AdS transitory asymptotically to the massivly higher-spin string phase coupled to the Einstein gravity on the flat background with zero cosmological constant. The above scenario is schematically illustrated in *fig. 1*.

It should be mentioned that one can look at the above scenario in two different ways. Firstly, it may be treated in a straightforward way as a scenario for the fundamental unified theory, i.e. the unified hypothetical lagrangian, or its spontaneously broken versions,

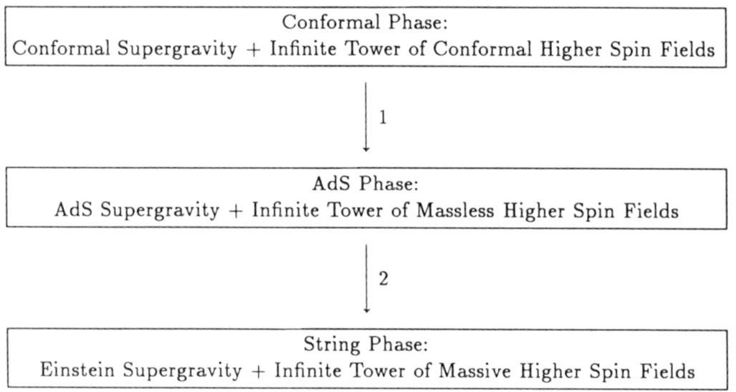

FIGURE 1: Phases of Unified Theory. The arrows 1,2 denote the symmetry breakings (see in the text).

actually describes this picture at all energies. Then, apparently, the scenario should be played in high dimensions. Secondly, it may be considered as a hierarchy of effective theories, each working effectively only in its own energy domain; and their lagrangians, generally speaking, are not connected in a straightforward way with each other. Subsequently it also might be considered in four dimensions.

To have such a scenario of phase transitions it is necessary first to demonstrate **a)** that a critical AdS string (superstring) model does exist not only in 26(10) dimensions; **b)** that an AdS higher spin theory exists in the same space-time dimension with the same degrees of freedom as in AdS string model. Such a family of critical strings in $D < 26(D < 10)$ described by anti-de Sitter non-compact caset models $SO(D -$

$1,2)/SO(D-1,1)$ (euclidian version $SO(D,1)/SO(D)$) and their (Kazama-Suzuki) superextension has been recently introduced *(Bars & Nemeshansky (1991); Fradkin & Linetsky (1991c))*.

It was demonstrated that these models are free of conformal anomaly in any $D < 26$ and the anomaly cancellation condition actually fixes the value of the "effective radius of the universe" at a given D. The crucial features of these vacuum solutions in string theory are: **1)** the space-time is non-flat and **2)** there is a space-time dependence of the background fields (metric, dilaton and, generally, also antisymmetric tensor field). For $D = 2$ such a calculation has been performed by *(Elitzur, Forge & Rabinovici (1990); Witten (1991); Dijkgraaf, H. Verlinde & E. Verlinde (1991); Tseytlin (1991))*. For $D = 3$ by *(Fradkin & Linetsky (1991); Crescimanno (1991); Bars & Sfetsos (1991); Chamseddine (1991))*. A preliminary cosmological interpretation of these results is proposed. Roughly, string theory seems to admit non-trivial cosmological (time-dependent) background solution where the universe created as an anti-de Sitter universe (after the phase transition from the higher spin phase to the string phase in AdS) and then under the influence of the fundamental fields, which acquires non-zero background value (dilaton, metric, Higgs particle, etc.) increasing in time, it begins to expand and the higher spin gauge fields become (generally step by step) massive. Let us note, that non-perturbative effects leading to the appearance of a non-trivial dilaton potential (with a mass term) are very essential.

In the case, when a selfconsistence theory exists only in $D > 4$, it is natural to expect a Kaluza-Klein version, when in string phase by the non-static solutions of the background, only three-space dimensions have the effective expansion (inflation) by time, which brings us to an asymptotical flat four space-time. (The remaining dimensions can contract or oscillate with a radius of Planck order.)

ACKNOWLEDGMENT

The author would like to thank Professor G. Scarpetta and the Department of Theoretical Physics of Salerno and Naples University for kind hospitaliy.

APPENDIX

We adopt here the notations and conventions. The Greek indices $\mu, \nu, \rho, \sigma = 0, 1, 2, 3$ are the indices of components of differential forms. The flat Minkowsky metric is $\eta_{\mu\nu}(+, -, -, -)$. The two-component dotted and undotted spinorial indices $\dot\alpha, \dot\beta, \cdots$; $\alpha, \beta, \cdots$ take on the values 1 and 2. They are raised and lowered by means of the symplectic metric $\varepsilon_{\alpha\beta} = -\varepsilon_{\beta\alpha}, \varepsilon^{\alpha\beta} = -\varepsilon^{\beta\alpha}, \varepsilon_{12} = \varepsilon^{12} = 1$ as

$$A^\alpha = \varepsilon^{\alpha\beta} A_\beta, \qquad A_\alpha = \varepsilon_{\beta\alpha} A^\beta \tag{A.1}$$

and analogously for dotted indices.

A symmetrization is implied separately for any set of upper or lower dotted or undotted spinorial indices denoted by the same letters. The ususal summation convention is understood for each pair of a lower and upper index denoted by the same letter. The number of indices is indicated in parantheses (except for a single index). After the symmetrization with the indices is carried out, the maximal possible number of upper and lower indices denoted by the same letter should be contracted. For instance,

$$A_{\alpha(n)} = \frac{1}{n!} \left(A_{\alpha_1 \cdots \alpha_n} + (n! - 1) \text{ permutations of } \alpha_1 \cdots \alpha_n \right), \tag{A.2}$$

$$C_{\alpha(n-m)} = A_{\alpha(n)} B^{\alpha(m)} = A_{(\beta_1 \cdots \beta_m \alpha_{m+1} \cdots \alpha_n)} B^{\beta_1 \cdots \beta_m}, \; n \geq m, \tag{A.3}$$

where brackets denote full symmetrization. $A_\mu = \sigma^{\alpha\dot\rho} A_{\alpha\dot\beta}; \; A_{\alpha\dot\beta} = \sigma^\mu_{\alpha\dot\beta} A_\mu$.

The flat vierbein is

$$\sigma_\mu^{\alpha\dot\beta} = (I, \sigma_1, \sigma_2, \sigma_3)^{\alpha\dot\beta}, \tag{A.4}$$

where I is the unit matrix and $\sigma_{1,2,3}$ are the Pauli matrices. The flat vierbein satisfies the following properties:

$$\sigma_{\mu\alpha\dot\beta} \sigma^{\mu\gamma\dot\delta} = 2\delta_\alpha^\gamma \delta_{\dot\beta}^{\dot\delta}, \qquad \sigma_{\mu\alpha\dot\beta} \sigma^{\nu\alpha\dot\beta} = 2\delta_\mu^\nu. \tag{A.5},(A.6)$$

The vierbein one-form is

$$\sigma_{\alpha\dot\beta} = \sigma_{\mu\alpha\dot\beta} \, dx^\mu, \tag{A.7}$$

and for its exterior product we have

$$2\sigma^{\alpha\dot{\beta}} \wedge \sigma^{\gamma\dot{\delta}} \;=\; \varepsilon^{\alpha\gamma}\,\bar{\sigma}^{\dot{\beta}\dot{\delta}} \;+\; \varepsilon^{\dot{\beta}\dot{\delta}}\,\sigma^{\alpha\gamma}, \tag{A.8}$$

where

$$\sigma^{\alpha\gamma} \;=\; \sigma^{\alpha\gamma}_{\mu\nu}\, dx^{\mu} \wedge dx^{\nu}, \qquad \bar{\sigma}^{\dot{\beta}\dot{\delta}} \;=\; \bar{\sigma}^{\dot{\beta}\dot{\delta}}_{\mu\nu}\, dx^{\mu} \wedge dx^{\nu} \tag{A.9}$$

and

$$\sigma_{\mu\nu}{}^{\alpha\gamma} \;=\; \sigma_{\mu}{}^{\alpha}{}_{\dot{\beta}}\,\sigma_{\nu}{}^{\gamma\dot{\beta}}, \qquad \bar{\sigma}_{\mu\nu}{}^{\dot{\beta}\dot{\delta}} \;=\; \overline{\left(\sigma_{\mu\nu}{}^{\beta\delta}\right)} \;=\; \sigma_{\mu\gamma}{}^{\dot{\beta}}\,\sigma_{\nu}{}^{\gamma\delta}. \tag{A.10}$$

The two-forms $\sigma^{\alpha(2)}$ and $\bar{\sigma}^{\dot{\beta}(2)}$ ($\sigma^{\alpha\gamma} = \sigma^{\gamma\alpha}$, $\bar{\sigma}^{\dot{\beta}\dot{\delta}} = \bar{\sigma}^{\dot{\delta}\dot{\beta}}$) are antiself-dual and self-dual respectively,

$$\star\,\sigma^{\alpha(2)} \;=\; -\,i\,\sigma^{\alpha(2)}, \qquad \star\,\bar{\sigma}^{\dot{\beta}(2)} \;=\; \subset\,\bar{\sigma}^{\dot{\beta}(2)}, \tag{A.11}$$

or in the components

$$\frac{1}{2}\,\varepsilon^{\mu\nu\rho\sigma}\,\sigma_{\rho\sigma}{}^{\alpha(2)} \;=\; -\,i\,\sigma^{\mu\nu\alpha(2)}, \tag{A.12a}$$

$$\frac{1}{2}\,\varepsilon^{\mu\nu\rho\sigma}\,\bar{\sigma}_{\rho\sigma}{}^{\dot{\beta}(2)} \;=\; i\,\bar{\sigma}^{\mu\nu\dot{\beta}(2)}. \tag{A.12b}$$

The quantities $\sigma_{\mu\nu}{}^{\alpha(2)}$ and $\bar{\sigma}_{\mu\nu}{}^{\dot{\beta}(2)}$ are, as a matter of fact, projectors on the antiself-dual part and self-dual part of the arbitrary two-form R ('t Hooft tensors)

$$R_{\mu\nu} \;=\; \frac{1}{4}\,\sigma_{\mu\nu}{}^{\alpha(2)}\,R_{\alpha(2)} \;+\; \frac{1}{4}\,\sigma_{\mu\nu}{}^{\dot{\beta}(2)}\,\bar{R}_{\dot{\beta}(2)}, \tag{A.13}$$

$$R_{\alpha(2)} \;=\; \frac{1}{4}\,i\,\varepsilon^{\mu\nu\rho\sigma}\,\sigma_{\mu\nu\alpha(2)}\,R_{\rho\sigma}, \qquad \bar{R}_{\dot{\beta}(2)} \;=\; \overline{\left(R_{\beta(2)}\right)}. \tag{A.14}$$

Evidently, the following relations hold:

$$\int R \wedge R \;=\; i \int d^{4}x\left(\bar{R}_{\dot{\beta}(2)}\,\bar{R}^{\dot{\beta}(2)} \;-\; R_{\alpha(2)}\,R^{\alpha(2)}\right), \tag{A.15}$$

$$\int R \wedge \star\, R \;=\; -\int d^{4}x\left(R_{\alpha(2)}\,R^{\alpha(2)} \;+\; \bar{R}_{\dot{\beta}(2)}\,\bar{R}^{\dot{\beta}(2)}\right), \tag{A.16}$$

due to the identities

$$\sigma^{\alpha(2)} \wedge \sigma_{\gamma(2)} \;=\; -\,16\,i\,\delta^{\alpha}_{\gamma}\delta^{\alpha}_{\gamma}\, d^{4}x, \tag{A.17a}$$

$$\bar{\sigma}^{\dot{\beta}(2)} \wedge \bar{\sigma}_{\dot{\delta}(2)} \;=\; 16\,i\,\delta^{\dot{\beta}}_{\dot{\delta}}\delta^{\dot{\beta}}_{\dot{\delta}}\, d^{4}x, \tag{A.18a}$$

$$\sigma_{\alpha(2)} \wedge \bar{\sigma}_{\dot{\beta}(2)} \equiv 0, \tag{A.19a}$$

or in the components

$$\varepsilon^{\mu\nu\rho\sigma} \, \sigma_{\mu\nu}{}^{\alpha(2)} \, \sigma_{\rho\sigma\gamma(2)} \;=\; -\,16\,i\,\delta^{\alpha}_{\gamma}\delta^{\alpha}_{\gamma}, \tag{A.17b}$$

$$\varepsilon^{\mu\nu\rho\sigma} \, \bar{\sigma}_{\mu\nu}{}^{\dot{\beta}(2)} \, \bar{\sigma}_{\rho\sigma\dot{\delta}(2)} \;=\; 16\,i\,\delta^{\dot{\beta}}_{\dot{\delta}}\delta^{\dot{\beta}}_{\dot{\delta}}, \tag{A.18b}$$

$$\varepsilon^{\mu\nu\rho\sigma} \, \sigma_{\mu\nu}{}^{\alpha(2)} \, \bar{\sigma}_{\rho\sigma}{}^{\dot{\beta}(2)} \;\equiv\; 0. \tag{A.19b}$$

REFERENCES

- Bars I. and Nemeshansky D., 1991, Nucl. Phys. **B348**, 89

- Bars I. and Sfetsos K., 1991, Prepr. USC-91/HEP-B5

- Beresin F.A., Method of Second Quantization (Acad. Press 1966)

- Chamseddine A.H., 1991, Preprint ZU-TH-31-1991

- Crescimanno M., 1991, Preprint LP2-30947

- Dijkgraaf R., Verlinde H. and Verlinde E., 1991, Princeton Univ. preprint PUPT-1252/91

- Elitzur E., Forge A. and Rabinovici E., 1990, Racah Institute preprint RI-143/90

- Fradkin E.S., 1989, The problem of unification of all interaction and self-consistency; lecture on the the occasion of awarding the Dirac medal 1988, ICPT Trieste.

- Fradkin E.S., Linetsky V.Ya., 1989a, Phys. Lett. **B231**, 97

    - 1989b, Mod. Phys. Lett. **A4**, 731, 2635, 2649

    - 1990a, Ann. Phys. **198**, 252, 293

    - 1990b, Mod. Phys. Lett. **A5**, 1167

    - 1991a, Nucl. Phys. **B350**, 274

    - 1991b, Phys. Lett. **B253**, 97, 107

    - 1991c, Phys. Lett. **B261**, 26

    - 1991d, Phys. Lett. **B277**, 73; Harvard preprint HUPT-91/1044

    - 1991e, Mod. Phys. Lett. **A6**, 2693

    - 1991f, Mod. Phys. Lett. **A6**, 617, 217

    - 1991g, An essay published in the Proceedings of the Capri Conference dedicated to P.G. Berman's seventyfifth birthday (1991: World Scientific)

- Fradkin E.S., Tseytlin A.A., 1984, Phys. Lett. **134B**, 187; 307

    - 1985a, Phys. Rep. **119**, 233

    - 1985b, Phys. Lett. **B162**, 245

- 1985c, Nucl. Phys. **B261**, 1

- 1985d, Phys. Lett. **B158**, 316

- 1985e, Phys. Lett. **B160**, 69

- 1985f, Phys. Lett. **B163**, 123

– Fradkin E.S., Vasiliev M.A., 1986, Dokl. Acad. Nauk USSR, **29**, 1100

- 1987a, Phys Lett. **B189**, 89

- 1987b, Nucl. Phys. **B291**, 141

- 1987c, Ann. Phys. (N.Y.) 177, 63

- 1988, Int. J. Mod. Phys. **A3**, 2983

– Green M.B., Schwarz J.H., 1984, Phys. Lett. **B149**, 117

– Green M.B., Schwarz J.H. and Witten E., String Theory, Cambridge 1987

– Konstein S.E., Vasiliev M.A., 1989, Nucl. Phys. **B312**, 402

- 1990, Nucl. Phys. **B331**, 475

– MacDowell S., Mansouri F., 1977, Phys. Rev. Lett. **38**, 739

– van Nieuwenhuizen P., 1987, Phys. Rep. **68**, 189

– Sohnius M.F. and West P.C., 1987, Phys. Lett. **B105**, 352

– Tseytlin A.A., 1991, Johns Hopkins Univ. preprint JHU-91009

– Vasiliev M.A., 1987, Fortschr. Phys. **35**, 741

- 1988, Fortschr. Phys. **36**, 33

- 1989a, Ann. Phys. **190**, 59

- 1989b, Nucl. Phys. **B324**, 503

- 1990, Phys. Lett. **B243**, 378

- 1991, Phys. Lett. **B257**, 110

– Witten E., 1991, preprint IASSNS-HEP-91/12

SPACE-TIME AND QUANTUM PHYSICS

Rudolf Haag

II. Institut für Theoretische Physik
Universität Hamburg

Before beginning with the subject of my talk it is fitting to think of the occasion. Congratulations and thanks to Eduardo who, by his perseverence in life and his achievements provided the cause for our meeting in this wonderful setting. I think back to the time in Copenhagen, almost fourty years ago, when I first met Eduardo. We both were then hopeful young scientists, fascinated by the international atmosphere and generosity of Niels Bohr's institute. From that time I want to recall one typical Caianiello saying: "If you happen to have a good idea, stop immediately working on it. Then you can enjoy it for a long time". I had often occasion to experience the truth of this by ignoring the warning.

Turning now to the topic of my talk, let us first take a quick look at the changes which our understanding of space and time has undergone in classical physics. There are three main stages.

1) Newton's mechanics with the notion of absolute space, absolute time. Space is a God given 3-dimensional continuum in which Euclidean geometry applies. It serves as a vessel into which pieces of matter, say particles, are cast. Time is a God given 1-dimensional metric continuum. The vessel (space) is unaffected by the flow of time. Only the configuration of its content changes. The laws governing this change are the subject of the theory.

2) Faraday-Maxwell electrodynamics and its offspring, the special theory of relativity. Every point of space at any time is now thought to be the carrier

of physical properties, the fields. Tied to this picture is the immensely fruitful principle of locality: The laws connect only physical quantities associated with neighboring points. This principle is still sharpened in special relativity. Space and time separately are no longer intrinsically meaningful. We must consider them together as forming a 4-dimensional space-time continuum in which all causal influences propagate within cones ("Minkowski space"). Note that a point particle is now no longer the simplest object. It is a trajectory in Minkowski space, a "world line". The elements of Minkowski space are space-time points, physically speaking they are point events. Let us keep this in mind for later reference.

3) General relativity with its removal of the last global features of the theory, namely the global, a priori given causal and metric structure of Minkowski space. These structures become now also dependent on the physical state and determined by local laws.

Comment. Leibniz rejected Newton's notion of absolute space and maintained that we should not regard space as a vessel with a reality of its own but only as a system of relations between objects such as particles. More pointedly: empty space is an empty concept. This criticism loses some but not all of its relevance in field theory. Only in stage 3 (general relativity) can we regard it as implemented, provided we interpret the diffeomorphisms of space-time together with the associated transformation of the fields not as symmetries, leading from one possible physical situation to a different one but only as an arbitrariness in the description of one and the same physical reality.

The unbroken line of success ensueing from the principle of locality and the field concept opened the vision of a universal field theory in which particles appear only in a secondary rôle as singularity lines or as regions of extremely high field strength. This was Einstein's dream in the last decades of his search. But it remained a dream. History followed a different course. Quantum physics dominated the scene. The study of atomic structure first led to quantum mechanics which brought particles again to the center of the stage. But the behaviour of the subatomic particles was markedly different from that of the billard balls of

42

Newton's mechanics. The principles of quantum mechanics consist first of all of a very specific mathematical formalism of which I only recall some catch words like "non commutative algebra", "Hilbert space". I said "first of all" because it was this structure which was found first by the work of Heisenberg, Schrödinger, Dirac, Born, Jordan. Then followed the long extended debate about its interpretation, its implications for our picture of the physical world. The central figure in shaping the language, the philosophy of quantum mechanics was Niels Bohr. This brings us back to Copenhagen. One talks about the Copenhagen interpretation of quantum mechanics. What is it? Much is subsumed nowadays under this heading to which Bohr might not have completely subscribed. But there are two central themes stressed by Bohr.

(i) the problems of divisibility. We must single out a part of the world, the "physical system" which we want to discuss. In doing so we introduce a subjective division. On the one side there is the "system", on the other the rest of the world which includes the observer and his instruments. The description of the two parts is necessarily different. On the observer side we are forced to use a realistic language, the language of classical physics "because we have to tell our friends what we have done and what we have learnt". We also accept there some features, unexplained by physical theory, such as free will and consciousness of the observer. On the side of the system we should avoid using any realistic picture (see (ii) below). Only by a measurement can we produce a real property related to the system. But this, the measuring result, is not an attribute of the system. It is an attribute of the interaction process between the system and the measuring apparatus. The lack of objective, real properties of an individual system at any particular time between its creation (isolation from the rest) and its response by an interaction with the rest relates again to a problem of divisibility. One cannot cut an "irreducible quantum process" into parts in any intrinsically meaningful way. It also relates to complementarity (see (ii)). Heisenberg discussed the consequences of shifting the cut between system and the rest of the world, considering parts of the measuring instruments as belonging to the system. This can be done to some extent but the cut (mostly called the Heisenberg cut) can not be completely eliminated. Drastically expressed: quantum theory of the universe is an empty notion.

(ii) Complementarity. Truth is like the veiled picture of Sais. We can never obtain full knowledge or give a complete description. To every aspect there is a complementary one. The more we focus on the one the more we loose sight of the other.

Remarks. Space and time are attached to the observer side of the Heisenberg cut and therefore classical concepts. Thus one might say that in quantum mechanics one gets along with Newton's concept of absolute space and time. But there is a difference. We can no longer imagine that a point in space is marked by the position of a particle. It is often said that "we cannot know the position of a particle unless it is measured" and that, due to complementarity such a measurement destroys all knowledge of the momentum. But we should express it in a sharper, more realistic form. It has nothing to do with "knowledge". Position is just no attribute of a particle alone in the world. It is an attribute of an interaction process of two particles, one being part of the system, the other part of the detection device. And then it marks a point (or small region) in space-time; it is an event. Moreover the space-time extension of the event as well as the significant assignment of the collision partner within the detection device involves the energy-momentum transfer. Depending on this we may interpret the event as a collision of the observed particle with a molecule, an atom, a nucleus, a quark... with correspondingly large or small extension in space-time. I find it noteworthy that quantum mechanics also suggests that the notion of an event (ideally a space-time point) is more basic than that of a point in space at some given time.

Now let us come to the incorporation of the principle of locality into quantum physics. It took only a few years from the establishment of quantum mechanics to the beginnings of quantum electrodynamics and, more generally, of relativistic quantum field theories. It was begun by subjecting a classical field theory to the same formal procedure, called "quantization", which was so successful in the step

44

from Newton's mechanics to quantum mechanics. In the simplest case, that of a scalar field satisfying the relativistic wave equation or Klein-Gordon equation, this procedure was a complete success and lead to the miraculous result that the field theory after "quantization" just describes a system of arbitrarily many identical, non-interacting, spinless particles obeying Bose statistics. Similarly, a slightly modified quantization procedure applied to the free electromagnetic field leads to the description of non interacting photons. Still more startlingly, if one considers the free Dirac equation as a classical field theory then a corresponding quantum theory of arbitrarily many non interacting electrons and positrons obeying Fermi statistics emerges by simply replacing in the quantization rules the commutators by anticommutators. The miracle is often regarded as a manifestation of complementarity, a field-particle duality: to every type of (elementary ?) particle one has an associated field and vice versa, to every field a particle type. From this point of view the quantized theory of a coupled Maxwell-Dirac field system is regarded as the theory of interacting photons, electron and positrons. But I do not think that this is a good way to think. Field-particle duality is a feature of simple models and does not catch the essence of quantum field theory.

Let me recall a remark by Heisenberg from a conversation in 1957: "The S-matrix is the roof of the theory, not its foundation". Instead of S-matrix we may equally well substitute "the types of physical particles". We do not know of any clear cut distinction between elementary and composite particles and - apart from the electromagnetic and gravitational field - we rarely encounter any "field aspect" in observations. The relevant point on the second level of quantum physics is the ordering of observables by their placement in space-time, the assignment of a set of observables to each space-time region in such a way that the relativistic causal structure is respected. Space-time is still considered as classical Minkowski space, given a priori without regard to the physical state. But the "physical system" is no longer materially defined as a collection of particles. Instead we might say that a space-time region is now regarded as a system; it is, of course, necessarily an open system. The quantum fields serve to establish the correspondence between regions in space-time and their associated algebras of observables. They are, if one wishes, a coordinatization of the system of local algebras. As a consequence of the principle of locality one finds that, closer to the foundations of the theory than the concept of particles is the charge structure. One has charge quantum numbers with a law of composition, conjugation and a "statistics parameter" determining their exchange symmetry. Besides locality (relativistic causality) there are two further essential properties needed to have a reasonable theory. The first may be called stability. It is usually formulated as the requirement that translations in space-time are a symmetry of the theory and that there exists a translation invariant "ground state", the vacuum, in terms of which energy may be defined. The other is called nuclearity. It demands that in the set of states with energy below some chosen finite value the linear combinations of a finite number of states suffice to describe the behaviour in a finite region of space-time arbitrarily well. It is gratifying that such a scheme (i.e. the assignment of algebras of observables to space-time regions satisfying the mentioned requirements) defines a complete theory including its physical interpretation. It determines the particle types which occur, their collision cross sections, the thermodynamic equilibrium states etc.

It also appears that all qualitative consequences and even a few quantitative ones agree with all present day experience in high energy physics and this will probably persist when the highest energies reached in accelerators are increased in the next decade.

It is, however, also clear that some fundamental change of the picture is necessary if we want to go to the third level and aspire to a synthesis between quantum physics and general relativity. Much work directed towards this goal has been done in the past fifteen years. There are extensive studies of "quantization" of Einstein's theory, starting from various choices of classical variables. There is the observation that a local version of the idea of supersymmetry leads naturally to a theory of spinorial matter fields with gravitational interaction, the so called supergravity. In these approaches on still presupposes a classical $(3+1)$-dimensional space and time. In common with Einstein's theory it is only the metric and causal structure of this continuum which is not given a priori; it is now subject to quantum fluctuations.

A step still further away from the classical picture follows the idea that a quantum structure of space-time should be described by a "non-commutative generalization of differential geometry". Finally we may remember Leibniz' criticism and maintain that we should no longer regard space-time and matter as separate entities but understand the former as a system of relations between possible events. The work on "superstrings" may perhaps be interpreted as an attempt in this direction. Another such scenario, closer to special relativistic quantum physics and the ideas of non-commutative geometry is based on a net of algebras, partially ordered by inclusion, in which the projectors in each algebra are interpreted as possible events. In a coarse grained description the inclusion relation between algebras must then be isomorphic to the inclusion of subsets in a 4-dimensional continuum.

I do not think that any one of the mentioned attempts has come very far yet and luckily my time is over so that I do not have to enlarge on speculations and on my personal opinions about their merits.

SIMPLE SPINORS AND NULL MOMENTUM SPACES

P. Budinich

Summary. Should space-time be conceived as an element of a Clifford algebra, of which simple spinors space S is a minimal left ideal, then (constant) spinor-component anticommutators are naturally obtained. In this frame Weyl and Maxwell, Dirac, pion-nucleon field equations in momentum space (for two-, four-, eight-component spinors, respectively) may be identified with the Cartan equations defining simple spinors and the fields are defined on compact manifolds. The stereographic projections on conformally flat spaces may produce form factors which are identical with the regularizing Pauli-Villars ones. These may be also obtained by subtraction of a mass term from spinor field Lagrangians in momentum space deduced from End $S \to \mathbb{R}$.

1. Introduction

Several of the recent attempts to overcome the difficulties encountered by the theory of elementary particles in recent decades (strings, superstrings, G.U.T., quantum groups, non commutative geometry [1]) have in common the idea of inserting the traditional conception of pseudo-euclidean space-time in a wider contest. The need to enlarge the algorithms adopted for the description of natural phenomena is not new and, in the history of science it has rendered good service for the discovery of new laws of nature. One needs only to remind the imbedding of the fields of numbers from integers to rational-, real-, complex-, quaternionic-numbers and, in geometry, from 3-dimensional Euclidean space to 4-dimensional pseudo-euclidean Minkowski space-time and then to pseudo-Riemannian and higher dimensional spaces. Now also these are being imbedded in wider manifolds and it is the turn of algebraic geometry. Algebras are not commutative in general and, Clifford algebras in particular, are anticommutative. This property is specially attractive for those who cherish the idea of the fundamental role of spinors in the elementary laws of nature. In fact the Clifford algebra $Cl(V, Q)$ corresponding to a vector space V endowed with a quadratic form Q over a field K contains not only the image $\overset{(i)}{(V)}$ of the vector space V and the one $\overset{(o)}{(V)}$ of the field K but also it contains the corresponding spinor space S (in the form of minimal left ideals). The mentioned anticommutativity in turn, if extended to S may then give rise to the hope to throw some light to the quantization problem of spinor fields to start with. Then, if the Cartan definition of simple spinors is adopted, one may envisage the possibility not only of deriving geometrically some of the basic and well established equations of physics like Weyl, Dirac and Maxwell ones in momentum space but also momentum space itself in the projective form of null quadrics, which, being obviously compact, do not admit in principle the embarassing concept of infinity.

In this paper we tentatively try to perform a few steps in this line of research.

In § 2 and 3 we remind some general properties of Clifford algebras and simple spinors relevant for our work. In § 4 we derive the basic anticommuting equations for (constant) simple spinor components. In § 5 we define the concept of spinor fields (on compact momentum space manifolds) based on Cartan equations defining simple spinors and subsequently identify Weyl and Maxwell equation obtained from 2-component spinors. In § 6 and 7 we present some considerations on $\mathbb{R}^{4,2}$ space and some heuristic conjecture based on the hypothesis that ordinary space-time and momentum space are both compactified. In § 8 we show that the minimal Clifford algebras needed to construct ordinary, real, compactified space-time from simple spinors (-bilinear polinomia) is Cl(5,3) for Weyl spinors and Cl(5,1), Cl(7.1) for Dirac spinors, for which, in the corresponding momentum space Dirac equations and pion-nucleon equations presenting SU(2) isotopic spin symmetry are naturally derived. In § 9 it is shown how, from the spinor formalism spinor lagrangians (in momentum space) are obtained which, after subtraction of a mass term (resembling the one adopted in mass renormalization), generate equations on compact manifold which, after stereographic projections, determine a conformal factor identical to the Pauli-Villars one.

2. <u>Clifford algebras</u>

Let V represent a vector space over a field K with a quadratic form Q. For u and v vectors of V a symmetric bilinear form $B(u, v) \in K$ may be defined [2]:

Def. 1 $\qquad B(u, v) = Q(u + v) - Q(u) - Q(v).$ $\hfill (2.1)$

The Clifford algebra Cl(V, Q) may be defined through the linear map f:

$$f : V \rightarrow Cl(V, Q)$$

such that for every vector $u \in V$:

Def. 2 $\qquad f^2(u) = Q(u).$ $\hfill (2.2)$

From these definitions it follows that for u and v vectors of V the anticommutator of their images in Cl(V, Q) equals the bilinear form $B(u, v) \in K$ defined above:

$$B(u, v) = f(u) f(v) + f(v) f(u) \tag{2.3}$$

Therefore, quite generally, to scalar products in V there correspond anticommutators in the corresponding Clifford algebra. In particular, orthogonal vectors in V have null anticommutator in $Cl(V, Q)$.

Let the field $K = \mathbb{R}$ and $V = R^{k,\ell}$ and let $e_1, e_2, \dots e_k, e_{k+1}, \dots e_{k+\ell}$ represent its orthonormal basis. If $\gamma_\mu = f(e_\mu)$, $\mu = 1, 2, \dots k + \ell$ represent their image in $Cl(V, Q) = Cl(k, \ell)$, then:

$$\frac{1}{2} [\gamma_\mu, \gamma_\nu]_+ = g_{\mu\nu} \qquad \mu, \nu = 1, 2, \dots k+\ell . \tag{2.4}$$

where $g_{\mu\nu}$ is diagonal (with k elements equal $+ 1$ and ℓ equal -1). The $k + \ell$ elements γ_μ, basis of $Cl(k,\ell)$ are said to "generate" $Cl(k, \ell)$ since any one of its elements $\omega \in Cl(k, \ell)$ may be expressed in the form of a sum of j-vectors:

$$\omega = \overset{(0)}{V} + \overset{(1)}{V} + \overset{(2)}{V} + \dots \overset{(j)}{V} + \dots \overset{(k+l)}{V} \tag{2.5}$$

where $\overset{(o)}{V} = a_{(o)}$, $\overset{(1)}{V} = a_\mu \gamma^\mu$, $\dots \overset{(j)}{V} = a_{\mu_1 \dots \mu_j} \gamma^{\mu_1} \dots \gamma^{\mu_j}$ with $\mu_1 < \mu_2 \dots < \mu_j$. We can identify V with its image V in $Cl(k,\ell)$.

We will assume $k = \ell = m$ by which $Cl(m, m)$ may be real[*] [3], and we will adopt also the isotropic basis defined by:

$$n_\alpha = \frac{1}{2} (\gamma_{2\alpha-1} - \gamma_{2\alpha}) \; ; \quad p_\alpha = \frac{1}{2} (\gamma_{2\alpha-1} + \gamma_{2\alpha}) , \qquad \alpha = 1, 2 \dots m , \tag{2.6}$$

where we assumed $g_{\mu\nu}$ neutral ($g_{\mu\nu} = +1, -1, +1, \dots -1$).

n_α, p_α obey:

$$[n_\alpha, n_\beta]_+ = 0 = [p_\alpha, p_\beta]_+$$

$$[n_\alpha, p_\beta]_+ = \delta_{\alpha\beta}. \tag{2.7}$$

---

[*] The algebra $Cl(k, \ell)$ is complex in general (remember $Cl(3,0)$ for which $\gamma_1\gamma_2\gamma_3 = i\,1$). The Clifford algebra $Cl(2m)$ of $\mathbb{C}^{2m}$, restricted to the real identifies with $Cl(m, m)$ of $V^{m,m}$.

The null vectors $\{n_1, n_2, \dots n_m\}$ span a maximal totally null (MTN) subspace N of V, while $\{p_1, p_2, \dots p_m\}$ span another MTN subspace P of V and since $N \cap P = \{0\}$ every vector u of V admits a unique decomposition

$$u = n + p$$

where $n \in N$, $p \in P$. If we define

$$n = n_1 \, x_1 + \dots n_m \, x_m \quad ; \quad p = y_1 \, p_1 + \dots y_m \, p_m$$

where x_α, y_α are arbitrary numbers (they may be real), we have

$$Q(u) = x_1 \, y_1 + \dots x_m \, y_m \, . \tag{2.8}$$

The "volume element":

$$\Gamma = \gamma_1 \, \gamma_2 \, \cdots \, \gamma_{2m} = [n_1, p_1]_- \, [n_2, p_2]_- \, \cdots \, [n_m, p_m]_- \tag{2.9}$$

anticommute with every γ_μ, n_α, p_α and, since $\Gamma^2 = 1$, it may be considered representing in $Cl(m, m)$ a unit vector "orthogonal" to V.

3. Simple spinors

Clifford algebras $Cl(V, Q)$ may be represented in linear spaces [2]. Spinors are vectors of the representation space of $Cl(V, Q)$. Let $u \equiv f(u)$ represent the image[*] in $Cl(V, g)$ of a vector $u \in V$. The equation

$$u \, \phi = 0 \tag{3.1}$$

where ϕ is a spinor, defines a totally null subspace $M\,(\phi)$ of V:

$$M\,(\phi) = \{u \in V \mid u \, \phi = 0\}, \tag{3.2}$$

in fact from (3.1) follows $Q(u) = 0$ for $\phi \neq 0$ but also $B(u, v) = 0$ if $v \in V$ is another vector of V satisfying $v \, \phi = 0$.

Def. 3   The spinor ϕ is <u>simple</u> if $M\,(\phi)$ defined by (3.2) is a MTN subspace of V.

According to Cartan [2] the simple spinor-direction (which means a spinor up to an arbitrary, non zero, multiplicative constant factor) is equivalent to the corresponding MTN subspace of V.

---

[*]   In the orthonormal basis (2.4) $u \equiv u_\mu \, \gamma^\mu$ while in the isotropic one (2.6) $u \equiv x_\alpha \, n^\alpha + y_\alpha \, p^\alpha$.

In the case of Cl(m, m) the m-dimensional MTN subspace $M(\phi) \in V$ corresponds to ϕ simple, which must obey

$$\Gamma \phi_{\pm} = \pm \phi_{\pm} \tag{3.3}$$

since Γ is orthogonal to V. $\phi_{\pm}$ are called Weyl spinors of opposite chirality. For Cl(m, m) they may be real (Majorana) [3].

Spinors may be also represented as elements of minimal left ideals *(mli)* of Cl(m, m) [4]. In particular the m-vector

$$\omega = n_m \, n_{m-1} \, \cdots \, n_2 \, n_1 \tag{3.4}$$

represents a MTN plane of $R^{m,\,m}$, and is an element of a *mli* of Cl(m, m), therefore it may be thought as a simple spinor whose defining equation (3.1) is

$$(x_1 \, n_1 + x_2 \, n_2 + \dots x_m \, n_m) \, \omega = 0 \tag{3.5}$$

for arbitrary $x_1, x_2, \dots x_m$ (real or complex). The spinor ω (with positive chirality: $\Gamma = + 1$) was called "standard" by Cartan [2] and may be represented by a 2^{m-1}-component one-column matrix with all zero elements but one (generally the first) equal one.

Let us consider the collection of $2^m \, m \, l \, i$ s :

$$\omega_0 = \omega \; ; \quad \omega_1 = p_1 \omega, \quad \omega_2 = p_2 \, \omega , \dots \omega_m = p_m \, \omega \; ;$$

$$\tag{3.6}$$

$$\omega_{12} = p_1 \, p_2 \, \omega, \dots \quad ; \quad \dots \; \omega_{12 \dots m} = p_1 \, p_2 \cdots p_m \omega.$$

They are all MTN subspaces and may be considered as a basis of spinor space, named Fock basis in ref. [4]. Every spinor may be obtained by acting with the elements of Cl(m, m) on the basis (3.6), it will therefore have the form:

$$\phi = \sum_{k=0}^{m} \sum \zeta_{\alpha_1 \dots \alpha_k} \, \omega_{\alpha_1 \dots \alpha_k} \tag{3.7}$$

For a simple spinor however its 2^{m-1} components may not be independent. In fact, as known, the MTN plane of $R^{m,m}$ depends on m(m-1)/2 parameters, which, for m > 3 is smaller than $2^{m-1}-1$ (number of parameters defining the direction of ϕ).

The constraint equations may be easily obtained by observing that from every element of the basis (3.6), say $\omega_0 = \omega$, every simple spinor ϕ with the same helicity may be obtained by acting on

52

it with the operator s derived from the spin group (spin group divided by the stability group) given by [4]:

$$s = \zeta_0 \prod_{j<k} (1 + \frac{\zeta_{jk}}{\zeta_0} \, p_j \, p_k) \, . \tag{3.8}$$

comparing $\phi = s\omega_0$ with eq.(3.7) we get easily the constraint equations. As an example for m = 4 we obtain the constraint equation:

$$\zeta_0 \, \zeta_{1234} = \zeta_{12} \, \zeta_{34} + \zeta_{13} \, \zeta_{42} + \zeta_{14} \, \zeta_{23} \tag{3.9}$$

which is the Pfaffian of the antisymmetric matrix $\zeta_{\alpha\beta}$ as first underlined by E. Caianiello [5].

4.   <u>Anticommuting spinor components</u>

We have seen how, given a vector space V over a field K endowed with a quadratic form Q, we may associate to it a Clifford algebra Cl(V, Q) containing the images of both K and V, represented by $\overset{(o)}{V}$ and $\overset{(1)}{V}$ of eq.(2.5) respectively. But also spinor space S is contained in Cl(V, Q) in the form $m \, l \, i \, s$, represented by eq.(3.7), say. Now since the map from V to K, or a scalar product in V, is represented in Cl(V, Q) by anticommutators, we may expect the same for inner product in S when $S \subset Cl(V, Q)$. But spinor anticommutators are what is postulated ad hoc when quantizing (II quantization) spinor field theories and, from these anticommutators, both Pauli exclusion principle and, through bilinear spinor polinomia, the standard quantization commutator $[p, q] = i\hbar$ is then deduced. The possibility of deriving spinor anticommutators directly, and quite generally, from Cl(V, Q), would then throw some new light on the possible role of Clifford algebras and spinors in quantum physics.

Let us go back to the Fock basis $\omega_0, \omega_1, \ldots \omega_{jk} \ldots \omega_{12\ldots m}$ of spinor space S given by eqs.(3.6). As matrices its elements may be represented by normal matrices E_{ab} defined by $(E_{ab})_{\alpha\beta} = \delta_{a\alpha} \, \delta_{b\beta}$. As an example for $V^{m,m}$ we have in the Cartan representation [3]:

$$\omega_0 = E_{1,n}$$

with:

$$n = 1 + \sum_{j=0}^{I(m)} 2^{2j} \tag{4.1}$$

and I(m): = I [1/2 (m-1)] is the maximal integer contained in 1/2 (m-1).

Therefore the other elements of the Fock basis (3.6) are represented by:

$$\omega_\rho = E_{\rho+1, n} \tag{4.2}$$

where ρ enumerates the 2^m indices appearing in (3.6), and the general spinor ϕ given by eq.(3.7) is represented by a 2^m x 2^m matrix with only the nth-column with non zero elements.

The concept of inner product may be defined for Clifford algebras and spinor spaces [3]. Noting that $n_\alpha{}^\dagger \equiv {}^t n_\alpha = p_\alpha$ (since n_α have real elements) we may now define the basis of the dual spinor space S^*. We have only to start from the MTN plane $P = p_1 \, p_2 \, \cdots \, p_m = \Pi_0 = {}^t\omega_0$.

$$\Pi_0 = P \; ; \; \Pi_1 = n_1 \, P, \; \Pi_2 = n_2 \, P, \, \ldots \, \Pi_m = n_m \, P;$$

$$\Pi_{12} = n_1 \, n_2 \, P, \, \ldots \; ; \, \ldots \qquad \Pi_{12..m} = n_1 \, n_2 \cdots n_m P. \tag{4.3}$$

The normal matrices will be:

$$\Pi_\rho = E_{n, \rho+1} \tag{4.4}$$

and the dual spinor ϕ^* will be represented by a 2^m x 2^m matrix with only the nth-line with non zero elements.

Let us now consider the "standard" spinor ω_0 and its dual Π_0. Their inner product $<\Pi_0 \, \omega_0>$ equals 1. As $m \, l \, i$ s of $Cl(m, m)$ their anticommutator is

$$[\omega_0 \, \Pi_0]_+ = E_{11} + E_{nn}$$

Therefore

$$\mathrm{Tr} \, \frac{1}{2} \, [\Pi_0, \omega_0]_+ \; = \; <\Pi_0 \, \omega_0> = 1 \tag{4.5}$$

This may be easily generalized to every pair of spinors ϕ and ψ (and to $Cl(k, l)$) (through the matrix B defined by $B \, \gamma_a = {}^t\gamma_a \, B$)

$$\mathrm{Tr} \, \frac{1}{2} [B \, \phi, \, \psi]_+ = B \, (\phi, \psi) \tag{4.6}$$

where $B(\phi, \psi)$ is the standard definition of inner product in spinor space [3] (in the particular case above $B = 1$).

We may now adopt a Lemma due to Chevalley (III, 24 of ref. [2]) by which if $M(\phi)$ and $M(\psi)$ represent the MTN planes corresponding to the simple spinor ϕ and ψ we have that:

$$M(\phi) \cap M(\psi) \neq 0 \text{ if and only if } B \, (\phi, \psi) = 0. \tag{4.7}$$

In our Fock basis (3.6) and its dual (4.3) we have that each pair of elements of each one of the basis have some common element. Therefore:

$$\omega_\rho \cap \omega_\tau \neq \{0\} \neq \Pi_\rho \cap \Pi_\tau \ ,$$

where the MTN planes $M(\omega_\rho)$, $M(\omega_\tau)$, $M(\pi_\rho)$, $M(\pi_\tau)$ are meant for ω_ρ, ω_τ, π_ρ, π_τ respectively.

Therefore because of (4.7) and (4.6)[(*)]

$$[\omega_\rho, \omega_\tau]_+ = 0 = [\Pi_\rho, \Pi_\tau]_+ \tag{4.8}$$

We have instead:

$$\omega_\rho \cap \Pi_\rho = \{0\}$$

while

$$\omega_\rho \cap \Pi_\tau \neq \{0\} \ \text{ for } \rho \neq \tau$$

where again the corresponding MTN planes are meant, therefore:

$$\mathrm{Tr}\,\frac{1}{2}[\omega_\rho, \Pi_\tau]_+ = \delta_{\rho\tau} \tag{4.9}$$

which are the needed anticommutators.

The procedure may easily be extended to $V = R^{m+1,\,m-1}$ including Minkowski space-time; one needs only to substitute γ_2 with $i\gamma_2$ and $^\tau\omega$ with $\omega^\dagger$ in the above computations.

<u>5.</u>   <u>Spinor fields. Wave equations. The example of space-time</u>

Simple spinors defined above are to be considered as constant vectors of spinor space S or constant elements of *mli* s of $Cl(V,Q)$ equivalent to MTN subspaces of V. As an example eq.(3.5) associates to the standard (constant) simple spinor $\omega = n_m\, n_{m-1} \dots n_2\, n_1$ the m-dimensional null plane N. A generic vector of N is a linear combination:

$$v_{(\omega)} = x_1\, n_1 + x_2\, n_2 + \dots x_m\, n_m \ ; \tag{5.1}$$

with arbitrary coefficient $x_1, x_2 \dots x_m$, of the m independent null vectors $n_1, n_2, \dots n_m$.

[(*)]   All the elements of the Fock basis (normal matrices $E_{\alpha,n}$) are nil potent but one normal matrix $E_{n,n}$ which is idempotent, this corresponds to the fact that spinor directions are determined by all but one of the spinor components and the one may have arbitrary value.

We may however transform equation (3.5) written in the form $v_{(\omega)} \, \omega = 0$ with the operator s given in (3.8), representing the result of a rotation in V (which determines the parameters ζ_{jk}/ζ_0), and considering ω a spinor of S, we obtain after rotation:

$$s \, v_{(\omega)} \, s^{-1} \, s \, \omega \; = \; p_\mu \, \gamma^\mu \, \psi_+ = 0 \tag{5.2}$$

where p_μ is a null vector ($\in$ MTN plane $M(\psi_+)$) of V, and $\psi_+ = 1/2 \, (1 + \Gamma) \, \psi$ is a simple Weyl spinor of positive chirality (ψ is a Dirac spinor [2]) whose components will be ζ_0, ζ_{jk} (subject, for $m > 3$ to constraint equations).

We may now introduce also another interpretation of simple spinors. Consider the null vector p with components p_μ as an arbitrary null vector and then eq.(5.2) defines a space of spinor fields ψ_+ (p) as functions of p; that is as simple-spinor <u>fields</u> on the null cone of V. The null vector p is the common intersection of all possible MTN planes $M(\psi_+(p))$.

As an example in $\mathbb{R}^{3,1}$ we have

$$s \, (x_1 n_1 + x_2 n_2) \, s^{-1} \, s \, n_2 n_1 = p_\mu \, \gamma^\mu \, (1 + \gamma_5) \, \psi = p_\mu \, \sigma^\mu \, \psi_+ = 0 \tag{5.3}$$

where $\sigma_\mu = \{\boldsymbol{\sigma}, 1\}$, $\boldsymbol{\sigma} = \{\sigma_1, \sigma_2, \sigma_3\}$ are Pauli matrices and $\psi_+ = \begin{pmatrix} \zeta_0 \\ \zeta_{12} \end{pmatrix}$ is a space-time Weyl spinor. Eq.(5.3) may be solved and we have:

$$\frac{\zeta_{12}}{\zeta_0} = \frac{p_1 + ip_2}{p_0 - p_3} = \frac{\pi_1 + i\pi_2}{1 - \pi_3} \; , \tag{5.4}$$

(where $\pi_j = p_j/p_0$) which means that the spinor direction $\zeta_{12}/\zeta_0 = \tau$ is determined by the stereographic projection of the S_2 sphere

$$S_2 \; : \qquad \pi_1^2 + \pi_2^2 + \pi_3^2 = 1 \tag{5.5}$$

on the complex plane where τ represents the complex variable. If we interpret $V = \mathbb{R}^{3,1}$ as Minkowski space then the sphere S_2 is the celestial sphere as well known [6].

However we could as well interpret $V = \mathbb{R}^{3,1}$ as momentum space (and p_μ as momentum of a massless neutrino or as Poynting vector of a light ray), and in this case eq.(5.3) acquires the physical meaning of the Weyl equation for a massless neutrino of positive chirality in momentum space[*] and $\psi_+ = \psi_+$ (p) e^{ipx} represent spinor field, elementary solution of the standard Weyl

---

[*] Eq.(5.3) in the form $\vec{p} \cdot \vec{\sigma} \, \psi_+ = p_0 \, \psi_+$ may be interpreted as an eigenvalue equation. This could allow to extend the anticommutators (4.9) to $\mathrm{Tr} \, 1/2 \, [\omega_\rho \, (p), \, \Pi_\tau(p')]_+ = \delta_{\rho\tau} \, \delta_{pp'}$, which is the starting point for spinor fields in II quantization.

56

equation $\partial \psi_+ = 0$. $\psi_+(p)$ is a spinor field, whose direction is determined at each point τ of the complex plane, stereographic projection of S_2.

Both interpretations are legitimate. The first gives rise to optical geometry [7], the second to standard field equations.

Observe that in both cases equation (3.1) or (5.2) refers only to a subspace of V represented by its light cone and this subspace is invariant for the automorphism groups of V e.g. for rotations which in the subspace induce conformal transformations.

We could even conceive these subspaces as "defined" by the simple spinor fields. In fact we have (see proposition 7 of ref. [4]):

<u>Proposition 5.1:</u>   The vector

$$p_\mu = B \, (\phi, \, \gamma_\mu \, \psi) \tag{5.6}$$

is a null-vector of V if and only if ψ is simple (while ϕ arbitrary).

Therefore given a simple spinor ψ we may generate the null vector p_μ of the null cone of V in the form of bilinear spinor polinomia. It may be shown [4] that in this way all vectors of the null cone of V may be generated, and through eq.(3.1) or (5.2) the simple spinor field $\psi(p)$ on the manifold.

<u>5.1   Maxwell equations</u>

Let us now consider $\omega = n_m \, n_{m-1} \, ... \, n_2 \, n_1$ as an m-vector element of a *mli* of Cl(V, g), then the result of transformation of eq.(3.5) with s will be:

$$s \, v_{(\omega)} \, s^{-1} \, s \, \omega \, s^{-1} = p_\mu \, \gamma^\mu \, F_m^+ \tag{5.3'}$$

where $F_m^+ = (1 + \Gamma) \, \Sigma \, F_{\mu_1 \mu_2 .. \mu_m} \, \gamma^{[\mu_1} \gamma^{\mu_2} ... \gamma^{\mu_m]}$, sum extended to all antisimmetrized products of $\gamma^{\mu_1} ... \gamma^{\mu_m}$, represents the rotated m-vector ω in Cl(V, Q) whose components will depend on $\zeta_0 \, \zeta_{jk}$ appearing in s (and components of ψ_+ in eq.(5.3)).

In the considered space-time example, after a few easy steps one may uniquely derive from (5.3') two of the vacuum-Maxwell equations in momentum space [8]:

$$p_\mu \, F_+^{\mu\nu} = 0$$

the other two would be obtained from $s\,\omega_1\,s^{-1} = F_2^-$. Again, if p_μ is a null vector of momentum space, F_2^+ and F_2^- represent the e.m. field on the light cone (plane wave solutions of $DF = 0$)[(*)].

<u>6.</u>   <u>The space $\mathbb{R}^{4,2}$</u>

Equation (3.1) adopted by Cartan [2] for the definition of simple spinors allows an interpretation which brings to the concept of spinor fields on null cones. The fact that for space-time eq.(3.1) implies both Weyl and Maxwell field-equations (in momentum space) induces to explore the extension of the above study to $\mathbb{R}^{4,2}$.

Again, as before, both coordinate space and "momentum" spaces should be considered on equal footing. However, while in space-time $\mathbb{R}^{3,1}$ the meaning of momentum space is obvious, for the conformal extensions $\mathbb{R}^{4,2}$ (and $\mathbb{R}^{5,3}$) of space-time momentum space definition deserves a further analysis. For the moment we adopt the concept of "wave-number space" which derives from the definition of Fourier transforms of functions taking values in these spaces.

Let us start from $\mathbb{R}^{4,2}$, and let $\eta_1, \dots \eta_6$ represent the components of a null ray in an orthonormal reference system. It is known that on the projective quadric defined by Q: $\{\eta \in \mathbb{R}^{4,2}/$ $\eta^2 = 0\}$, modulo rescaling, Minkowski space-time $\mathbb{R}^{3,1}$ is densely represented; its coordinates x_μ may be represented e.g. [9] by:

$$x_\mu = \frac{2\eta_\mu}{\eta_5 + \eta_6}\,R \quad ; \quad \eta_5 + \eta_6 \neq 0 \tag{6.1}$$

where R is a length. The projective quadric PQ is compact, diffeomorphic to $S^3 \otimes S^1/Z_2$. Adding to $M^{3,1}$ represented by (6.1) the 3-dimensional intersections ζ of PQ with $\eta_5 + \eta_6 = 0$ (light cone at infinity [6]) compactified space-time $\overline{M}^{3,1}$ is obtained. S0(4,2) transformations in $\mathbb{R}^{4,2}$ induce in $M^{3,1}$ conformal transformations which constitute the largest automorphism group preserving the light cone [10], in $\mathbb{R}^{3,1}$ [(**)].

[(*)]   It would be interesting to explore the possible physical meaning of eq.(5.3') for m = 3, 4.

[(**)]   Observe that Q is invariant for S0(4,2) therefore, if our space-time is effectively $\overline{M}^{3,1}$ we would not dispose of any transformation to go out of it and to explore the two extra dimensions. In particular Maxwell equations in vacuum, which are strictly conformally covariant have to show on Q their manifestely covariant form. The study of their interaction with matter fields, and the requirement of its self consistency, could teach us somethig on the role of the conformal group in physics in general, similarly to what happened with the discovery of special relativity.

58

Among the conformal transformations of special interest is the conformal inversion:

$$x_\mu \rightarrow \frac{x_\mu}{x^2} \, R^2 \tag{6.2}$$

corresponding to the $\eta_5 \rightarrow -\eta_5$ in $R^{4,2}$. Then it maps every point inside of the hyperboloid $x^2 = 1$ with a point outside it. In particular it maps the origin to infinity, and, in $\overline{M}^{3,1}$, to the light cone at infinity ζ.

In the definition represented by (6.1), $M^{3,1}$ results as the stereographic projection of $S^3 \otimes S^1/Z_2$ on the pseudoeuclidean flat space $\mathbb{R}^{3,1}$. Consequently we may derive from (6.1) the conformal factor to appear in front of the invariant metric form $dx_\mu \, dx^\mu$:

$$ds^2 = \Omega_x^2 \, dx_\mu \, dx^\mu$$

where

$$\Omega_x^2 = \frac{R^2}{(R^2 + x^2)^2 + R^2 x_0^2}$$

and x_0 is the time component of x. Therefore the volume element dV in $\mathbb{R}^{3,1}$ will be:

$$dV_x = \Omega_x^4 \, d^4 x$$

Observe that $\Omega_x \rightarrow 0$ for $x_\mu \rightarrow \infty$ therefore a field theory in such a space should be infrared (but not ultraviolet) convergent even for massless fields. Alternatively, a quantized system in $S^3 \otimes S^1/Z_2$ may be expected to present a non-zero lower limit for linear momentum but not an upper. It is difficult to reconcile such assymmetrical behaviour with the general properties of conformal covariance which implies dilatation and scale invariance.

This difficulty may be overcome if we impose, as suggested by the spinor approach, that also momentum space is obtained from the null cone of $\mathbb{R}^{4,2}$ "wave number" space. The problem of deriving ordinary momentum space from Fourier transforms in $\mathbb{R}^{4,2}$ restricted to the null cone, and define accordingly field theories is under study [11], for the moment we will present some heuristic considerations on the subject.

7. <u>Compactified ordinary and momentum space from $\mathbb{R}^{4,2}$</u>

Let $\mathbb{R}^{4,2}$ represent "wave number" space and $\pi_1, \dots \pi_6$ represent the component of a null vector building up the quadric:

$$Q': \{\pi \in \mathbb{R}^{4,2} \; ; \; \pi^2 = \vec{\pi}^2 - \pi_0^2 + \pi_5^2 - \pi_6^2 = 0\} \; , \tag{7.1}$$

in the projecting version of it a (compactified) $\Pi^{3,1}$ "wave number-space" defined by:

$$\rho_\mu = \frac{2\pi_\mu}{\pi_5 + \pi_6} \; , \quad \pi_5 + \pi_6 \neq 0 \tag{7.2}$$

is densely contained.

Again PQ' is compact and diffeomorphic to $S^3 \otimes S^1/Z_2$.

Let us now suppose that η_a ($a = 1, 2, 3, 0, 5, 6$) have the dimension of a length l, therefore π_a have the dimension of l^{-1}. Suppose the two $S^3 \otimes S^1/Z_2$ to have the form:

$$X : \{\eta \in \mathbb{R}^{4,2} \; ; \; \eta_1^2 + \eta_2^2 + \eta_3^2 + \eta_5^2 = R^2 = \eta_0^2 + \eta_6^2\} \tag{7.3}$$

$$Y : \{\pi \in \mathbb{R}^{4,2} \; ; \; \pi_1^2 + \pi_2^2 + \pi_3^2 + \pi_5^2 = \Pi^2 = \pi_0^2 + \pi_6^2\} \tag{7.4}$$

where R and Π are the common radiuses of the spheres S^3 and S^1.

For R and Π very large we may expect that the Fourier transform induced on X and Y will be quasi identical with the standard one in ordinary Minkowski space-time almost everywhere except in the neighbourhood of the origin and infinity. In fact because of the maximal radius R of the compact manifold X we may expect that Y, the wave number space, will present a minimal radius $\pi_0 \sim 1/R$ (in this region the Fourier integral will be substituted by a series) and viceversa the manifold X will have a minimal radius $r_0 \sim 1/\Pi$.

The conformal inversion (6.2) in $\mathbb{R}^{3,1}$ spaces, implying:

$$x^2 \to \frac{x^2}{x^4} = \frac{1}{x^2}$$

(x dimensionless) will in particular induce in coordinate space

$$\frac{r_0}{R} \to \frac{R}{r_0}$$

or, since $r_0 \sim 1/\Pi$ and $R \sim 1/\pi_0$

$$r_0 \, \pi_0 \to R \, \Pi \qquad (7.5)$$

and the same in wave number space. In this way the scale-symmetry between X and Y is re-established. We have that compactification eliminates both the infinity and the infinitesimal in both spaces.

Obviously we have to expect[*] $r_0 \, \pi_0 \ll 1$ and $R \, \Pi \gg 1$.

For physical applications we need linear momenta which are obtained by multiplying π_a by an action H:

$$p_a = H \, \pi_a \qquad (7.6)$$

therefore the conformal inversion represented by (7.5) will become:

$$\frac{r_0 \, p_0}{H} \to \frac{R \, P}{H} \qquad (7.7)$$

where $p_0 = H \, \pi_0$ and $P = H \, \Pi$.

As mentioned the conformal inversion maps the points internal to a 2-sphere of radius 1 ($x^2 = 1$ for $x_0 = 0$) to the external ones. Radius one, in our case, is defined by

$$\frac{r \, p}{H} = 1 \qquad (7.8)$$

or if we identify p with m c, and H with the Planck action h then (7.8) defines $r = \lambda$ Compton wave length:

$$\frac{\lambda \, m \, c}{h} = 1 \qquad (7.9)$$

and the internal sphere means: when distances r are smaller than the Compton wave length $r \ll \lambda = h/mc$ (and quantum mechanics is valid) while external means $r \gg \lambda = h/mc$ (when classical

---

[*]   If we take R = radius of the universe ($\sim 10^{10}$ light years) and Π = Planck mass/h then: $R\Pi \sim 10^{62}$, $r_0 \, \pi_0 \sim 10^{-62}$.

mechanics is valid). In this way the conformal inversion would acquire a suggestive physical meaning. In particular Fourier transforms in $\mathbb{R}^{4,2}$ will induce in $\mathbb{R}^{3,1}$ spaces densely contained in X and Y transforms quasi-identical to the standard ones adopted in physics, with the exception of a neighborhood of the origin ($r \ll \lambda$) where the Fourier integral will have to be substituted by the discrete Fourier series defined by the compact manifolds[*].

If we admit that $h\,\pi_a = p_a$ represent momenta then we may interpret eq.(7.2) to represent ordinary linear momentum space $P^{3,1}$:

$$p_\mu = \frac{2\,\pi_\mu}{\pi_5 + \pi_6}\; M \; ; \quad \pi_5 + \pi_6 \neq 0 \tag{7.10}$$

where M (times c) represent a unit of mass. The resulting space will be conformally flat with conformal factor[**]

$$\Omega_p^2 = \frac{M^2}{(M^2 + p^2) + M^2 p_0^2}\quad , \tag{7.11}$$

where p_0 is the fourth component of p . Ω^4 will appear in front of the volume element in momentum space:

$$d\,V_p = \Omega_p{}^4\; d^4\,p \tag{7.12}$$

In this way, if we assume that, as suggested by conformal covariance of Maxwell equations, space-time is conformally flat, and that, as suggested by simple spinor theory, both ordinary and momentum space have to be taken on equal footing and therefore that both are conformally flat, then we should expect both infrared and ultraviolet regularization for field theories in these spaces[***], which is reasonable for a property originating from conformal covariance where, in principle, there is no distinction between large and small.

---

[*] It is interesting to observe that 0(4,2) in momentum space generates the dynamical group of the H atom [12] therefore its quantum numbers will presumably label the indices of the Fourier series in x-space.

[**] Observe that for $\pi_6 \neq 0$ we may also express p_μ given by (7.10) in the form: $p_\mu = 2t_\mu\,(1+ t_5)^{-1} M \; ; \quad 1 + t_5 \neq 0$ where $t_j = \pi_j/\pi_6$; ($j = 1, 2, 3, 0, 5$). In this case the conformal factor takes the form $\Omega_p = M^2\,(M^2 + p^2)^{-1}$ which is identical with the Pauli-Villar regularization factor.

[***] If this is an intrinsic property of these spaces, than any field theory, even if only relativistically covariant, should present these properties, since it would be a property affecting all integrations performed in these spaces.

A conformally flat space contains the light cone which, by itself, as projective manifold, is compact.

In turn, the conformally flat space (or the corresponding compact manifold) may be considered as light cone (projective) of a higher-dimensional space which in turn can be conformally flat. And so on.

It will be of interest to investigate the structure of Cartan- and Dirac-equations in these higher-dimensional spaces since their properties might manifest themselves even at lower dimensions and in particular in space-time[*].

8. Higher-dimensional spaces

8.1 Weyl spinors

According to Proposition 5.1, for $\psi_\pm = 1/2 \, (1 \pm \gamma_7) \, \Psi$, Weyl spinors or twistors, associated with $Cl_0(4,2)$, the 6-vector

$$p_a = B \, (\Psi^c \, \gamma_a \, \psi_\pm) \qquad a = 1, 2, 3, 0, 5, 6 \tag{8.1}$$

where Ψ^c is charge conjugate of the Dirac spinor Ψ, is null in $\mathbb{R}^{4,2}$.

However p_a are complex [13] in fact we have the corollary to Prop. 5.1:

<u>Corollary 8.1</u>:   The null vectors of $\mathbb{R}^{m+1, \, m-1}$ with components:

$$p_A^{(\pm)} = B(\Psi^c, \gamma_A \, \psi_\pm) \qquad A = 1, 2 \dots 2m \tag{8.2}$$

where Ψ is a Dirac spinor associated with $Cl(m+1, m-1)$ and Ψ^c its charge conjugate, and $\psi_\pm = 1/2 \, (1 \pm \gamma) \, \Psi$ simple, is real (or imaginary) for m even and complex for m odd. In the latter case $p_A^{(+)}$ and $p_A^{(-)}$ are complex conjugate. Therefore we have:

<u>Proposition 8.1</u>   The minimal conformal extension $\mathbb{R}^{m+1, \, m-1}$ of space-time to allow the spinorial generation of compactified $\mathbb{R}^{3,1}$ in the form of bilinear spinor polinomia is $\mathbb{R}^{5,3}$.

Therefore p_A given by (8.2) for m = 4 are real and null in $\mathbb{R}^{5,3}$. They are rays of a light cone where $\mathbb{R}^{4,2}$ is densely contained. $\mathbb{R}^{3,1}$ is contained in turn in its projective light cone.

---

[*]   In the same way as Dirac equation gives rise to spin in the Schroedinger non-relativistic theory of the electron.

This conformal-spinorial generation of $\mathbb{R}^{3,1}$ from Cl(5,3)-simple spinors has consequences of interest for physics since, as known, $\mathbb{R}^{5,3}$ presents interesting symmetry properties as triality discovered by Cartan. They will be analyzed in a separate paper.

These constructions priviledges Weyl spinors which imply parity violation. And in fact they seem to have a central role in elementary physical phenomena as weak interactions- and neutrino physics seem to indicate. However Dirac spinors also seem to have important roles when mass is involved. They allow another genesis of null rays and corresponding compact manifolds.

8.2   <u>Dirac spinors</u>

We have another corollary to proposition 5.1:

<u>Corollary 8.2</u>:   Let Ψ represent a Dirac spinor associated with Cl(2m-1, 1) such that $\psi_\pm = 1/2$ $(1 \pm \Gamma)\Psi$ are simple. Then

$$p_A = \Psi^\dagger \gamma_0 \gamma_A \Psi \qquad A = 1, 2 \ldots 2(m+1) \qquad (8.3)$$

with $\gamma_A = \{\gamma_1, \gamma_2 \ldots \gamma_{2m}, \gamma, i1\}$ are the real components of a null vector of $\mathbb{R}^{2m+1,\,1}$, the corresponding compact manifold is S^{2m}.

This corollary is based on the isomorphisms $Cl_0(2m) \sim Cl(2m-1)$ and $Cl_0(2m-1) \sim Cl(2m-2)$ [3]

As an example for $m = 2$ we have the real null vector in $\mathbb{R}^{5,1}$:

$$p_a = \{\tilde{\psi}\,\gamma_\mu\,\psi,\ \tilde{\psi}\,\gamma_5\,\psi,\ i\tilde{\psi}\,\psi\} \qquad (8.4)$$

where ψ is a Dirac spinor of Minkowski space-time and $\tilde{\psi} = \psi^\dagger \gamma_0$.

For ψ Majorana spinor $\tilde{\psi}\,\psi = 0$ and the Cartan-eq. reduces to

$$(p_\mu \gamma^\mu \gamma_5 + p_5)\psi = 0 \qquad (8.5)$$

with $p_\mu p^\mu = -p_5^2$ which may be interpreted as the standard Dirac equation in momentum space with $p_5 = m$.

For $m = 4$ we obtain:

64

$$(p_\mu \, 1 \otimes \gamma_\mu + \vec{\pi} \cdot \vec{\sigma} \otimes \gamma_5 + iM)\, N \; = \; 0 \tag{8.6}$$

where $N = \begin{pmatrix} \psi_1 \\ \psi_2 \end{pmatrix}$ is a doublet of Dirac spinors, and:

$$\vec{\pi} = \frac{1}{4}\, N^\dagger\, \vec{\sigma} \otimes \gamma_5\, N$$

$$iM \; = \; i(\tilde{\psi}_1\, \psi_1 + \tilde{\psi}_2\, \psi_2)$$

with $N^\dagger = |\tilde{\psi}_1\ \tilde{\psi}_2|$.

Eq.(8.6) represents the standard Dirac equation for the nucleon doublet N in SU(2) covariant interaction with the pseudo-scalar pion π in momentum space[(*)].

9. Lagrangians

The geometrical structure of Lagrangian densities (in momentum space) may also be derived from simple spinor geometry. In fact given two spinors ψ and ϕ associated with $Cl(2m) = End\ S$, there is the decomposition [4]:

$$\psi \otimes B\, \phi \; = \; \sum_{j=0}^{m} \overset{(j)}{V} \tag{9.1}$$

where the j-vectors are those appearing in eq.(2.5) and, for $\phi = \psi$ simple, the sum reduces to one term: $\overset{(m)}{V}$.

In particular for $Cl(2m-1, 1)$ and $B\phi = \psi^\dagger \gamma_0 = \tilde{\psi}$ and we may derive from (9.1)

$$\frac{1}{2^{m-2}}\, (\tilde{\psi}\,\psi)^2 = L = \tilde{\psi}\, (P_a\, \gamma^a + P_{2m-1}\, \gamma^{2m-1} + i\, M)\, \psi, \quad a = 1, 2, \dots 2m-2 \tag{9.2}$$

where

$$P_A \; = \; \frac{1}{2^{m-1}}\, \tilde{\psi}\, \gamma_A\, \psi \qquad A = 1, 2, \dots 2m \tag{9.3}$$

[(*)] The γ_5-matrix appearing in eq.(8.6), and the consequent pseudoscalar nature of $\pi_\pm = 1/2\,(\pi_1 \pm i\pi_2)\, \pi_0$, building up $\vec{\pi}$, derive from the construction of N as a doublet of Dirac spinors which imply the following choice for the $Cl(5,1)$ generators $\Gamma_\mu = \gamma_\mu \otimes 1$, $\Gamma_5 = \gamma_5 \otimes \sigma_1$, $\Gamma_6 = \gamma_5 \otimes \sigma_2$. .

and ψ is a Dirac spinor associated with Cl(2m-3,1).

Eq. (9.2) has the structure of a spinor Lagrangian density (in momentum space). Through the subtraction of a term δm having the form of a mass term:

$$\delta m = \alpha \, \tilde{\psi} \, \psi$$

with $\alpha = 2 \, iM$ we obtain a Lagrangian density for a spinor field on a compact manifold implying, as we have seen, ultraviolet convergence. This reminds the procedure of mass renormalization.

9.1 Example of space-time

Let ψ represent a Dirac spinor of Cl(3,1) and let us set in (9.1) $B\phi = \psi^\dagger = \tilde{\psi}$. We obtain from it the real number:

$$\tilde{\psi} \, (\psi \otimes \tilde{\psi}) \, \psi = \tilde{\psi} \, [iM + P_\mu \, \gamma^\mu + P_5 \, \gamma^5 + f_{\mu\nu} \, [\gamma^\mu, \gamma^\nu]_- + P_{\mu 5} \, \gamma_\mu \, \gamma^5] \, \psi \, , \qquad (9.4)$$

where $iM = 1/4 \, \tilde{\psi} \, \psi$, which has the form of a Lagrangian density in momentum space. We easily obtain in accordance with (9.2):

$$L = \frac{1}{2} \, (\tilde{\psi} \, \psi)^2 = \tilde{\psi} \, (iM + P_\mu \, \gamma^\mu + P_5 \, \gamma^5) \, . \qquad (9.5)$$

If we now operate the subtraction:

$$L - 2 \, \tilde{\psi} \, iM \, \psi = L_0 = \tilde{\psi} \, (-iM + P_\mu \, \gamma^\mu + P_5 \, \gamma^5) \, \psi \, , \qquad (9.6)$$

L_0 defines the equation for the spinor field $\psi(P)$:

$$(P_\mu \, \gamma^\mu + P_5 \, \gamma^5 - iM) \, \psi \, (P) = 0. \qquad (9.7)$$

As we have seen $\psi(P)$ is defined on the null cone $P_\mu P^\mu + P_5^2 + M^2 = 0$ diffeomorphic to S^4. Through the stereographic projection:

$$p_\mu = \frac{2 P_\mu}{P_5 + iM} \, M$$

we obtain, as before, the conformally flat momentum space $\Pi^{3,1}$ with conformal factor[*]:

$$\Omega = \frac{iM^2}{M^2 + p^2}$$

(9.8)

of the Pauli-Villars form, which will induce ultraviolet convergence.

10. Conclusion

We have seen how simple-spinor-genesis of null-vectors and corresponding compact manifolds may give rise, through the identification of the equations defining simple spinors with some of the well-known equations of physics (Weyl, Maxwell, Dirac, etc.) to the explanation of some of their properties, which then would simply be the manifestation of the intimate non trivial structure of space-time and of the larger algebraic geometries in which it is immersed and from which our intuitive, but not exhaustive, conception of pseudo-euclidean Minkowski space-time (non compact) may be derived. Other aspects of simple-spinor geometry in eight-dimensional spaces of possible relevance for physics and the role of Dirac versus Weyl spinor geometry will be exposed elsewhere.

Acknowledgements

The author wishes to thank Professors A. Barut, L. Dabrowski, G.F. Dell'Antonio, and R. Raczka for helpful discussions.

---

[*]　Alternatively, the conformally flat momentum space is $\Pi^{1,3}$ and the conformal factor $M^2/(M^2 - p^2)$.

R E F E R E N C E S

[1]   A. Connes, Geometrie non Commutative; Inter Editors, Paris, 1990.

[2]   There are excellent expositions on Clifford algebras and spinors in many books; see, for example:
E. Cartan, Lecons sur la Théorie des Spineurs; Hermann, Paris, 1938.
C. Chevalley, The Algebraic Theory of Spinors; Columbia U.P. New York, 1954.
I.R. Porteous, Topological Geometry; Cambridge U.P. Cambridge, 1981.
R. Penrose and W. Rindler, Spinors and Space-Time; Cambridge U.P. Cambridge, 1986.
I.M. Benn and R.W. Tucker, An Introduction to Spinors and Geometry; Hilyer, Bristol, 1987.

[3]   P. Budinich and A. Trautman, The Spinorial Chessboard; Trieste Notes in Physics, Springer, Berlin, 1988.

[4]   P. Budinich and A. Trautman, J. Math. Phys. **30**, 2125, 1989.

[5]   E.R. Caianiello and A. Giovannini, Lett. Nuovo Cimento **34**, 301, 1982.

[6]   R. Penrose and W. Rindler, Spinors and Space-time; Cambridge U.P. Cambridge, 1986.

[7]   I. Robinson and A. Trautman, Proceedings of the Conference on New Theories in Physics, Kazimiers 1988; World Scientific, Singapore, 1989.

[8]   P. Budinich, Symmetry in Nature (in honour of L.A. Radiacati di Brozolo) pg. 141; Scuola Normale Superiore, Pisa, 1989. A. Barut article in this volume.

[9]   P.A.M. Dirac, Annals of Mathematics, **37**, 429, 1936.

[10]   M. Flato and D. Sternheimer, Comptes Rendus Acad. Sc., Paris, **263**, A936, 1963.

[11]   P. Budinich and R. Raczka, in preparation.

[12]   A.O. Barut and G.L. Bornzin, J. Math. Phys. 15, 1000 (1974) and references therein.

[13]   P. Budinich, The Running Seminar on Spinors; Lett. n. 14, SISSA 1991.

Construction of Pure Spinors and Neutrinos from the Electromagnetic Field

A.O. BARUT

Department of Physics, University of Colorado, Boulder, CO 80309

We construct from the square root of certain combinations of the electromagnetic fields E and B pure Cartan spionors ζ_L and ζ_R satisfying the Weyl equations. Thus neutrinos are in some sense the "square roots" of light.

Eduardo Caianiello has always been deeply interested in the spinor structure of fundamental particles.[1] In this spirit I consider the connection between the Maxwell and Weyl fields and Cartan's pure spinors.

§1. There are many discussions of the spinor form of Maxwell's equations, beginning with the early work of Darwin[2] in 1928. In all these studies[3-14] the components of E and B (or, of $E + iB$) are related to the components of ψ_α of a spinor. This is not what we shall consider in this work, because this connection seems to be somewhat formal: The equations have the form of a Dirac equation, but the transformation properties of ψ are determined by those of E and B, their spinor character is not clear. Rather I analyse another connection here in which the spinors appear as the "square root" of the electromagnetic field, in fact automatically as Cartan's pure spinors.[15]

§2. Physically we shall apply this connection to a relationship between light and neutrinos. Again this will be different than the longstanding idea of the "neutrino theory of light" going back to de Broglie,[16] right after the neutrino hypothesis of Pauli, and later discussed by many others.[17-24] Here photons are thought to be the bound states of ν and $\bar{\nu}$. This idea has not been conclusively developed in to a theory, as there are difficulties with spin-statistics, and with the fact that photons interact strongly with electrons and neutrinos do not. Rather, in our case, the neutrinos appear as some very special configurations of the electromagnetic field which we shall now determine.

§3. Recall that according to Cartan,[25] isotropic or null vectors in a 3-dimensional complex vector space $\not\!\!C^3$ can be related to 2-dimensional pure spinors as follows. From

$$z_1^2 + z_2^2 + z_3^2 = 0 \; , \; z_i \varepsilon \; \not\!\!C^3 \tag{3}$$

the pure spinor $\zeta = \begin{pmatrix} \zeta_0 \\ \zeta_1 \end{pmatrix}$ is obtained by

$$z_1 = \frac{1}{2}\left(\zeta_1^2 - \zeta_0^2\right), z_2 = -\frac{i}{2}\left(\zeta_1^2 + \zeta_0^2\right), z_3 = \zeta_0\zeta_1 \tag{2}$$

or,

$$\zeta_0 = \pm\sqrt{-z_1 + iz_2} \; , \; \zeta_1 = \pm\sqrt{z_1 + iz_2} \tag{3}$$

The components ζ_0 and ζ_1 are linearly independent. The spinor ζ is oriented, i.e., signs must be chosen in (3) to specify it completely. The above correspondence can be expressed succinctly by the spinor equation

$$(\boldsymbol{\sigma} \cdot \boldsymbol{z})\zeta = 0 \tag{4}$$

Now it is clear that for free (plane) electromagnetic fields $\boldsymbol{E}$ and $\boldsymbol{B}$ we have $\boldsymbol{E} \cdot \boldsymbol{B} = 0$ and $|\boldsymbol{E}| = |\boldsymbol{B}|(c = 1)$, hence the vectors

$$\boldsymbol{z}_{\substack{L \\ R}} = \boldsymbol{E} \mp i\boldsymbol{B} \tag{5}$$

are null vectors in $\not\!\!C^3$. Consequently the equations

$$(\boldsymbol{\sigma} \cdot \boldsymbol{z}_{\substack{L \\ R}})\zeta_{\substack{L \\ R}} = 0 \tag{6}$$

hold. Maxwell's equations, on the other hand, can be written as

$$\sigma_\mu \partial^\mu (\boldsymbol{\sigma} \cdot \boldsymbol{z}_L) = 0 \; , \; \tilde{\sigma}_\mu \partial^\mu (\boldsymbol{\sigma} \cdot \boldsymbol{z}_R) = 0 \tag{7}$$

where

$$\sigma_\mu = (I, \boldsymbol{\sigma}), \; \tilde{\sigma}_\mu = (I, -\boldsymbol{\sigma}) \equiv \sigma^\mu$$

But the ζ's have the spinorial property, and not the z's, and are given by

$$\begin{aligned} \zeta_{0\,\substack{L \\ R}} &= \pm\left[-(E_1 \mp iB_1) + i(E_2 \mp iB_2)\right]^{1/2} \\ \zeta_{1\,\substack{L \\ R}} &= \pm\left[(E_1 \mp iB_1) + i(E_2 \mp iB_2)\right]^{1/2} \end{aligned} \tag{8}$$

keeping in mind that $E_3 \mp iB_3 = \zeta_{0\,\substack{L \\ R}}\zeta_{1\,\substack{L \\ R}}$.

We wish to obtain an equation for ζ. For this purpose we apply the Weyl operators $\sigma\partial, \tilde{\sigma}\partial$ to eq.(6)

$$\sigma_\mu \partial^\mu \left[(\boldsymbol{\sigma} \cdot \boldsymbol{z}_L)\zeta_L\right] = 0$$
$$\tilde{\sigma}_\mu \partial^\mu \left[(\boldsymbol{\sigma} \cdot \boldsymbol{z}_R)\zeta_R\right] = 0 \tag{9}$$

One might hastily conclude that the spinors satisfy the Weyl equations because of (7). However, the two operators in (9) do not commute and we obtain

$$\left[\sigma_\mu\partial^\mu(\boldsymbol{\sigma} \cdot \boldsymbol{z}_L)\right]\zeta_L + \boldsymbol{\sigma} \cdot \boldsymbol{z}_L \sigma_\mu\partial^\mu\zeta_L + 2i\boldsymbol{\sigma} \cdot (\boldsymbol{z}_L \nabla \zeta) = 0 \tag{10}$$

and a similar equation for ζ_R with $\tilde{\sigma}_\mu\partial^\mu$. The last term in (10) is due to the commutator. The first term in (10) is zero due to Maxwell's equations (7). Consequently, the spinors $\zeta_{L,R}$ satisfy the Weyl-neutrino equation.

$$\sigma_\mu\partial^\mu\zeta_L = 0, \quad \tilde{\sigma}_\mu\partial^\mu\zeta_R = 0 \tag{11}$$

provided that

$$\boldsymbol{\sigma} \cdot (\boldsymbol{z}_L \wedge \nabla\zeta) = 0 \tag{12}$$

The spinor ζ can be written in the following equivalent forms

$$\zeta = \pm \begin{pmatrix} \sqrt{-z_1 + iz_2} \\ \sqrt{z_1 + iz_2} \end{pmatrix} = \pm \frac{i}{\sqrt{z_1 - iz_2}} \begin{pmatrix} z_1 - iz_2 \\ z_3 \end{pmatrix} = \pm \frac{i}{\sqrt{z_1 + iz_2}} \begin{pmatrix} z_3 \\ z_1 + iz_2 \end{pmatrix} \tag{13}$$

Without loss of generality, we can take the fields $\boldsymbol{E}$ and $\boldsymbol{B}$ to be in the plane perpendicular to x_3, the direction of propagation. Then $z_3 = 0$ $(E_3 = B_3 = 0)$ and we have

$$\frac{\partial\zeta}{\partial t} = \pm\frac{1}{2\sqrt{z_1 + iz_2}} \begin{pmatrix} 0 \\ \frac{\partial}{\partial t}(z_1 + iz_2) \end{pmatrix}$$

$$\boldsymbol{\sigma} \cdot \nabla\zeta = \pm\frac{1}{2\sqrt{z_1 + iz_2}} \begin{pmatrix} i\left(\frac{\partial z_2}{\partial x_1} + \frac{\partial z_1}{\partial x_2}\right) \\ -\frac{\partial}{\partial x_3}(z_1 + iz_2) \end{pmatrix}$$

The Weyl equation is satisfied, if $\frac{\partial z_2}{\partial x_1} = \frac{\partial z_1}{\partial x_2}$, since $\frac{\partial}{\partial t}(z_1 + iz_2) = -\frac{\partial}{\partial x_3}(z_1 + iz_2)$ by virtue of Maxwell's equations. And the above condition imply that the components of $\boldsymbol{E}$ (and $\boldsymbol{B}$) have the same dependence on x_1 and x_2, e.g. they are independent of x_1 and x_2 (as in the case of plane waves moving in the x_3-direction).

It is easy to see that if we parametrize the spinor as

$$\zeta = \begin{pmatrix} -\sin\frac{\theta}{2} & e^{-i\varphi/2} \\ \cos\frac{\theta}{2} & e^{i\varphi/2} \end{pmatrix}$$

then the corresponding electromagnetic fields are

$$E = \frac{1}{2} \begin{pmatrix} -\cos\theta\cos\varphi \\ \cos\theta\sin\varphi \\ -\sin\theta \end{pmatrix} \ , \ B = \frac{1}{2} \begin{pmatrix} \sin\varphi \\ -\cos\varphi \\ 0 \end{pmatrix}$$

In particular, we can obtain the basis spinors $\zeta_L \sim \binom{0}{1}$ and $\zeta_R \sim \binom{1}{0}$. With the metric $C = \begin{pmatrix} 0 & 1 \\ -1 & 0 \end{pmatrix}$ the spinors (13) have zero norms:

$$\zeta^T C \zeta = 0.$$

In conclusion, the construction of Cartan spinors statisfying the Weyl equations puts the photon-neutrino connection in a new light. Because, on the other hand, we can construct in principle all particle states (heavy lepton, baryons, intermediate bosons) from the two basic particles electron and neutrino interacting via the electromagnetic field,[26] the present considerations may bring us further towards unification in particle physics.

References

1. E. R. Caianiello, *Physics Reports*, **137**, 21 (1986) and references therein
2. C. G. Darwin, *Proc. Roy. Soc. London*, **118**, 654 (1928)
3. O. Laporte and G. E. Uhlenbeck, *Phys. Rev.*, **37**, 1380 (1931)
4. J. R. Oppenheimer, *Phys. Rev.*, **38**, 725 (1931)
5. H. E. Moses, *Nuovo Cim. Suppl.*, **7**, 1 (1958)
6. J. S. Lomont, *Phys. Rev.*, **111**, 1720 (1958)
7. A. O. Barut, *Electrodynamics and Classical Theory of Fields and Particles*, 1964 and 1980 (Dover, N.Y.)
8. R. Mignani, E. Recami and M. Baldo, *Lett. Nuovo Cim.*, **11**, 568 (1974)
9. A. A. Frost, *Found. of Phys.*, **5**, 619 (1975)
10. H. Sallhofer, *Z. Naturf.*, **33a**, 1378 (1978)
11. A. da Silveira, *Z. Naturf.*, **34a**, 646 (1979)
12. W. E. Baylis, *Am. J. Phys.*, **68**, 918 (1980)
13. V. M. Simulik, *Theor. of Math. Physics*, **87**, 386 (1991)
14. S. Thattey, *Ann. der Physik*, **48**, 353 (1991)
15. P. Budinich, *Phys. Reports*, **137**, 35 (1986)
16. L. De Broglie, *Compt. Rend.*, **195**, 862 (1932); **199**, 813 (1984)
17. P. Jordan, *Z. Phys.*, **93**, 464 (1935)
18. R. Kronig, *Physica*, **3** 1120 (1936)
19. M. H. L. Price, *Proc. Roy. Soc. London*, **165**, 247 (1938)
20. I. M. Barbour, A. Bietti and B. F. Touschek, *Nuovo Cim.*, **28**, 453 (1963)
21. B. Ferretti, *Nuovo Cim.*, **33**, 265 (1964)
22. W. A. Perkins, *Phys. Rev.*, **137**, B 1291 (1965)
23. P. Bandyopadhyay and P. R. Chandhuri, *Phys. Rev.*, **D3**, 1378 (1971)
24. H. Inoue, T. Tajima and S. Tanaka, *Propr. Theor. Phys.*, **48**, 1338 (1972)
25. E. Cartan, *Bull. Soc. Math. de France*, **41**, (1913); *La theorie des spineurs*, (Hermann, Paris, 1936)
26. A.O. Barut, *Ann. der Physik*, **43**, 83 (1986)

*To Eduardo who tought us that
one good question may be more
useful than several answers.*

SYMMETRIES AND REDUCTION:
INTERACTING SYSTEMS OUT OF FREE ONES [1]

G. Marmo[#,@] and G. Vilasi[*,@]

[#]*Dipartimento di Scienze Fisiche - Università di Napoli,
Mostra d'Oltremare, Pad.19 - I-80125 Napoli — Italy.*

[*]*Dipartimento di Fisica Teorica e S.M.S.A. - Università di Salerno,
Via S. Allende, I-84081 Baronissi (SA) — Italy.*

[@]*Istituto Nazionale di Fisica Nucleare - Sezione di Napoli
Mostra d'Oltremare, Pad.20 - I-80125 Napoli — Italy.*

Abstract

We reconsider the reduction procedure for the number of degrees of freedom
in the Hamiltonian formalism and extend it to the Lagrangian, Newtonian and
Poisson formalisms. With help of few examples, we show that interacting systems
may be obtained by reducing free ones.

---

[1]Supported in part by the Italian Ministero dell'Università e della Ricerca Scientifica (MURST)

1. Introduction

It is well known that a constant of the motion can be used to "reduce the number of degrees of freedom" of a Hamiltonian dynamical system [1], [2], [3].

An adeguate number of constants of motion may give rise to action-angle variables which are used to handle completely integrable systems.

In the Hamiltonian or Lagrangian formalism constants of the motion are associated with symmetries of the equations of the motion, therefore the reduction procedure relies on both concepts. Constants of the motion provide us with invariant submanifols and symmetries, acting on them, provide us with equivalence classes. Our reduced dynamics becomes a dynamics on equivalence classes. Very often the "reduced space", quotient manifold of equivalence classes, can be imbedded as a submanifold of the starting space and we get a dynamical evolution of data with a physical interpretation provided already by our initial model system.

An interesting feature of the reduced system is that it may turn out to be non linear even if the starting one was linear and therefore integrable via exponentiation. Many completely integrable systems actually turn out to be reduced system of free or simple ones [4]. One may be tempted to conjecture that any completely integrable system should arise as reduction of a simple system. By simple here we mean a system which can be integrated via exponentiation (i.e. like going from a Lie algebra to a corresponding Lie group).

Nowdays one knows that various descriptions of dynamical systems are available. One starts with a vector field Γ on a manifold M (carrier space) and by further qualification of M and Γ one finds a Lagrangian description on $M = TQ$ and Γ second order [2], [3], or a symplectic description on (M,ω), with ω a symplectic structure and Γ a symplectic vector field. When (M,Γ) is thought of as a classical limit of a quantum system, M will be required to be a Poisson manifold and Γ to be a Hamiltonian vector field with respect to the given Poisson structure.

When dealing with the reduction of systems using additional structures one is obliged to worry also about them, i.e. the reduction procedure should be compatible with the additional structure on the carrier space.

At this point, it should be clear that the usual reduction procedure we are familiar with in the Hamiltonian formalism shall be reconsidered to extract the main ingredients which can be used also in other formalisms. In the first few sections we shall give some indications on how to generalize the induction procedure in various instances; afterwards we apply these reduction procedures to simple systems and show how we can get interacting ones.

2. The General Reduction Procedure [5]

Let us consider the mains steps involved in any reduction procedure. We consider a carrier space M for our dynamical system Γ, a vector field on M. We shall indicate by ϕ_Γ^t the local flow of Γ.

A general reduction procedure requires two ingredients:
i) A submanifold Σ of M to which Γ is tangent, i.e.

$$\Gamma(m) \in T_m\Sigma, \quad \forall m \in \Sigma$$

ii) An equivalence relation $\mathcal{R}$ on Σ which is compatible with Γ. Compatibility here means that

$$m\mathcal{R}m' \iff \phi_\Gamma^t(m)\mathcal{R}\phi_\Gamma^t(m') \quad \forall m,m' \in \Sigma$$

The reduced system $\tilde{\Gamma}$ will be defined on $\tilde{\Sigma}_R = \frac{\Sigma}{\mathcal{R}}$ and it is the projection of Γ along the natural projection: $\pi_R : \Sigma \to \tilde{\Sigma}_R$

A dual view point can be given on the algebra of observables, the smooth functions on M, say $\mathcal{F}(M)$. The dynamics here is defined by $\dot{f} = L_\Gamma f, \quad \forall f \in \mathcal{F}$

A reduction procedure will require a projection $\pi_\Sigma : \mathcal{F} \to \mathcal{F}_1$ This projection can be visualized in terms of the identification map $i_\Sigma : \Sigma \to M$, by setting :
$\pi_\Sigma(f) = i_\Sigma^*(f)$

The tangency condition in this dual view is translated in requiring the existence of a derivation Γ_1 on $\mathcal{F}_1$ such that:

$$\pi_\Sigma(L_\Gamma f) = L_{\Gamma_1}\pi_\Sigma(f)$$

condition ii) is translated by requiring the existence of a subalgebra $\tilde{\mathcal{F}} \in \mathcal{F}_1$ such that $L_{\Gamma_1}\tilde{\mathcal{F}} \subset \tilde{\mathcal{F}}$. The restriction of Γ_1 to $\tilde{\mathcal{F}}$ can be denoted by $\tilde{\Gamma}$ and provides us with the analog of the reduced dynamics we had an $\tilde{\Sigma}_R$. This dual view point is useful when dealing with Poisson dynamics.

Remark: Our two steps reduction procedure give rise to the following diagrams

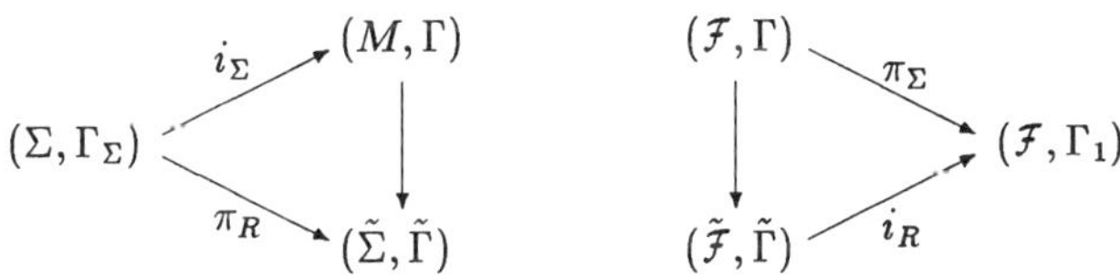

where $i_R(\tilde{\mathcal{F}}) = \pi_R^*(\mathcal{F}(\tilde{\Sigma}))$ and $\pi_\Sigma(\mathcal{F}) = i_\Sigma^*(\mathcal{F})$. We shall now consider what happens when we deal with additional structures.

3. Reduction with Additional Structures.

a) Lagrangian Formalism

In the Lagrangian formalism we have $M = TQ$ and Γ is a Lagrangian vector field [3], i. e.

$$i_\Gamma d\theta_\mathcal{L} = -dE_\mathcal{L}$$

where in local coordinates

$$\theta_\mathcal{L} = \frac{\partial\mathcal{L}}{\partial\dot{q}^i}dq^i \qquad E_\mathcal{L} = \dot{q}^i\frac{\partial\mathcal{L}}{\partial\dot{q}^i} - \mathcal{L}$$

and $d\theta_{\mathcal{L}}$ is non degenerate. Here the reduction procedure should give rise to $\tilde{\Sigma} = T\tilde{Q}$ otherwise there is no reduced second order dynamics and a fortiori no Lagrangian description. However even if $\tilde{\Sigma} = T\tilde{Q}$ and $\tilde{\Gamma}$ is second order, there is no simple way to get a Lagrangian description for $\tilde{\Gamma}$. To put it differently, the restriction $\mathcal{L}_{\Sigma}$ of $\mathcal{L}$ to Σ, will not be related to $\tilde{\pi}_R(\tilde{\mathcal{L}})$ even if $\tilde{\mathcal{L}}$ does exist.

To find a Lagrangian description for $\tilde{\Gamma}$ on the reduced space we have to solve an inverse problem [6]. If we impose more stringent condictions to get a Lagrangian $\tilde{\mathcal{L}}$ directly related to the starting $\mathcal{L}$, we find that such stringent conditions will not be met even for the usual reduction of central potential. Thus the recourse to the inverse problem seems to be unavoidable.

b) Symplectic Formalism.

Here we have a symplectic structure ω on M such that

$$i_{\Gamma}\omega = -dH$$

On the submanifold Σ we get a 2-form $\omega_{\Sigma} = i_{\Sigma}^{*}\omega$ which in general will be degenerate.

A condition for ω to be compatible with the equivalence relation $\mathcal{R}$ can be given in terms of the following vector fields. We consider the algebra of functions $\pi_{R}^{*}(\mathcal{F})(\tilde{\Sigma}))$ and find all solutions of the algebraic equations on Σ

$$i_X\omega_{\Sigma} = \pi_{R}^{*}f \tag{1}$$

Now ω_{Σ} will be compatible with the equivalence relation $\mathcal{R}$ if local flows associated with solutions of (1) are compatible in the standard way :

$$\phi_X^s(m)\,\mathcal{R}\,\phi_X^s(m') \iff m\mathcal{R}m'$$

Remark: As ω_{Σ} is degenerate there will be functions $f \in \mathcal{F}(\tilde{\Sigma})$ for which no solution exists. If the set of solutions is empty ω_{Σ} will be trivially compatible with the equivalence relation.

It should be noticed that the usual reduction procedure of the symplectic formalism goes along the following lines: the submanifold Σ is a level set of constants of the motion for Γ (or, more generally, defined by invariant relations for Γ).

The equivalence relation is provided by the involutive distribution associated with ker ω_{Σ} with the assumption that rankω_{Σ} is constant on Σ. In this restricted situation ω_{Σ} will be obviously compatible with the equivalence relation and will give rise to a non degenerate 2-form $\tilde{\omega}$ on the quotient space $\tilde{\Sigma}$.

The reduced dynamics $\tilde{\Gamma}$ will be symplectic with respect to $\tilde{\omega}$ with Hamiltonian function $\tilde{H}$ being the projection of $H_{\Sigma} = i_{\Sigma}^{*}(H)$. Unlike the Lagrangian formalism, in this case there is a definite connection between the reduced Hamiltonian and the starting one.

c) The Poisson Formalism.

We recall very briefly that, in this case, the Poisson structure on M is given by saying that $\mathcal{F}(M)$ is a Lie algebra with respect to some Lie bracket

$$\{\,,\,\} \;:\; \mathcal{F} \times \mathcal{F} \;\to\; \mathcal{F},$$

with the additional property that

$$f \longmapsto X_f$$

is a Lie algebra homomorphism from $\mathcal{F}$ into the Lie algebra of derivations $Der(\mathcal{F})$. this last requirement means that $\{f \cdot g, h\} = f\{g \cdot h\} + \{f, h\}g$ and $\{f, g\} \mapsto X_{\{f,g\}}$. The derivation X_f is defined by setting $X_f \cdot g := \{f, g\}$.

Now our dynamical vector field Γ is required to be defined by the Poisson bracket according to

$$L_\Gamma \cdot f = \{H, f\}$$

for any function $f \in \mathcal{F}$. A reduction procedure in this formalism would require something like $\pi_\Sigma : \mathcal{F} \to \mathcal{F}_1$ is a Poisson projection ,i. e.

$$\pi_\Sigma\left[\{f, g\}\right] =: \{\pi_\Sigma(f), \pi_\Sigma(g)\}_1$$

defines a Poisson structure $\{\,,\,\}_1$ on $\mathcal{F}_1$. This requirement will not be satisfied in general. For instance consider $T^*\mathbb{R}^2$, with standard Poisson brackets

$$\{p_i, q_j\} = \delta_i{}_j \qquad i, j \in \{1, 2\}$$

By taking Σ to be thew level set determined by p_2 equal to a constant we would get $\pi_\Sigma\left(\{p_2, q_2\}\right) = \{\pi_\Sigma p_2, \pi_\Sigma q_2\}$. The second side should be zero because $\pi_\Sigma p_2$ is a constant while the first side should be one. Therefore the reduction as stated would not be possible. The reader familiar with the Dirac-Bergman constraint formalism easily realizes that one is forced into distinguishing first class Σ from second class ones. There by the reduction procedure would be much more involved.

4. Relation with complete integrability.

Up to now, in our discussion of the reduction procedure we have not taken into consideration how to use these generalized procedures.

In this connection we recall that many known completely integrable systems arise from reduction of free systems. These starting free system however are known to exbihit alternative Lagrangian, Hamiltonian or Poisson descriptions. Therefore by using different routes to reduce the starting system we might be able to give rise to different non linear dynamical systems which turn out to be completely integrable.

Our interest in the Poisson formalism arises because of its predominance in many infinite dimensional situation. These aspects for field theories will be dealt with elsewhere. Here we shall proceed by considering specific examples.

5. Interacting systems out of free ones.[4]

For simplicity we shall illustrate some of our general considerations by using a simple and well known example .

Reduction of a free motion on $\mathbb{R}^3$.

We start with the equations of motion on $T\mathbb{R}^3$, say $\ddot{\vec{r}} = 0$. We shall obtain a non linear system on $\mathbb{R}^+$. The submanifold Σ will be defined as the level set of the angular momentum; Γ will be tangent to Σ because angular momentum is a constant of the motion. The equivalence relation will be provided by the rotation group which is a symmetry group for Γ and acts on Σ.

The reduced dynamics on $T\mathbb{R}^+$ turns out to be given by $\ddot{r} = \ell^2/r^3$ where ℓ^2 is the fixed value of the square of the angular momentum, the value ℓ behaves like a coupling constant.

By using a different Σ, say the one determined by the value E of the energy instead of the angular momentum we find, by using again the equivalence relation associated with the rotation group, the non linear equation $\ddot{r} = \left(\frac{2E}{r}\right) - \left(\frac{\dot{r}^2}{r}\right)$.

If one fixes a value of a convex linear combination of energy and angular momentum to determine a new Σ while keeping the rotation group for the equivalence relation we find a non linear equation of motion given by:

$$\ddot{r} = \frac{\alpha\ell^2 + (1 - \alpha)2E + (\alpha - 1)^2\dot{r}^2}{\alpha r^3 + (1 - \alpha)r}.$$

For additional details on this example in the Lagrangian formalism see [5].

6. Reduction in the Symplectic formalism.

On $T^*\mathbb{R}^3$ we consider the standard Hamiltonian description provided by

$$H = \frac{1}{2m}\left(p_r^2 + \frac{L^2}{r^2}\right) = \frac{p^2}{2m}$$

where

$$p_r = \frac{\vec{p}\cdot\vec{r}}{r} \quad , \quad \vec{L} = \vec{p}\wedge\vec{r}.$$

The submanifold is given by

$$\Sigma_\ell = \left\{(\vec{r},\vec{p}) \vdash p^2 r^2 - r^2 p_r^2 = \ell^2\right\}$$

The equivalence relation is provided by the rotation group and the reduced dynamics on $T^*\mathbb{R}^+$ is described by $\tilde{H} = \frac{1}{2m}\left(p_r^2 + \frac{\ell^2}{r^2}\right)$ with $\tilde{\omega} = dp_r \wedge dr$.

If we determine Σ_E by fixing $p^2 = 2mE$ with equivalence relation given by the rotation group we get a reduced dynamics given by

$$\tilde{\Gamma} = \frac{p_r}{m}\frac{\partial}{\partial r} + \left(\frac{2E}{r} - \frac{p_r^2}{mr}\right)\frac{\partial}{\partial p_r}$$

which is Hamiltonian with respect to the symplectic structure on $T^*\mathbb{R}^+$ $\tilde{\omega} = r^2 dp_r \wedge dr$. The Hamiltonian function is

$$H = Er^2 - \frac{r^2 p_r^2}{2m}$$

This time ker ω_E is generated by the starting dynamics itself therefore the symplectic description of the reduced dynamics is not related to the starting symplectic description.

7. Reduction in the Poisson formalism

A simplified reduction procedure that takes into account the existing additional structure from the beginning, could proceed along the following lines. We fix a set of constants of the motion, say $f_1, f_2, \ldots, f_k$ such that $\{f_i, f_j\} = 0$ and look for the Lie subalgebra of $\mathcal{F}(M)$, say

$$\mathcal{F}_f \equiv \{g \in \mathcal{F} \vdash \{g, f_i\} = 0 \qquad \forall i \in (1, \ldots, k).$$

Then we consider the dynamical evolution restricted to this subalgebra, i. e.

$$\dot{g} = \{H, g\} \qquad \text{for} \qquad g \in \mathcal{F}_f$$

Let us apply this procedure to our example. We consider on $T^*\mathbb{R}^3$ with the standard Poisson bracket the free dynamical system described by

$$H = \frac{1}{2m} \left(p_r^2 + \frac{L^2}{r^2} \right) = \frac{p^2}{2m}$$

We take $f_1 = L^2$ and see very easily that $x = (\vec{p} \cdot \vec{r})/r$, $y = r$, $z = p^2$ and the three function represented by $\vec{L} = \vec{p} \wedge \vec{r}$ are a basis for $\mathcal{F}_{L^2}$.

As for the dynamics we find

$$\frac{d}{dt} x = \frac{1}{my}(z - x^2)$$
$$\frac{d}{dt} y = \frac{x}{m}$$
$$\frac{d}{dt} z = 0 \quad , \quad \frac{d\vec{L}}{dt} = 0,$$

described by the Poisson bracket

$$\{x, y\} = 1 \quad , \quad \{y, z\} = -2x \quad , \quad \{z, x\} = \frac{2}{y}(z - x^2)$$

with Hamiltonian function

$$H = \frac{z}{2m}$$

A more general reduction procedure for Poisson structures is the following one.

On the Poisson algebra $(\mathcal{F}(M), \{\ ,\ \})$, we select a Lie subalgebra $A \subset \mathcal{F}(M)$.

We consider the normalizer of A in $\mathcal{F}(M)$, i. e. the Lie subalgebra $E \subset \mathcal{F}(M)$ such that $\{E, A\} \subset A$. We get a short exact sequence of Lie algebras by construction:

$$0 \to A \to E \to B \to 0$$

The Lie algebra B is defined by the sequence. Now any Hamiltonian function in E gets reduced by our sequence, indeed we find

$$\frac{d}{dt} f_i = \{H, f_i\} = a_i(f)$$

and on each equivalence class say $[b_a] \in B$, we find

$$\frac{d}{dt}[b_a] = \{[H], [b_a]\}_B \in B.$$

In our previous example A was generated by L^2 while $(p^2, r^2, (\vec{p}\cdot\vec{r}), \vec{p}\wedge\vec{r})$ generated B. The dynamics along A is trival, $\dot{L}^2 = 0$ and we restricted ourselves to the dynamics along B. The sequence was a split sequence with

$$z = p^2 \qquad y = r \qquad x = \frac{\vec{p}\cdot\vec{r}}{r} \qquad \vec{L} = (L_x, L_y, L_z)$$

and

$$\{z, x\} = 2\left(\frac{z - x^2}{y}\right) \qquad \{z, y\} = 2x \qquad \{x, y\} = 1$$

$$\{\vec{L}, z\} = \{\vec{L}, y\} = \{\vec{L}, x\} = 0 \qquad \{L_i, L_j\} = \epsilon_{ijk} L_k$$

We could have started with A identified as the angular momentum algebra. The subalgebra E would be identified with the algebra generated by $(p^2, r^2, \vec{r}\cdot\vec{p}, \vec{r}\wedge\vec{p})$. These six functions are linked by $p^2 r^2 - (\vec{r}\cdot\vec{p})^2 = (\vec{r}\wedge\vec{p})^2$. The subalgebra B will be generated by $(p^2, r^2, \vec{r}\cdot\vec{p})$. It is clear that by choosing appropriate functions to generate B we might be able to have a finite dimensional Lie algebra which generates B, in this way the quantization procedure for the quotient dynamics would be much more viable for we could look for unitary representation of the corresponding Lie group.

8. Geodetical Motions out of Simple Ones. [7]

On $\mathbb{R}^6$ we consider the following Hamiltonian

$$H = \frac{1}{2}(\vec{p}\wedge\vec{q})^2$$

with standard symplectic structure

$$\omega = dp_i \wedge dq^i.$$

Equation on motion can be integrated by exponentation, indeed by fixing $\vec{p}\wedge\vec{q} = \vec{k}$ on such submanifold the dynamics becomes an element of the Lie algebra of the rotation group acting in a canonical way on $T^*\mathbb{R}^3$. Of course such an element is not a fixed elements but changes with $\vec{k}$. Our system is then a simple system. We fix now the submanifold $\Sigma = \{(p,q) \vdash q^2 = 1\}$ as an equivalence relation on Σ we take $(p_1, q_1)\mathcal{R}(p_2, q_2)$ if $q_1 = q_2$ and $p_1 = p_2 + sq$, for some $s \in \mathbb{R}$. It is possible to imbedd the quotient manifold as a submanifold of $\mathbb{R}^6$ by selecting in every equivalence class an element $\bar{p}$ such that $\bar{p} \cdot \vec{q} = 0$, i.e. in the equivalence class of $(\vec{p}, \vec{q})$ we consider $\bar{p} = \vec{p} - (\vec{p} \cdot \vec{q})\vec{q}$ with $q^2 = 1$. In this way our submanifold is T^*S^2. The dynamical vector field turns out to be

$$\Gamma = (q^2\vec{p} - \vec{q} \cdot \vec{p}\vec{q})\frac{\partial}{\partial \vec{p}} + (\vec{q} \cdot \vec{p}\,\vec{p} - p^2\vec{q})\frac{\partial}{\partial \vec{p}}$$

On the submanifold T^*S^2 it becomes

$$\tilde{\Gamma} = \vec{p}\frac{\partial}{\partial \vec{q}} - p^2\vec{q}\frac{\partial}{\partial \vec{p}}$$

with equations of motion in the second order form given by

$$\ddot{\vec{q}} = -(\dot{q})^2\vec{q}$$

which represent the geodetical flow on the unitary sphere.

Remark: Had we started with Lagrangian $\mathcal{L} = \frac{1}{2}(m\dot{\vec{q}} \wedge \vec{q})^2$ on the tangent bundle of $\mathbb{R}^3$ we would have found a dynamical system with constraints in the sense of Dirac-Bergman. This example was investigated in a pionering paper by R. Haag [8].

It is interesting to observe that our starting dynamical system admits another Hamiltonian description on $T^*\mathbb{R}^3$ with a different symplectic structure. It is quite easy to see that

$$H = \frac{1}{2}\left(\vec{p} \wedge \vec{q} + n\frac{\vec{q}}{q}\right)^2$$

gives the same equations of motion with

$$\omega = dp_i \wedge dq^i + n\varepsilon_{ij}k\frac{q_k}{q^3}dq^i \wedge dq^k$$

This symplectic structure can be restricted to T^*S^2, the submanifold in $T^*\mathbb{R}^3$ realizing $\tilde{\Sigma}$, therefore our geodetical motion on S^2 admits a bihamiltonian description.

9. Conclusions

We have shown that interacting systems can be obtained by reduction procedure from free ones. Reduction procedures can be used also in quantum mechanics and field theory. As a matter of fact our examples can also be treated as quantum systems and the reduction procedure carried on. However in several examples it appears that quantization and reduction do not commute, i.e.first quantize and then reduce or first reduce and then quantize give different results. Then one has to develop the reduction procedure directly in the quantum framework.

In field theory, Kaluza-Klein type theories offer already an instance where vacuum Einstein equations, in higher dimensions, by reduction give rise to gravitational and Yang-Mills fields on space time. By carrying on this procedure in quantum field theory one should be able to get interacting theories out of free ones. We shall come back to these issues elsewhere.

References

[1] Whittaker E.T. *A treatise on the analytical dynamics of particles and rigid bodies* Cambridge University Press, London (1904, 4th edn, 1959)

[2] Abraham R, Marsden J.E. *Foundations of Mechanics*, Benjamin, Reading Mass (1978)

[3] Marmo G., Saletan E.J., Simoni A. and Vitale B. *Dynamical Systems, a Differential Geometric Approach to Symmetry and Reduction* John Wiley, Chichester UK 1985.

[4] Kazdan D., Kostant B. and Sternberg S. , Comm. Pure Applied Math. **31**, 481 (1978) Olshanetsky M.A., Perelomov A.M., Phys. Rep. **71**, 313 (1981);Phys. Rep. **94**, 313 (1983)

[5] Landi G., Marmo G., Sparano G. and Vilasi G. *A generalized reduction procedure for dynamical systems* Modern Phys. Lett **6A**, 3445-3453 (1991)

[6] Morandi G., Ferrario C., Lo Vecchio G., Marmo G. and Rubano C., *The Inverse problem in the Calculus of variations and the geometry of the tangent bundle*, Phys. Rep. **188**, 147-284 (1990)

[7] Moser J., *Various Aspects of Integrable Hamiltonian Systems*, in Dynamical Systems *Progr. Math.* vol.8, Birkhäuser, Boston (1980).

[8] Haag R., *Der kanonische Formalismus in antarteten Fäller*, Z. Angew. Math. Mech. **32**, 197 (1952)

ON THE INVARIANCE PROPERTIES OF REDUCTION THEORIES*

ALBERTO RIMINI

Dipartimento di Fisica Nucleare e Teorica, Università di Pavia, Pavia

1. Introduction.

It is common experience of all physicists that the interpretation of the wave function, up to a factor, as the state of an individual physical system works perfectly well in ordinary situations, even though it brings into evidence some problematic aspects of quantum mechanics with systems consisting of distant microscopic parts. Such an interpretation is made possible by the reduction principle, which amounts to suspend from time to time the validity of the general principle of evolution (the Schrödinger equation) and implies some kind of splitting of the relevant part of the world into conceptually different parts (observed–observer, system–apparatus, microscopic–macroscopic, quantum–classical). To get rid of these unsatisfactory aspects of the reduction principle one tries to do the so–called theory of quantum measurement, which consists in including in the description, besides the system, the apparatus and possibly other parts of the world, and in applying to such an enlarged system universally valid principles, first of all the Schrödinger equation. One should either deduce from the description of the enlarged system the standard formulation or, at least, show that the two pictures are consistent. In doing the theory of quantum measurement, however, one meets difficulties such that the interpretation of the wave function as the state of an individual physical system cannot be maintained. One can then give up the program of the theory of quan-

---

* Supported in part by Ministero dell'Università e della Ricerca Scientifica e Tecnologica and Istituto Nazionale di Fisica Nucleare.

tum measurement, which is essentially the position of the original Copenhagen interpretation, or insist in the description of the enlarged system but renouncing in any case to pretend that the elements of the quantum–mechanical description be real properties of the described objects.

In the frame of a realistic conception of our description of physical systems, the only open possibilities are either adopting a concept of state which includes additional variables besides the wave function or accepting the idea of introducing modifications to the Schrödinger equation. We repeat this statement using Bell's words — "Either the wave function, as given by the Schrödinger equation, is not everything, or it is not right. ... If, with Schrödinger, we reject extra variables, then we must allow that his equation is not always right. I do not know that he contemplated this conclusion, but it seems to me inescapable".[1]

The modification of the Schrödinger equation, if this way is taken, is aimed at describing reduction as a real physical process, so that it must have the form of a definite mathematical recipe which is not "disregarded from time to time on the basis of supplementary, imprecise, verbal prescriptions".[1] Since the reduction process is nonlinear and stochastic, the modified Schrödinger equation will have such features. In order to repeat the quantitative success of quantum mechanics in the description of physical systems, the modification must have practically un-observable consequences in all ordinary situations and become effective only when the emergence of superpositions of macroscopically distinguishable states requires reduction. The possibility of carrying out such a program has been demonstrated recently by the theory of spontaneous localization.[2] According to this proposal, each particle of a system of distinguishable particles undergoes, at uniformly distributed random times with a definite frequency, a sudden spontaneous localization within a region of a definite size located in a random place with a probability distribution depending on the wave function in accordance with the square–modulus rule. It turns out that it is possible to choose the frequency and size parameters in such a way that microscopic systems and the internal structure of macroscopic systems are practically unaffected, that the stochastic behaviour bestowed by the process on macroscopic objects is negligible and that, nevertheless, superpositions of macroscopically distinguishable states are quickly reduced to one of their components. Subsequently, the spontaneous localization model was combined with previous proposals of describing reduction as a continuous stochastic process in the Hilbert space[3,4,5] and was formulated in nonrelativistic second–quantized form.[6] A complete quantitative discussion of the physical consequences of the resulting continuous spontaneous localization theory was also given.[6,7]

The just mentioned theory of reduction is based on the Itô stochastic differential equation

$$
\begin{aligned}
\mathrm{d}|\psi(t)\rangle = & -iH\mathrm{d}t|\psi(t)\rangle \\
& + \int \mathrm{d}\vec{x}\left[\gamma^{1/2}\left(N(\vec{x}) - \langle N(\vec{x})\rangle_{\psi(t)}\right)\mathrm{d}W_{\vec{x}}(t) \right. \\
& \left. - \tfrac{1}{2}\gamma\left(N(\vec{x}) - \langle N(\vec{x})\rangle_{\psi(t)}\right)^2 \mathrm{d}t\right]|\psi(t)\rangle,
\end{aligned}
\tag{1.1}
$$

where $W_{\vec{x}}(t)$ are independent Wiener processes such that

$$
\overline{\mathrm{d}W_{\vec{x}}(t)} = 0,
$$

$$
\overline{\mathrm{d}W_{\vec{x}}(t)\,\mathrm{d}W_{\vec{x}'}(t)} = \delta(\vec{x} - \vec{x}')\,\mathrm{d}t
\tag{1.2}
$$

and

$$
\langle N(\vec{x})\rangle_{\psi} = \langle\psi|N(\vec{x})|\psi\rangle.
\tag{1.3}
$$

Eq. (1.1), which preserves the norm of the state vector, is interpreted as the evolution equation in the Schrödinger picture. The dynamical variable $N(\vec{x})$ is the particle density around $\vec{x}$ defined through a suitable average procedure. In ref. 6 it was chosen

$$
N(\vec{x}) = \left(\frac{\alpha}{2\pi}\right)^{3/2}\sum_s \int \mathrm{d}\vec{x}'\exp\left(-\tfrac{1}{2}\alpha(\vec{x}' - \vec{x})^2\right)a^+(\vec{x}',s)\,a(\vec{x}',s),
\tag{1.4}
$$

where $a^+(\vec{x},s)$ and $a(\vec{x},s)$ are the creation and annihilation operators, respectively, of a boson or fermion at point $\vec{x}$ with spin component s. The orders of magnitude of the parameters α and γ are chosen in such a way that the stochastic term in eq. (1.1) has, as its practically unique effect, that of inducing a very fast reduction of superpositions of states having macroscopically different densities of particles to one of their components.[6,7]

For invariant H, eq. (1.1) has all invariance properties which are usually demanded from a nonrelativistic theory (with the obvious exception of time–reversal invariance), as one can easily check. Of course, since it is a stochastic equation, invariance means that, for two different observers, subjectively identical evolutions *have the same probabilities*, rather than *are both solutions*. The problem of getting a Lorentz invariant formulation was first raised by Bell.[1,8] He suggested that the Tomonaga–Schwinger approach to relativistic quantum field theory, where the Schrödinger equation is written in an invariant form by replacing time by an arbitrary space–like surface, could provide a framework in which instantaneous jumps of the wave function can be described. "But there is an immense difficulty in doing this. To what quantities are you going to address the jumps? ... The way of constructing a sensible quantity in quantum field theory is to take a little integral of the local quantity over a finite volume of space and time. But although that is useful for some purposes, you cannot do that in a Lorentz invariant way, and if you want your theory to be Lorentz invariant, I think the jumping has to be addressed to a local quantity, and the trouble is that all local quantities, to my knowledge, in ordinary quantum field theory fluctuate infinitely."[8] This is essentially the difficulty met by some attempts at a relativistic reduction theory that have been made recently.[9]

The aim of the present discussion is to propose a new way towards a Lorentz invariant formulation of the continuous spontaneous localization theory. We will study the invariance of eq. (1.1) for space and time translations. The idea is to repeat for the proper orthochronous Poincaré group, in a Tomonaga–Schwinger framework, what is done in the following sections for translations.

2. Invariant construction of the stochastic equation.

88

In this section we will invent eq. (1.1) by means of a procedure such that the invariance for space–time translations is automatic.

Given an observer O using space time coordinates $\vec{x}, t$ we consider the set of all observers (which we call beaters) $B_\tau = \tau O$ obtained from O by a translation τ, i. e. using coordinates

$$(2.1) \qquad \vec{x}_\tau = \vec{x} - \vec{a}_\tau, \quad t_\tau = t - b_\tau.$$

Let T_τ be the application which transforms the state vector $|\psi\rangle$ used by O into the state vector $T_\tau|\psi\rangle$ used by B_τ to describe the system in the same objective situation. T_τ can be taken bijective, linear and isometric. It is time independent because the transformation is time independent. For each beater B_τ we consider a process–defining dynamical variable A_τ in such a way that all A_τ's are subjectively identical relative to the corresponding B_τ. We then associate to each beater a stochastic change of the state vector given by

$$
(2.2) \qquad T_\tau^+ \left[\gamma^{1/2}\, \delta_\epsilon^{1/2}(t_\tau) \left(A_\tau - \langle\psi(t)|T_\tau^+ A_\tau T_\tau|\psi(t)\rangle \right) dw_\tau(t_\tau) \right.
$$
$$
\left. - \tfrac{1}{2}\gamma\, \delta_\epsilon(t_\tau) \left(A_\tau - \langle\psi(t)|T_\tau^+ A_\tau T_\tau|\psi(t)\rangle \right)^2 dt_\tau \right] T_\tau|\psi(t)\rangle,
$$

where $w_\tau(t_\tau)$ are independent Wiener processes such that

$$
(2.3) \qquad \overline{dw_\tau(t_\tau)} = 0,
$$
$$
\overline{dw_\tau(t_\tau)\, dw_{\tau'}(t_\tau)} = \delta(\vec{a}_\tau - \vec{a}_{\tau'})\, \delta(b_\tau - b_{\tau'})\, dt_\tau.
$$

The function $\delta_\epsilon(t_\tau)$ which modulates in time the strength of the process is taken to be

$$(2.4) \qquad \delta_\epsilon(t_\tau) = \frac{1}{2\epsilon}\, \chi_{(-\epsilon,\epsilon)}(t_\tau).$$

On the whole, the stochastic evolution of the state vector obeys the Itô equation

$$
(2.5) \qquad d|\psi(t)\rangle = - iH\,dt|\psi(t)\rangle
$$
$$
+ \int d\vec{a}_\tau \int db_\tau \left[\gamma^{1/2}\, \delta_\epsilon^{1/2}(t_\tau) \left(A_\tau^o - \langle A_\tau^o \rangle_{\psi(t)} \right) dw_\tau(t_\tau) \right.
$$
$$
\left. - \tfrac{1}{2}\gamma\, \delta_\epsilon(t_\tau) \left(A_\tau^o - \langle A_\tau^o \rangle_{\psi(t)} \right)^2 dt_\tau \right] |\psi(t)\rangle,
$$

where

$$(2.6) \qquad A_\tau^o = T_\tau^+ A_\tau T_\tau$$

is A_τ in the description of O.

Eq. (2.5) is invariant for space–time translations by construction. In fact, each process appearing in the integral is associated to a beater in a subjectively identical way. If we change from the observer O to a space–time translated observer $\bar{O}$, for each beater $B_\tau = \tau O$ in a given relation to O there is a beater $\bar{B}_\tau = \tau\bar{O}$ in the same relation to $\bar{O}$. Going from O to $\bar{O}$ (and from $|\psi(t)\rangle$ to a subjectively identical $|\bar{\psi}(\bar{t})\rangle$) simply amounts to replacing the contribution of each B_τ to the integral with that of the corresponding $\bar{B}_\tau$. The invariance property of eq. (2.5) is explicitly demonstrated in the next section.

Eq. (1.1) is easily derived from (2.5). Let A_τ be the density around the origin of the space frame of B_τ. In such a case A_τ^o is simply $N(\vec{a}_\tau)$ and is independent of b_τ. Note further that, because of (2.1), $\mathrm{d}t_\tau = \mathrm{d}t$ and $\delta_\epsilon(t_\tau) = \delta_\epsilon(t - b_\tau)$. Then the $\mathrm{d}b_\tau$–integral of the second term in the square brackets in eq. (2.5) reduces to

$$(2.7) \qquad \int \mathrm{d}b_\tau\, \delta_\epsilon(t - b_\tau) = 1$$

and that of the first term defines a new stochastic process

$$(2.8) \qquad \int \mathrm{d}b_\tau\, \delta_\epsilon^{1/2}(t - b_\tau)\, \mathrm{d}w_\tau(t - b_\tau) = \mathrm{d}W_{\vec{a}_\tau}(t),$$

which, due to eqs. (2.3), has the properties (1.2). Therefore, with the above choice of A_τ, eq. (2.5) reduces to eq. (1.1).

3. Proof of invariance.

Given two observers or beaters O and B, let $[B \overset{T}{\leftarrow} O]$ and $[B \overset{S}{\leftarrow} O]$ be the bijective, linear and isometric applications which transform the state vector $|\psi\rangle$ used by O into the state vectors $[B \overset{T}{\leftarrow} O]|\psi\rangle$ and $[B \overset{S}{\leftarrow} O]|\psi\rangle$ used by B to describe the system in the same objective and subjective, respectively, situations. In this notation the application T_τ of sect. 2 is $[B_\tau \overset{T}{\leftarrow} O]$. The applications of the two types satisfy the identity[10]

$$(3.1) \qquad [O \overset{T}{\leftarrow} \tau O][\tau O \overset{S}{\leftarrow} O'] = [O \overset{S}{\leftarrow} \tau^{-1} O'][\tau^{-1} O' \overset{T}{\leftarrow} O'].$$

The unitary operator

$$(3.2) \qquad [O \overset{T}{\leftarrow} \tau O][\tau O \overset{S}{\leftarrow} O] = [O \overset{S}{\leftarrow} \tau^{-1} O][\tau^{-1} O \overset{T}{\leftarrow} O]$$

performs the active transformation of the system. We assume the invariance of H

$$(3.3) \qquad H = [O \overset{S}{\leftarrow} \bar{O}][\bar{O} \overset{T}{\leftarrow} O] H [O \overset{T}{\leftarrow} \bar{O}][\bar{O} \overset{S}{\leftarrow} O],$$

and the subjective identity of A_τ's

$$(3.4) \qquad A_{\check{\tau}} = [B_{\check{\tau}} \overset{S}{\leftarrow} B_\tau] A_\tau [B_\tau \overset{S}{\leftarrow} B_{\check{\tau}}].$$

Consider, besides O, a new observer $\bar{O} = \bar{\tau} O$ using coordinates

$$(3.5) \qquad \bar{\vec{x}} = \vec{x} - \bar{\vec{a}}, \quad \bar{t} = t - \bar{b}.$$

Let observer O apply eq. (2.5) to the state vector $[O \overset{T}{\leftarrow} \bar{O}][\bar{O} \overset{S}{\leftarrow} O]|\psi(t)\rangle$. Putting $|\bar{\psi}(\bar{t})\rangle = [\bar{O} \overset{S}{\leftarrow} O]|\psi(t)\rangle$, taking into account that $d\bar{t} = dt$ and changing the name of the integration variable from τ to $\check{\tau}$, the equation becomes

$$
\begin{aligned}
d|\bar{\psi}(\bar{t})\rangle = {} & -i[\bar{O} \overset{T}{\leftarrow} O] H [O \overset{T}{\leftarrow} \bar{O}] d\bar{t} \, |\bar{\psi}(\bar{t})\rangle \\
& + \int d\vec{a}_{\check{\tau}} \int db_{\check{\tau}} \left[\gamma^{1/2} \delta_\epsilon^{1/2}(t_{\check{\tau}}) \left([\bar{O} \overset{T}{\leftarrow} O] A_{\check{\tau}}^O [O \overset{T}{\leftarrow} \bar{O}] \right. \right. \\
& \qquad\qquad\qquad\qquad \left. - \langle [\bar{O} \overset{T}{\leftarrow} O] A_{\check{\tau}}^O [O \overset{T}{\leftarrow} \bar{O}]\rangle_{\bar{\psi}(\bar{t})} \right) dw_{\check{\tau}}(t_{\check{\tau}}) \\
& \qquad\qquad - \tfrac{1}{2} \gamma \delta_\epsilon(t_{\check{\tau}}) \left([\bar{O} \overset{T}{\leftarrow} O] A_{\check{\tau}}^O [O \overset{T}{\leftarrow} \bar{O}] \right. \\
& \qquad\qquad\qquad\qquad \left. \left. - \langle [\bar{O} \overset{T}{\leftarrow} O] A_{\check{\tau}}^O [O \overset{T}{\leftarrow} \bar{O}]\rangle_{\bar{\psi}(\bar{t})} \right)^2 dt_{\check{\tau}} \right] |\bar{\psi}(\bar{t})\rangle
\end{aligned}
$$

$$(3.6)$$

where $t_{\check{\tau}} = t - b_{\check{\tau}}$ is the time coordinate used by $B_{\check{\tau}} = \check{\tau} O$. Due to the invariance property (3.3), the operator appearing in the Schrödinger term in eq. (3.6) can be written

$$(3.7) \qquad [\bar{O} \overset{T}{\leftarrow} O] H [O \overset{T}{\leftarrow} \bar{O}] = [\bar{O} \overset{S}{\leftarrow} O] H [O \overset{S}{\leftarrow} \bar{O}].$$

In the stochastic term, we transform the integration variable from $\hat{r}$ to τ related to $\hat{r}$ by $\hat{r} = \tau\bar{\tau}$. Then one can write the chain of equalities

$$
\begin{aligned}
[\bar{O} &\overset{T}{\leftarrow} O]A^o_{\hat{r}}[O \overset{T}{\leftarrow} \bar{O}] \\
&= [\bar{O} \overset{T}{\leftarrow} O][O \overset{T}{\leftarrow} B_{\hat{r}}]A_{\hat{r}}[B_{\hat{r}} \overset{T}{\leftarrow} O][O \overset{T}{\leftarrow} \bar{O}] \\
&= [\bar{O} \overset{T}{\leftarrow} B_{\hat{r}}]A_{\hat{r}}[B_{\hat{r}} \overset{T}{\leftarrow} \bar{O}] = [\bar{O} \overset{T}{\leftarrow} \bar{B}_r]\bar{A}_r[\bar{B}_r \overset{T}{\leftarrow} \bar{O}] \\
&= [\bar{O} \overset{T}{\leftarrow} \bar{B}_r][\bar{B}_r \overset{S}{\leftarrow} B_r]A_r[B_r \overset{S}{\leftarrow} \bar{B}_r][\bar{B}_r \overset{T}{\leftarrow} \bar{O}] \\
&= [\bar{O} \overset{S}{\leftarrow} O][O \overset{T}{\leftarrow} B_r]A_r[B_r \overset{T}{\leftarrow} O][O \overset{S}{\leftarrow} \bar{O}] \\
&= [\bar{O} \overset{S}{\leftarrow} O]A^o_r[O \overset{S}{\leftarrow} \bar{O}]
\end{aligned}
\tag{3.8}
$$

having defined $\bar{B}_r = \tau\bar{O} = \hat{r}O = B_{\hat{r}}$ and having taken into account eqs. (3.4) and (3.1). Note that $t_{\hat{r}} = \bar{t}_r$. From the relations

$$
\vec{a}_{\hat{r}} = \vec{a}_r + \bar{\vec{a}}, \quad b_{\hat{r}} = b_r + \bar{b}
\tag{3.9}
$$

it follows that the Jacobian is 1. Furthermore, the properties of the stochastic process $w_{\hat{r}(\tau)}(\bar{t}_r)$ are given by

$$
\begin{aligned}
&\overline{\mathrm{d}w_{\hat{r}(\tau)}(\bar{t}_r)} = 0 = \overline{\mathrm{d}w_\tau(\bar{t}_r)}, \\
&\overline{\mathrm{d}w_{\hat{r}(\tau)}(\bar{t}_r)\,\mathrm{d}w_{\hat{r}(\tau')}(\bar{t}_r)} \\
&\quad = \delta\big((\vec{a}_\tau + \bar{\vec{a}}) - (\vec{a}_{\tau'} + \bar{\vec{a}})\big)\,\delta\big((b_\tau + \bar{b}) - (b_{\tau'} + \bar{b})\big)\,\mathrm{d}\bar{t}, \\
&\quad = \delta(\vec{a}_\tau - \vec{a}_{\tau'})\,\delta(b_\tau - b_{\tau'})\,\mathrm{d}\bar{t}_r = \overline{\mathrm{d}w_\tau(\bar{t}_r)\,\mathrm{d}w_{\tau'}(\bar{t}_r)},
\end{aligned}
\tag{3.10}
$$

so that eq. (3.6) is equivalent, as a stochastic equation, to the equation

$$
\begin{aligned}
\mathrm{d}|\bar{\psi}(\bar{t})\rangle = {} &-i[\bar{O} \overset{S}{\leftarrow} O]H[O \overset{S}{\leftarrow} \bar{O}]\mathrm{d}\bar{t}\,|\bar{\psi}(\bar{t})\rangle \\
&+ \int \mathrm{d}\vec{a}_r \int \mathrm{d}b_r \Bigg[\gamma^{1/2}\,\delta_\epsilon^{1/2}(\bar{t}_r)\Big([\bar{O} \overset{S}{\leftarrow} O]A^o_r[O \overset{S}{\leftarrow} \bar{O}] \\
&\qquad - \langle[\bar{O} \overset{S}{\leftarrow} O]A^o_r[O \overset{S}{\leftarrow} \bar{O}]\rangle_{\bar{\psi}(\bar{t})}\Big)\mathrm{d}w_\tau(\bar{t}_r) \\
&\qquad - \tfrac{1}{2}\gamma\,\delta_\epsilon(\bar{t}_r)\Big([\bar{O} \overset{S}{\leftarrow} O]A^o_r[O \overset{S}{\leftarrow} \bar{O}] \\
&\qquad - \langle[\bar{O} \overset{S}{\leftarrow} O]A^o_r[O \overset{S}{\leftarrow} \bar{O}]\rangle_{\bar{\psi}(\bar{t})}\Big)^2\mathrm{d}\bar{t}_r\Bigg]|\bar{\psi}(\bar{t})\rangle.
\end{aligned}
\tag{3.11}
$$

This is precisely the equation which is written by $\bar{O}$ if invariance is there.

92

4. Conclusions.

The individual stochastic processes (2.2) are not invariant, they are associated in an invariant way to translations. This fact is formally expressed by the property (3.4) of the process–defining quantities A_r. In spite of the lack of invariance of the individual processes, the construction of sect. 2 and the proof of sect. 3 show that it is possible to build an invariant stochastic process by integrating them on the group manifold.

In order that the procedure outlined above can be repeated to get a Poincaré invariant stochastic process, the individual processes must be translatable from one element of the group to another one. This condition should be fulfilled within a Tomonaga–Schwinger approach to relativistic quantum field theory. Furthermore, invariant integration must be possible and this should be obtained using the Haar measure. We think that there are good reasons to hope that the program can be carried out.

References

1. J.S. Bell, in *Schrödinger — Centenary celebration of a polymath*, edited by C.W. Kilmister (Cambridge University Press, Cambridge, 1987), p. 41.

2. G.C. Ghirardi, A. Rimini, T. Weber, in *Quantum Probability and Applications II (Heidelberg, 1984)*, edited by L. Accardi and W. von Waldenfels, Lecture Notes in Mathematics, vol. 1136 (Springer, Berlin, 1985), p. 223; Physical Review D **34** (1986), 470; Physical Review D **36** (1987), 3287.

3. L. Diósi, Physics Letters A **129** (1988), 419.

4. P. Pearle, Physical Review A **39** (1989), 2277.

5. N. Gisin, Helvetica Physica Acta **62** (1989), 363.

6. G.C. Ghirardi, P. Pearle and A. Rimini, Physical Review A **42** (1990), 78.

7. G.C. Ghirardi and A. Rimini, in *Sixty-Two Years of Uncertainty*, edited by A.I. Miller (Plenum Press, New York, 1990), p. 167

8. J.S. Bell, in *Themes in Contemporary Physics II*, edited by S. Deser and R.J. Finkelstein (World Scientific, Singapore, 1989), p. 1.

9. P. Pearle, in *Sixty-Two Years of Uncertainty*, edited by A.I. Miller (Plenum Press, New York, 1990), p. 193; G.C. Ghirardi, R. Grassi and P. Pearle, Foundations of Physics **20** (1990), 1271; G.C. Ghirardi, R. Grassi and P. Pearle, in *Symposium on the Foundations of Modern Physics 1990*, edited by P. Lahti and P. Mittelstaedt (World Scientific, Singapore, 1991), p. 109.

10. A.S. Wightman, Supplemento al Nuovo Cimento **14** (1959), 81.

QUANTUM MECHANICS AND BUBBLES IN WEYL SPACETIME

G. Papini and W. R. Wood

Department of Physics, University of Regina

Regina, Saskatchewan, S4S 0A2 Canada

1. Introduction

The present work represents a variation of a theme due to Caianiello [1] and as such it is only fitting that it be contributed to a conference in his honour. The theme in question is that of maximal acceleration, approached here from the purely geometrical viewpoint of Weyl space.

Caianiello's starting point is the eight-dimensional line element [2]

$$d\tau^2 = \left(1 - \frac{\hbar^2}{m_0^2 c^2}|\ddot{x}_\mu \ddot{x}^\mu|\right) ds^2 \tag{1}$$

from which maximal acceleration follows if one requires invariance of $d\tau$ under Lorentz transformations. Equation (1) bears a strong resemblance to a conformal transformation of the classical line element in Weyl space. This is particularly significant as Weyl space has at times been linked to quantum mechanics.

Weyl's geometry is characterized by a transport law whereby not only the direction of a vector changes, but also its length. Accordingly, if the vector length is ℓ at the beginning of the transport operation, it will have changed by

$$\delta\ell = \ell\kappa_\mu dx^\mu \tag{2}$$

after an infinitesimal displacement dx^μ and by

$$\delta\ell = \ell \oint \kappa_\mu dx^\mu = \ell \int f_{\mu\nu} d\sigma^{\mu\nu} \tag{3}$$

after displacement along a closed path. In (3), the κ_μ represent additional field quantities (usually associated with the electromagnetic field) and $f_{\mu\nu} = \kappa_{\nu,\mu} - \kappa_{\mu,\nu}$, where a comma denotes the partial derivative. In such a geometry, the comparison of lengths at two points which are not infinitesimally close is functionally dependent on the path that is chosen to link the points, and different paths will, in general, yield different results. This ambiguity may be removed by setting up a standard of length (a gauge) at each point so that a comparison of lengths can be made, independently of the path chosen, by referring the vector length to the local standard. Weyl's geometry, however, leaves the choice of the gauge, and consequently, the length of a vector, completely arbitrary. In fact, a change of gauge can be effected by means of a factor $\lambda(x)$ that changes from point to point. Then ℓ is transformed to $\ell' = \lambda(x)\ell$ and, to first order, $\delta\ell$ is transformed to $\delta\ell' = \ell'\kappa'_\mu dx^\mu$, where $\kappa'_\mu = \kappa_\mu + (\ln\lambda)_{,\mu}$. In particular, $ds^2 = g_{\mu\nu}dx^\mu dx^\nu$ becomes $(ds')^2 = \lambda^2 ds^2$, of which (1) is a particular example. In general, a tensor T is said to have a power (or weight) n if under a gauge transformation it transforms as $T' = \lambda^n T$. A gauge-covariant calculus, which derives a connection and leads to an operation of gauge-covariant differentiation, can be developed [3–5] without difficulty. This approach leads to a geometrical interpretation of gravitation and electromagnetism still unsurpassed in its beauty and simplicity.

The suggestion of a link between quantum mechanics and Weyl space was made long ago by London [6]. If, in fact, one considers the H-atom and assumes, for simplicity, that the electron orbits are circular, one obtains from classical mechanics and electromagnetism the relationship

$$\frac{mv^2}{r} = \frac{e^2}{r^2} \tag{4}$$

from which the orbital period $T = 2\pi r/v$ and velocity $v = e/\sqrt{mr}$ follow. However, during the time $\tau = cT$, the scale of length changes according to

$$\ell = \ell_0 \exp\left(\int_0^\tau \varphi_0 d\tau\right) = \ell_0 \exp(\varphi_0 \tau) \tag{5}$$

and vanishes on those orbits for which $\exp(\varphi_0 \tau) = 1$ or

$$\varphi_0 \tau = 2\pi i n, \tag{6}$$

where n is an arbitrary integer. Equation (5) implies a connection between the phase of the wave function and the change in length scale due to the unusual transport properties of Weyl space. Unfortunately, Bohr's formula for the radii of the orbits,

$$r_n = \frac{\hbar^2}{e^2 m} n^2, \tag{7}$$

follows from (4), (6) and $\varphi_0 = \alpha_e/r$ only if the fine structure constant α_e is imaginary. This dismaying situation was subsequently remedied by Caianiello [7] with the introduction of complex gauge transformations that return α_e to its real value. Ehlers, Pirani and Schild [8] arrive at a Weyl geometry by starting from primitive concepts such as event, light ray and freely falling particles. Since quantum mechanics should agree with classical mechanics in the classical limit, Weyl space may be regarded as an appropriate space in which to describe quantum phenomena. In fact, Santamato [9] has shown that a theory, fully equivalent to quantum mechanics, can be constructed where the "quantum force" of Bohm can be precisely related to the vector transport law of Weyl's geometry. Santamato assumes the law of transport to be determined by the (stochastic) motion of the particle itself, while in Bohm's interpretation only the initial conditions are random. These links provide, in the authors' opinion, sufficient motivation to explore further the geometrical aspects of quantum mechanics and the related question of maximal acceleration.

2. A particular theory of the Weyl type.

In 1973, Dirac [4] introduced the new action (the notation is that of ref. [5])

$$I = \int \left\{ -\frac{1}{4} f_{\mu\nu} f^{\mu\nu} - \beta^2 \bar{R} + k\beta_{*\mu}\beta^{*\mu} + c\beta^4 \right\} \sqrt{-g}\, d^4 x, \tag{8}$$

where

$$\bar{R} = R + 6\kappa^\alpha{}_{;\alpha} - 6\kappa^\alpha \kappa_\alpha, \tag{9}$$

$\beta(x)$ is a real scalar field that acts as a constraint and k and c are arbitrary constants. The importance of (8) is twofold. First, unlike Weyl's original action, (8) is linear in the curvature invariants. This property will prove particularly useful in the boundary value problem to be studied below. Second, (8) bears a strong resemblance [5] to the action leading to the Landau-Ginzburg theory of superconductivity, which is a well-known example of quantum theory. Unfortunately, a wave equation, which is a desired feature in the present work, cannot follow from the the action (8) with β real. This particular feature can, however, be obtained by starting from the action [5]

$$I_c = \int \left\{ -\frac{1}{4} f_{\mu\nu} f^{\mu\nu} + |\beta|^2 \bar{R} + k|\beta_{*\mu}\beta^{*\mu}| + \lambda|\beta|^4 \right\} \sqrt{-g} d^4 x +$$

$$+ \int \rho g^{\mu\nu} \gamma_\mu (\rho_{*\nu} - \varepsilon\rho\varphi_{,\nu}) \sqrt{-g} d^4 x, \tag{10}$$

where k and λ are arbitrary constants, $\beta = \rho e^{i\varphi}$, ρ and φ are real scalar fields, γ_μ is a Lagrange multiplier vector field and $\varepsilon = \pm 1$. The independent variation of the fields ρ, φ, $g_{\mu\nu}$, κ_μ and γ_μ leads to the equations

$$\bar{R} + 2\lambda\rho^2 + 2\varepsilon k\varphi_{,\mu}\varphi^{,\mu} = 0, \tag{11}$$

$$\left(\rho\rho^{*\mu}\right)_{*\mu} = 0, \tag{12}$$

$$\frac{1}{2}\rho^2 \left(\bar{R}^{\mu\nu} + \bar{R}^{\nu\mu} - g^{\mu\nu}\bar{R} \right) = \frac{1}{2} \left(f^{\mu\beta} f^\nu{}_\beta - \frac{1}{4} g^{\mu\nu} f_{\alpha\beta} f^{\alpha\beta} + \lambda g^{\mu\nu}\rho^4 \right) +$$

$$+ \rho^2 \left[\varepsilon k g^{\mu\nu}\varphi_{,\alpha}\varphi^{,\alpha} + 2(2 - \varepsilon k)\varphi^{,\mu}\varphi^{,\nu} + 2\varepsilon\varphi^{*(\mu\nu)} \right], \tag{13}$$

$$f^{\mu\nu}{}_{*\nu} = 4(k - 3\varepsilon)\rho^2\varphi^{,\mu}, \tag{14}$$

$$\rho_{*\mu} - \varepsilon\rho\varphi_{,\mu} = 0 \tag{15}$$

and, to ensure self-consistency, the equation

$$\left(\rho^2\gamma^\mu\right)_{*\mu} = 3\left(\rho\rho^{*\mu}\right)_{*\mu} = 0 \tag{16}$$

is imposed. Equation (16) has the particular solution $\rho^2\gamma^\mu = \varepsilon k\rho\rho^{*\mu}$. The theory also contains the wave equation [10]

$$(\nabla^\lambda + i\kappa^\lambda)(\nabla_\lambda + i\kappa_\lambda)\psi - \frac{\lambda}{3}|\psi|^2\psi - \frac{1}{6}R\psi = 0, \tag{17}$$

98

where ∇_λ is the covariant derivative associated with the Riemannian connection. In fact, if one writes $\psi = \rho e^{i\chi}$ and defines $\chi_{,\mu}$ according to

$$\sqrt{\frac{1}{3}(\varepsilon k - 3)}\varphi_{,\mu} \equiv \alpha\varphi_{,\mu} \equiv \chi_{,\mu} + \kappa_\mu, \tag{18}$$

the real part of (17) coincides with (11), while the imaginary part yields (12).

3. Motion of the Madelung fluid.

The motion of the Madelung fluid associated with the wave equation (17) can be calculated as follows. Equations (11) and (18) yield the equation

$$(\chi_{,\mu} + \kappa_\mu)(\chi^{,\mu} + \kappa^\mu) = -\frac{\varepsilon k - 3}{\varepsilon k}\rho^2\left(\frac{\lambda}{3} + \frac{\bar{R}}{6\rho^2}\right) \tag{19}$$

which, in Bohm's interpretation of quantum mechanics, is the Hamilton-Jacobi equation for a system of momentum $\chi^{,\mu} + \kappa^\mu = Mu^\mu$. From (19) one easily finds

$$M^2 = \frac{\varepsilon k - 3}{\varepsilon k}\rho^2\left(\frac{\lambda}{3} + \frac{\bar{R}}{6\rho^2}\right). \tag{20}$$

Since M is a scalar field of weight -1, the product Mu_μ is a gauge-invariant quantity and its gauge-covariant derivative with respect to the parameter s along the curve can be re-expressed as follows:

$$\frac{\Delta}{ds}(Mu_\mu) = \alpha\frac{\Delta}{ds}(\varphi_{,\mu}) \equiv \alpha\varphi_{*\mu\nu}u^\nu. \tag{21}$$

From $\varphi_{*\mu} = \varphi_{,\mu}$, one obtains

$$\varphi_{*\mu\nu} = \varphi_{*\mu,\nu} - \bar{\Gamma}^\alpha_{\mu\nu}\varphi_{,\alpha} = \varphi_{,\mu\nu} - \bar{\Gamma}^\alpha_{\mu\nu}\varphi_{,\alpha} \tag{22}$$

and, with the help of (15),

$$\varphi_{*\mu\nu} - \varphi_{*\nu\mu} = \varphi_{,\mu\nu} - \varphi_{,\nu\mu} = \varepsilon f_{\nu\mu}. \tag{23}$$

Substituting (23) into (21), one finds

$$\begin{aligned}
\frac{\Delta}{ds}(Mu_\mu) &= \alpha(\varphi_{*\mu\nu} - \varphi_{*\nu\mu})u^\nu + \alpha\varphi_{*\nu\mu}u^\nu \\
&= -\varepsilon\alpha f_{\mu\nu}u^\nu + \alpha\varphi_{*\nu\mu}\left(\frac{\alpha\varphi^{,\nu}}{M}\right) \\
&= -\varepsilon\alpha f_{\mu\nu}u^\nu + \frac{\alpha^2}{2M}(\varphi_{*\nu}\varphi^{*\nu})_{*\mu} \\
&= -\varepsilon\alpha f_{\mu\nu}u^\nu - M_{*\mu},
\end{aligned} \tag{24}$$

since $\varphi_{*\nu}\varphi^{*\nu} = -M^2/\alpha^2$. The equation of motion (24) also follows by differentiating (13) and imposing the Bianchi identities [11]; as well, its structure coincides with that of the corresponding equation of motion of a test particle [12]. Consider now the gauge-invariant and coordinate-invariant quantity

$$\Omega \equiv -\frac{1}{M^2}\Delta(Mu_\mu)\Delta(Mu^\mu)$$

$$= \frac{1}{M^2}\left(\frac{\Delta M}{ds}\right)^2\left[1 - M^2\left(\frac{\Delta M}{ds}\right)^{-2}\left(\frac{\Delta u_\mu}{ds}\right)^2\right]ds^2, \qquad (25)$$

where the conformal factor in the square brackets will henceforth be denoted by ξ^2. It defines a gauge-invariant proper time along a fluid element worldline. This invariant is the result of a conformal transformation applied to the Riemannian line element and characterizes the motion as the continuous unfolding of a conformal transformation in Weyl space. During the motion, Ω remains invariant in form and value, though the acceleration of the fluid does not. In order to study the problem of the maximal acceleration, one must deal with particles of finite size embedded in the Madelung fluid. In the next section, such a particle is constructed as a bubble of Riemannian space embedded in Weyl space. This obviously breaks the conformal invariance in the bubble's interior and fixes its scale. The coexistence rules for the two spaces are rather complicated, but may be obtained by extending the Gauss-Mainardi-Codazzi (GMC) formalism to Weyl space.

4. The Gauss-Mainardi-Codazzi Formalism.

Consider a timelike hypersurface Σ that divides spacetime into two four-dimensional regions V^I and V^E, both having Σ as their common boundary. Σ represents the history of an infinitesimally thin shell of matter. The intrinsic (2+1)-dimensional Weyl space on Σ is defined by the condition that the spacetime metrics in V^I and V^E induce the same metric $h_{\mu\nu}$ on Σ according to the equation

$$h_{\mu\nu} = g_{\mu\nu} - n_\mu n_\nu, \qquad (26)$$

where $n^\mu = g^{\mu\nu} n_\nu$ is the unit spacelike ($n_\mu n^\mu = 1$) vector field normal to Σ in the direction from V^I to V^E. In order to define the bending of Σ in V^I and V^E, one defines the three-dimensional extrinsic curvature tensor (or second fundamental form)

$$\bar{K}_{\mu\nu} = -h_\mu{}^\alpha \bar{\nabla}_\alpha n_\nu, \tag{27}$$

where $\bar{\nabla}_\alpha$ indicates gauge-covariant differentiation. The GMC equations relating the curvature in V^E to the intrinsic curvature are [13]

$$h_\beta{}^\sigma h_\alpha{}^\rho h_\nu{}^\delta h^\mu{}_\lambda \bar{R}^\lambda{}_{\delta\rho\sigma} = {}^3\bar{R}^\mu{}_{\nu\alpha\beta} + \bar{K}_{\alpha\nu} \bar{K}_\beta{}^\mu - \bar{K}_{\beta\nu} \bar{K}_\alpha{}^\mu, \tag{28}$$

$$n^\mu h_\alpha{}^\nu \bar{R}_{\mu\nu} = \bar{D}_\alpha \bar{K} - \bar{D}_\mu \bar{K}^\mu{}_\alpha \tag{29}$$

and

$$n^\alpha n^\beta h_\nu{}^\gamma h^\mu{}_\delta \bar{R}^\delta{}_{\beta\gamma\alpha} = \mathcal{L}_n \bar{K}^\mu{}_\nu - \bar{K}^\alpha{}_\nu \bar{K}^\mu{}_\alpha, \tag{30}$$

where $\bar{D}_\alpha$ is the operator of intrinsic, gauge-covariant differentiation and $\mathcal{L}_n$ is the Lie derivative. In order to obtain relations that can be referred to a particular model of matter, it is useful to re-express (28)–(30) in terms of the Einstein tensor $G_{\mu\nu}$. One obtains [13]

$$n_\mu n^\nu G^\mu{}_\nu = -\frac{1}{2}\left({}^3R + K_{\mu\nu} K^{\mu\nu} - K^2\right) - D_\mu \kappa^\mu + 2 h_\mu{}^\nu \kappa^\mu \kappa_\nu + 2K n^\mu \kappa_\mu, \tag{31}$$

$$n_\mu h_\alpha{}^\nu G^\mu{}_\nu = D_\alpha K - D_\mu K^\mu{}_\alpha \tag{32}$$

and

$$h^\alpha{}_\mu h_\beta{}^\nu G^\mu{}_\nu = {}^3 G^\alpha{}_\beta + \left(K^\alpha{}_\beta - h^\alpha{}_\beta K\right)_{,n} - K K^\alpha{}_\beta + \frac{1}{2} h^\alpha{}_\beta \left(K_{\mu\nu} K^{\mu\nu} + K^2\right) -$$

$$-2\left(K^\alpha{}_\beta - h^\alpha{}_\beta K\right) n^\lambda \kappa_\lambda + 2 h^\alpha{}_\beta h_\mu{}^\nu \kappa^\mu \kappa_\nu. \tag{33}$$

The Einstein equations $G_{\mu\nu} = T_{\mu\nu}$ then specify the particle model through $T_{\mu\nu}$. While the form of the GMC equations depends only on the nature of the geometry itself, the analysis of the junction conditions across a thin shell depends on the field equations, and hence on the nature of the theory selected. The following junction conditions across Σ are imposed: The quantities $g_{\mu\nu}$, κ_μ, φ and ρ are all assumed smooth in V^I and V^E, including Σ; they are all assumed continuous across Σ and their derivatives, which are discontinuous across Σ, are taken to be sufficiently continuous in (V^I, V^E) for the usual equations to apply.

Define

$$S^\mu{}_\nu \equiv \lim_{\varepsilon \to 0} \int_{-\varepsilon}^{\varepsilon} T^\mu{}_\nu dn \tag{34}$$

as the intrinsic stress-energy tensor on Σ. The junction conditions for the gravitational field are obtained by integrating the Einstein equations in the normal direction across Σ from $-\varepsilon$ to $+\varepsilon$ and then taking the limit $\varepsilon \to 0$. One gets

$$\lim_{\varepsilon \to 0} \int_{-\varepsilon}^{\varepsilon} n_\mu n^\nu G^\mu{}_\nu dn = 0 = n_\mu n^\nu S^\mu{}_\nu, \tag{35}$$

$$\lim_{\varepsilon \to 0} \int_{-\varepsilon}^{\varepsilon} n_\mu h_\alpha{}^\nu G^\mu{}_\nu dn = 0 = n_\mu h_\alpha{}^\nu S^\mu{}_\nu \tag{36}$$

and

$$\lim_{\varepsilon \to 0} \int_{-\varepsilon}^{\varepsilon} h^\alpha{}_\mu h_\beta{}^\nu G^\mu{}_\nu dn = \gamma^\alpha{}_\beta - h^\alpha{}_\beta \gamma = h^\alpha{}_\mu h_\beta{}^\nu S^\mu{}_\nu, \tag{37}$$

where

$$\gamma^\alpha{}_\beta \equiv [K^\alpha{}_\beta] \equiv \lim_{\varepsilon \to 0}[K^\alpha{}_\beta(n = +\varepsilon) - K^\alpha{}_\beta(n = -\varepsilon)] \tag{38}$$

and $\gamma \equiv \gamma^\mu{}_\mu$. Equation (37) differs from the corresponding equation in Riemannian space because the existence of $S^\mu{}_\nu$ may be here related to the conformal invariance of the theory instead of being introduced in an *ad hoc* manner. For the electromagnetic field, the junction conditions are given by

$$[n_\mu h_\alpha{}^\nu f_\nu{}^\mu] = h_\alpha{}^\mu j_\mu \quad \text{and} \quad [h_\mu{}^\alpha h^\nu{}_\beta f^\mu{}_\nu] = 0. \tag{39}$$

Using the definition

$$\{\Phi\} \equiv \lim_{\varepsilon \to 0} \frac{1}{2}[\Phi(n = +\varepsilon) + \Phi(n = -\varepsilon)], \tag{40}$$

one can obtain from (31)–(33) and (34)–(37) the equations

$$\{K^\mu{}_\nu\}S_\mu{}^\nu + [n_\mu n^\nu T^\mu{}_\nu] = 0 \tag{41}$$

and

$$D_\mu(h^\mu{}_\alpha h_\nu{}^\beta S^\alpha{}_\beta) + [n_\alpha h_\nu{}^\beta T^\alpha{}_\beta] = 0. \tag{42}$$

Equations (41) and (42) can be used to describe the balance of stress-energy-momentum in the shell. Indeed, the requirement that the gravitational and electromagnetic fields join at Σ in accordance with the junction conditions places constraints on the dynamical behavior of the bubble since Σ represents the *history* of the thin shell. The first step in the analysis of the bubble dynamics is to obtain a solution to the Einstein equations. This problem is made tractable by considering a spherically symmetric bubble.

The most general static, spherically symmetric metric has the form

$$ds^2 = -e^{\nu(r)}dt^2 + e^{\mu(r)}dr^2 + r^2(d\theta^2 + \sin^2\theta d\phi^2), \tag{43}$$

where $\nu(r)$ and $\mu(r)$ are functions that must be determined from the Einstein equations for specific expressions of $T_{\mu\nu}$. As energy-momentum tensor in V^I, one may take $T^I_{\mu\nu} = \lambda g_{\mu\nu}\rho_0^2/2$ with ρ_0 a constant, where, for λ less than, equal to, or greater than zero, the interior space is then de Sitter, Minkowski or anti-de Sitter, respectively. $T^I_{\mu\nu}$ can, in fact, be obtained from (13) for $\rho = \rho_0$ and $\varphi_{,\mu} = \kappa_\mu = 0$. The solution of the Einstein equations is represented in this case by $e^{-\mu_I} = 1 + \lambda\rho_0^2 r^2/6 = e^{\nu_I}$. For $T^E_{\mu\nu}$, we take

$$T^E_{\mu\nu} = \frac{1}{2\rho^2}E_{\mu\nu} + I_{\mu\nu} + \frac{1}{2}\lambda g_{\mu\nu}\rho^2 + H_{\mu\nu} \tag{44}$$

which coincides with the right-hand-side of (13) divided by $\rho^2/2$, where $E_{\mu\nu}$ is the usual Maxwell tensor,

$$I_{\mu\nu} = \frac{2}{\rho}\left(\rho_{;\mu\nu} - g_{\mu\nu}\rho_{;\alpha}{}^\alpha\right) - \frac{1}{\rho^2}\left(4\rho_{,\mu}\rho_{,\nu} - g_{\mu\nu}\rho_{,\alpha}\rho^{,\alpha}\right) \tag{45}$$

and

$$H_{\mu\nu} = -2(\varepsilon k - 3)\left(\varphi_{,\mu}\varphi_{,\nu} - \frac{1}{2}g_{\mu\nu}\varphi_{,\alpha}\varphi^{,\alpha}\right). \tag{46}$$

For small values of the current in V^E, the term $H_{\mu\nu}$ may be neglected and the solution has the form [13]

$$e^{-\mu_E} = \left(1 + r\frac{\rho'}{\rho}\right)^{-2}\left[1 - \frac{2m}{\rho r} + \frac{q^2}{4\rho^2 r^2} + \frac{1}{6}\lambda\rho^2 r^2\right], \tag{47}$$

$$e^{\nu_E} = (\ell_0\rho)^{-2}\left[1 - \frac{2m}{\rho r} + \frac{q^2}{4\rho^2 r^2} + \frac{1}{6}\lambda\rho^2 r^2\right],\tag{48}$$

where the field ρ is an arbitrary function of r and m, q and ℓ_0 are integration constants. By substituting $T_{\mu\nu}$ into (34) one finds [13]

$$h^{\alpha}{}_{\mu}h_{\beta}{}^{\nu}S^{\mu}{}_{\nu} = -2\sigma h^{\alpha}{}_{\beta},\tag{49}$$

where $\sigma \equiv [n^{\mu}(\ln\rho)_{,\mu}]$. This result, which also agrees with (35)–(37), indicates that for $\sigma > 0$, the bubble is under a surface tension that opposes the Coulomb repulsion due to the surface charge. The intrinsic stress-energy tensor (49) is characteristic of a domain wall of surface energy 2σ. One can also show from (42) that $\sigma = $ constant and from (41) that the normal forces acting on the respective sides of Σ are not equal.

5. Maximal acceleration

The spherical bubble constructed in the preceding section by employing the GMC formalism is capable of sustaining small radial oscillations whose period and amplitude can be determined without great difficulty. For a purely spherical shell, however, the symmetry requires that the tangential components of the acceleration vanish: $h_{\mu}{}^{\nu}a_{\nu} = 0$. As well, the fields from which the particle is constructed have the interesting property that the divergence of $T^{E}_{\mu\nu}$ (in the absence of the Madelung fluid tensor $H_{\mu\nu}$) vanishes identically, so that the tangential motion remains indeterminate.

An essential feature of Bohm's interpretation of quantum mechanics is the manner in which the particle is viewed as being inseparable from the Madelung fluid. Taking $\varphi_{,\mu} = 0$ in V^{I} and discontinuous across Σ leads to the possibility of embedding the bubble in the Madelung fluid in accordance with (41) and (42), while the surface stress-energy tensor (45) remains unchanged. In this manner, the bubble acquires a new dynamical nature as it is guided in its motion by the fluid in V^{E}. The realization of this guidance process can be seen by considering the dynamical behavior of an element of fluid at a point P on the exterior surface of the bubble. By construction, the metric tensor is continuous across Σ at P.

104

The motion of the fluid along its worldline from P to a subsequent point P', at which the fluid element and the bubble are still in contact, can be viewed as the result of a conformal transformation induced by the factor ξ^2, as indicated by (25). The metric tensor at P' in V^E can therefore be obtained from its corresponding value at P by applying a conformal transformation with the same factor ξ^2 and, by continuity, the intrinsic metric at P' is also determined. The resulting identity, $h_{\mu\nu}(P') = \xi^2 h_{\mu\nu}(P)$, leads to the conclusion that the bubble and fluid must move in step. Consequently, equations (19) and (25) may be applied to the particle itself.

On Σ, the tangential component of the acceleration must satisfy (25) which, when written in detail, reads

$$\Omega = \left[\frac{1}{M^2} \left(\frac{dM}{ds} \right)^2 - |h_{\mu}{}^{\nu} a^{\mu} a_{\nu}| \right] ds^2. \tag{50}$$

Since the factor in square brackets must be positive if Ω is to be interpreted as a gauge-invariant proper time, it follows that

$$|h_{\mu}{}^{\nu} a^{\mu} a_{\nu}| < \frac{1}{M^2} \left(\frac{dM}{ds} \right)^2. \tag{51}$$

The intrinsic equation (50) follows without any assumption about the symmetry; hence, equation (51) is valid for a bubble of arbitrary shape. From the equation of continuity one obtains

$$D_{\mu} u^{\mu} = -\frac{1}{M} M_{,\mu} u^{\mu} = -\frac{1}{M} \frac{dM}{ds}; \tag{52}$$

while, for a spherically symmetric shell,

$$D_{\mu} u^{\mu} = h_{\mu}{}^{\nu} \nabla_{\nu} u^{\mu} = h_{\mu}{}^{\nu} (u^{\mu}{}_{,\nu} + \Gamma^{\mu}{}_{\alpha\nu} u^{\alpha}) = g_{\mu}{}^{\nu} (\Gamma^{\mu}{}_{0\nu} u^0 + \Gamma^{\mu}{}_{1\nu} u^1). \tag{53}$$

The connection coefficients in (53) can be determined from the interior metric and are $\Gamma^{\mu}{}_{0\mu} = 0$ and $\Gamma^{\mu}{}_{1\mu} = 2/r$. With these, (53) yields

$$D_{\mu} u^{\mu} = 2\frac{\dot{R}}{R} \tag{54}$$

at $r = R$. Since $\dot{R} \leq 1$, one obtains from (54), (52) and (51)

$$|h_{\mu}{}^{\nu}a^{\mu}a_{\nu}| < \frac{1}{M^2}\left(\frac{dM}{ds}\right)^2 \leq \frac{4}{R^2}, \tag{55}$$

provided the shell is deformed during the motion from its original spherical shape. The maximal acceleration required by (55) is consistent with Caianiello's particle-dependent expression. This result provides further evidence of the quantum nature of Weyl space.

Acknowledgments

This research was supported by the Natural Sciences and Engineering Research Council of Canada.

References

[1] E. R. Caianiello, Lett. Nuovo Cim. **32** (1981) 65; E. R. Caianiello, S. De Filippo, G. Marmo and G. Vilasi, Lett. Nuovo Cim. **34** (1982) 112; E. R. Caianiello, Lett. Nuovo Cim. **41** (1984) 370; G. Scarpetta, Lett. Nuovo Cim. **41** (1984) 51; W. Guz and G. Scarpetta, in Quantum Field Theory, ed. F. Mancini (North Holland, Amsterdam, 1986), p. 233; M. Gasperini and G. Scarpetta, in Proc. of the Fifth M. Grossmann Meeting, Perth, eds. D. G. Blair and M. J. Buckingham (World Scientific, Singapore, 1989), p. 771; E. R. Caianiello, M. Gasperini and G. Scarpetta, Nuovo Cim. **B105** (1990) 259.

[2] E. R. Caianiello, A. Feoli, M. Gasperini and G. Scarpetta, Int. J. Theor. Phys. **29** (1990) 131.

[3] H. Weyl, Sitzugsber. Preuss. Akad. Wiss. (1918) 465; Raum. Zeit. Materie. 4th E. (Springer, Berlin, 1921; English translation: Space, Time, Matter, Dover Publications, New York, 1950).

[4] P. A. M. Dirac, Proc. Roy. Soc. London **A333** (1973) 403.

[5] D. Gregorash and G. Papini, Nuovo Cim. **B63** (1981) 487; G. Papini, in High-Energy Physics, Proc. of the 20th Orbis Scientiae Conf. Dedicated to P. A. M. Dirac's 80th Year, eds. B. Kursunoglu, S. L. Mintz and A. Perlmutter (Plenum Publishing Corp., New York, 1985), p. 179.

[6] F. London, Z. Physik **42** (1927) 375.

[7] E. R. Caianiello, Lett. Nuovo Cim. **25** (1979) 225.

[8] J. Ehlers, F. A. E. Pirani and A. Schild, in General Relativity, ed. L. O'Raifeartaigh (Oxford Univ. Press, 1972), p. 63.

[9] E. Santamato, Phys. Rev. **D29** (1984) 216.

[10] G. Papini, in Proc. of the Fifth M. Grossmann Meeting, Perth, eds. D. G. Blair and M. J. Buckingham (World Scientific, Singapore, 1989), p. 787.

[11] W. R. Wood and G. Papini, In preparation.

[12] D. Gregorash and G. Papini, Nuovo Cim. **B70** (1982) 259.

[13] W. R. Wood and G. Papini, Submitted for publication.

ACCELERATED STRINGS WITH LIMITED PROPER ACCELERATION

A. Feoli   G. Scarpetta

Dipartimento di Fisica Teorica e S. M. S. A. dell' Universitá di Salerno -

84081 Baronissi (Salerno), Italia

Istituto Nazionale di Fisica Nucleare - Sezione di Napoli

1 Introduction

Nearly ten years ago, interpreting quantum physics in the light of the principles governing information theory and modern system theory, E.R. Caianiello [1, 2, 3] introduced a new fundamental constant of nature, the maximal proper acceleration of a massive particle in arbitrary motion, a quantity that is deeply interconnected with the extended nature of particles, whose finite extension cannot be neglected without introducing in quantum field theory troubles and divergencies connected with the point-like approximation.

The maximal proper acceleration has been also derived starting from the first principles of quantum mechanics and relativity [4, 5]; let $|\psi> \in \mathcal{H}$ be the normalized ket describing a quantum particle and let $\hat{A}$ and $\hat{B}$ be two operators corresponding to given physical observables; it is well known that, for arbitrary non-commuting observables, it follows, from the Heisenberg uncertainty relations, that the product of uncertainties satisfies the following inequality

$$\Delta A \, \Delta B \geq \frac{1}{2} \left| <\psi|[\hat{A}, \hat{B}]|\psi> \right| \tag{1}$$

for $|\psi>$, $\hat{A}|\psi>$ and $\hat{B}|\psi> \in D(\hat{A}) \cap D(\hat{B})$. Choosing $\hat{B}$ as the hamiltonian operator $\hat{H}$ and $\hat{A}$ as the velocity operator of the particle, defined through the equation $i\hbar\hat{v} = [\hat{x}, \hat{H}]$, we have

$$\Delta v \,\Delta H \geq \frac{1}{2}\left|<\psi(t)|[\hat{v}, \hat{H}]|\psi(t)>\right| = \frac{1}{2}\hbar\left|\frac{d}{dt}<\psi(t)|\hat{v}|\psi(t)>\right| \qquad (2)$$

According to the Eherenfest theorem, the right hand side of eq. (2) represents the acceleration of the particle; therefore

$$|a| \leq \frac{2}{\hbar}\Delta v \,\Delta H \qquad (3)$$

Using the relativistic condition of the finite upper limit of the velocity, and, moreover, the condition that particles with negative rest energy are unphysical, one can derive that in the proper frame of the particle $\Delta v \leq c$ and $\Delta H \leq mc^2$, from which it follows

$$|a| \leq a_{max} = \frac{2mc^3}{\hbar} \qquad (4)$$

Eq. (3) implies that if the quantum state of the particle is an eigenfunction of the hamiltonian, then $\Delta H = 0$, and the acceleration of the particle is $|a| = 0$; this fact explains well known results of quantum mechanics, giving a clear explanation of the stability of atoms and of the absence of energy radiation from the electron, as long as it remains in an eigenstate of the hamiltonian.

Along this line of thinking, many papers have been published in the last years, each of one introducing the maximal proper acceleration starting from different motivations and from different theoretical schemes [6, 7, 8, 9, 13]; it is well to note that in some of these works the maximal acceleration is fixed by the Planck mass

$$m_p = \left(\frac{\hbar c}{G}\right)^{\frac{1}{2}}$$

It is well known that massive extended objects imply critical accelerations, determined from the extension of the particles and from the causal structure of the space–time manifold.

For instance in classical relativity [10] an object of proper length λ, in which one extreme point is moving with acceleration a with respect to the other, will develop a Rindler horizon at a proper distance a^{-1} from the accelerated extremity, so that all parts of the object can be causally connected only if $\lambda < a^{-1}$; this implies a proper critical acceleration $a_c \simeq \lambda^{-1}$ which depends on λ and diverges in the limit in which the objects reduces to a point–like particle.

In the quantum relativistic context, the analysis of string propagation in cosmological backgrounds revealed that accelerations higher than the critical one give rise to the onset of Jeans–like instabilities [11] in which the string oscillating modes develop imaginary frequencies and the string's proper length diverges. Gasperini [12] has given a very interesting cinematic interpretation of this string instability showing that it occurs when the acceleration induced by the background gravitational field is large enough to render the two string extremities causally disconnected because of the Rindler horizon associated to their relative acceleration. This critical acceleration a_c is determined by the string size λ and is given by $a_c = \lambda^{-1} = (m\alpha)^{-1}$ where m is the string mass and α^{-1} the usual string tension.

Frolov and Sanchez [13] derived that an universal critical acceleration $a_c \simeq \lambda^{-1}$ must be a general property of the strings; they analyze the dynamics of an open string of length λ in flat space with masses concentrated at its ends as "heavy" particles coupled to an external force, and prove that rigid equilibrium configurations of the accelerated string (moving as a rigid body without any excitation) exist only for acceleration less than the critical one, beyond which the string begins to grow indefinitely.

In previous cases the critical acceleration arises as a dynamical effect due to the interplay of the Rindler horizon with the finite extension of the string; in the Caianiello's proposal the maximal proper acceleration is a basic physical property of all the particles, so that it must be included from the outset in the physical laws and implies a modification of the metric structure of space-time.

The simplest theoretical framework which includes maximal proper acceleration, consists of considering as physical invariant, not the einsteinean four dimensional space time distance element, but a new one, more general, defined in an eight dimensional space–time tangent bundle TM, whose coordinates are [14]:

$$x^A = \left(x^\mu; \frac{\hbar}{mc}\dot{x}^\mu \right)$$

where m is the rest mass of the particle, $x^\mu = (ct, \vec{x})$ is the usual space–time four–vector, and

$$\dot{x}^\mu = \frac{dx^\mu}{ds}$$

110

are the relativistic four–velocity coordinates in the four-velocity fiber manifold. (Conventions: $A, B, \ldots = 0, 1, 2, \ldots 7$; $\mu, \nu, \ldots = 0, 1, \ldots 3$; from now on we use natural units $\hbar = c = 1$).

The four-dimensional space-time invariant

$$ds^2 = g_{\mu\nu} dx^\mu dx^\nu$$

has to be replaced now by a more general physical invariant representing the infinitesimal element of distance in TM

$$d\tau^2 = g_{AB} dx^A dx^B = g_{\mu\nu} dx^\mu dx^\nu + \frac{1}{m^2} g_{\mu\nu} d\dot{x}^\mu d\dot{x}^\nu. \tag{5}$$

thus introducing an upper limit on the proper acceleration which depends on the rest mass of the particle $a_{\max} = m$.

The introduction of an invariant interval in the eight-dimensional space–time tangent bundle TM can be also interpreted as a regularization procedure of field equations, alternative respect to methods which discretize the space-time with a lattice and introduce a fundamental length; the advantage is that in such a way one preserves the continuum structure of the space-time, expliciting, at the same time, the fundamental property of a higher limit of the proper acceleration of a particle.

In this paper we analyze the string's rigid motion [15] in a Rindler background, properly modified by the condition of finite maximal proper acceleration; the interest resides in exploiting the link between the finite extension of the string and the maximal acceleration; in particular we derive that for each value of the acceleration, there is a minimal value of the string length measured in the accelerated frame, that can be equal to zero only when the acceleration is maximal. This is an effect analogous to what happens in special relativity, where only particles with null rest mass move at the maximal velocity c; in maximal acceleration physics, only null length strings, i.e. point particles, can be accelerated to the maximal acceleration. In sect. 2 we discuss the embedding procedure, by which we reduce to an effective four–dimensional space–time geometry, and in sect. 3 we study the string motion in the Rindler space.

2 The embedding procedure

The construction of the dynamic laws can be realized defining a suitable action in the eight–dimensional space–time tangent bundle TM [2, 3]; or by an embedding procedure [16], the first step of a process of successive approximations, by which one reduces to an effective four-dimensional space-time geometry. The embedding procedure constructs an effective new space–time metric $\tilde{g}_{\mu\nu}(\xi)$, through the eight parametric equations $x^A = x^A(\xi^\mu)$ that correlate the coordinates x^A of TM to the coordinates ξ^μ chosen to parametrize the four–dimensional space–time manifold V_4:

$$\tilde{g}_{\mu\nu}(\xi) = g_{AB} \frac{\partial x^A}{\partial \xi^\mu} \frac{\partial x^B}{\partial \xi^\nu} = g_{\alpha\beta} \left(\frac{\partial x^\alpha}{\partial \xi^\mu} \frac{\partial x^\beta}{\partial \xi^\nu} + \frac{1}{m^2} \frac{\partial \dot{x}^\alpha}{\partial \xi^\mu} \frac{\partial \dot{x}^\beta}{\partial \xi^\nu} \right) \tag{6}$$

The first step approximation introduced by this procedure consists in defining the eight parametric equations $\{x^\mu = x^\mu(\xi^\alpha); \dot{x}^\mu = \dot{x}^\mu(\xi^\alpha)\}$ by the solutions to the ordinary relativistic equations of motion. For instance, in a pure gravitational case, the velocity field obeys to the classical geodesic equation of motion

$$d\dot{x}^\mu = -\Gamma^\mu_{\alpha\beta} \dot{x}^\alpha dx^\beta \tag{7}$$

and the "geodesic embedding" induces the following effective space-time metric $\tilde{g}_{\mu\nu}$:

$$d\tau^2 = \tilde{g}_{\mu\nu} dx^\mu dx^\nu = \left[g_{\mu\nu} + \frac{1}{m^2} \Gamma_{\alpha\mu\beta} \Gamma^\alpha_{\nu\gamma} \dot{x}^\beta \dot{x}^\gamma \right] dx^\mu dx^\nu \tag{8}.$$

This effective metric depends on the space-time x^μ on the four-velocity field $\dot{x}^\mu$ and on the connection Γ and thus represent a generalization of the metric of Finsler spaces.

The embedding procedure produces corrections to the given background four dimensional geometry, which disappear in the classical limit $m^{-1} \to 0$; however these corrections induce in general a non vanishing curvature even starting from an eight–dimensional space–time tangent bundle TM with a flat metric.

Applying the embedding procedure to the case of Rindler space, the classical four dimensional invariant

$$ds^2 = -\xi^2 d\eta^2 + d\xi^2 + dy^2 + dz^2 \tag{9}$$

(where the Rindler coordinates $\xi = 1/a$, $\eta = as$, and a is the acceleration) was replaced [17] by

$$d\tau^2 = - \left(\xi^2 - \frac{1}{m^2} \right) d\eta^2 + d\xi^2 + dy^2 + dz^2 \tag{10}$$

112

which defines a geometry of a curved manifold with scalar curvature

$$R = -\frac{2m^2}{(m^2\xi^2 - 1)^2} \tag{11}$$

This new metric provides in a natural way the horizon regularization introduced by H.J. De Vega and N. Sanchez [18] to quantize a string in an accelerated frame, so it seems the right background metric to study the dynamical behaviour of a uniformly accelerated string.

3. String motion in Rindler space

We consider an open string and suppose that there are two heavy particles (e.g. monopoles [19]) at its ends, numbered with the indices 1 and 2, on which some external force, for instance the gravitational one created by a static uniform infinite plane layer of matter, is applied in such a way that both particles are moving with the same constant proper acceleration a. In an inertial frame of reference the coordinates (t, x, y, z) are chosen in such a way that the x–axis coincides with the direction of acceleration, while the y–axis is parallel to the string of length L, whose ends have coordinate $y_1 = -y_2 = \frac{L}{2}$; correspondingly, the Rindler coordinates are (η, ξ, y, z) and in the accelerated Rindler frame the particles at the string ends obey the following boundary conditions:

$$\xi_1 = \xi_2 = \frac{1}{a} \qquad y_1 = -y_2 = \frac{L}{2} \qquad z_1 = z_2 = 0 \tag{12}$$

In this simplified model, all the informations regarding the interaction between the string and the external forces are contained in the boundary conditions. The string action in a curved manifold is given by [20]:

$$S = \int d\sigma d\theta \sqrt{\Lambda}\, \Lambda^{\alpha\beta}\, \tilde{g}_{\mu\nu} \partial_\alpha x^\mu \partial_\beta x^\nu \tag{13}$$

where the string tension has been normalized to one, $\Lambda_{\alpha\beta}$ is the world–sheet metric, $\tilde{g}_{\mu\nu}$ is the metric of the curved Rindler manifold, and θ and σ are the world–sheet time and space variables. As it is well known, this action is invariant under reparametrization of the world–sheet coordinates, and this allows us to choose $\Lambda_{\alpha\beta}$ in the so called conformal form

$$\Lambda_{\alpha\beta} = \Lambda(\sigma, \theta)\eta_{\alpha\beta}$$

Working in the conformal gauge, from the action (13) one derives the following equation of motion of the string:

$$\ddot{x}^{\mu} - x''^{\mu} + \Gamma^{\mu}_{\nu\delta}(\dot{x}^{\nu} + x'^{\nu})(\dot{x}^{\delta} - x'^{\delta}) = 0 \tag{14}$$

and the constraint equations:

$$\tilde{g}_{\mu\nu}(\dot{x}^{\mu}\dot{x}^{\nu} + x'^{\mu}x'^{\nu}) = 0 \qquad \tilde{g}_{\mu\nu}\dot{x}^{\mu}x'^{\nu} = 0 \tag{15}$$

where $\dot{x}^{\mu} = \partial x^{\mu}/\partial\theta$, $x'^{\mu} = \partial x^{\mu}/\partial\sigma$, and $\Gamma^{\mu}_{\nu\delta}$ are the Christoffel connections of the metric $\tilde{g}_{\mu\nu}$. Written explicitly, the equations (14) of motion of a string in the modified Rindler metric (10) and the constraint equations (15) become:

$$\left.\begin{aligned} \ddot{\eta} - \eta'' + \frac{2m^2\xi}{m^2\xi^2 - 1}(\dot{\xi}\dot{\eta} - \xi'\eta') &= 0 \\ \ddot{\xi} - \xi'' + \xi(\dot{\eta}^2 - \eta'^2) &= 0 \\ \ddot{y} - y'' &= 0 \end{aligned}\right\} \tag{16}$$

$$-(\xi^2 - m^{-2})(\dot{\eta}^2 + \eta'^2) + (\dot{\xi}^2 + \xi'^2) + (\dot{y}^2 + y'^2) = 0 \tag{17}$$

Now we are searching for a special solution of equations (16) (17) and (12), describing an uniformly accelerated string, moving as a rigid body without any excitation; this imply that η must be a linear function of θ, ξ a function of the only variable σ, and $z(\theta, \sigma) = 0$. We put $y = L\sigma$ so that the spatial parameter σ varies from $-1/2$ to $1/2$.

We obtain the following solution:

$$\xi = \sqrt{\frac{L^2}{\beta^2} + \frac{1}{m^2}}\cosh\left(\frac{\beta}{L}y\right) \tag{18}$$

$$\eta = \beta\theta \tag{19}$$

where β is a parameter that will be determined in the following.

The string configuration is described by the equation (18); it does not depend from the Rindler time η and hence the string as a whole is at rest in the Rindler frame; for $m \to \infty$ (18) reduces to the same solution found by Frolov and Sanchez [13]; moreover, because $\xi = 1/a$, we have

$$a = \frac{m\beta}{\sqrt{\beta^2 + m^2L^2}}\frac{1}{\cosh(\frac{\beta}{L}y)} \tag{20}$$

and the acceleration remain finite also in the limit $L \to 0$.

With regards to the β parameter, its value is fixed by the boundary condition that the string ends must move with the assigned acceleration a; this gives us the following equation:

$$\cosh\left(\frac{\beta}{2}\right) = \frac{\frac{\beta}{aL}}{\sqrt{1 + \frac{\beta^2}{m^2 L^2}}} \tag{21}$$

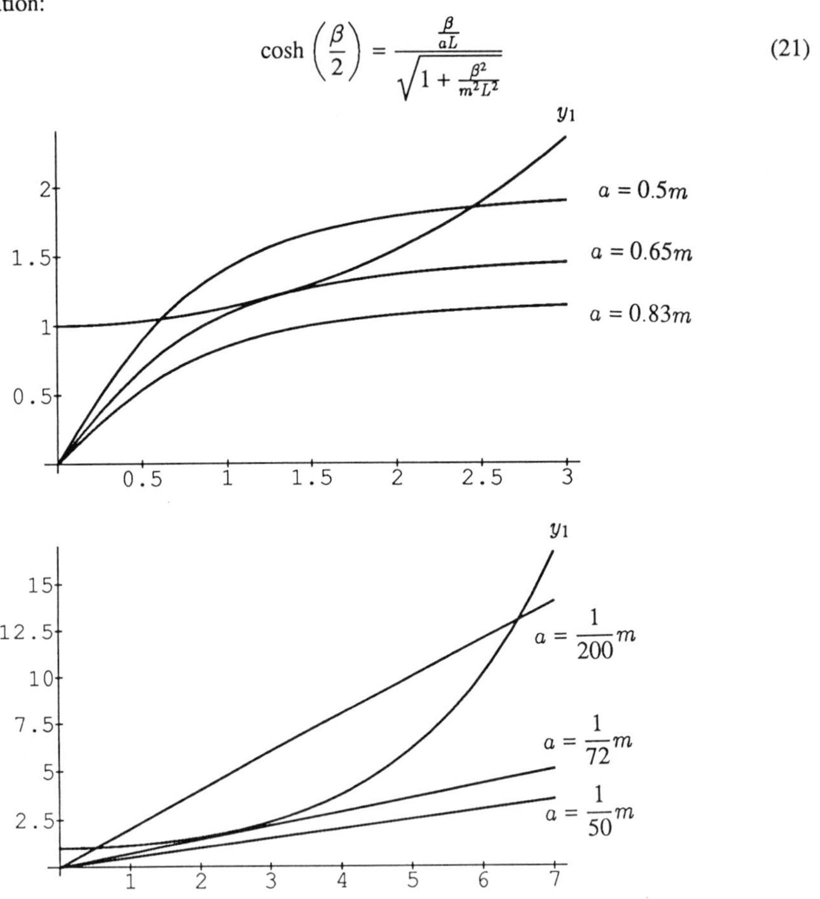

Fig. 1: plots of the functions $y_1 = \cosh(\frac{\beta}{2})$ and $y_2 = \frac{\beta}{aL}(1 + \frac{\beta^2}{m^2 L^2})^{-\frac{1}{2}}$ for two different values of the string length ($L = m^{-1}$ and $L = 100m^{-1}$)

As shown in fig. 1, where we plotted the left and right hand side of eq. (21) for different values of the acceleration a, this equation admits two, one, or no solution; for each value L, we have a corresponding critical value of the string acceleration a_c, above which there is no solution β to eq. (21), and viceversa for each value of the acceleration

a, there is a maximal value L_{max} above which again eq. (21) does not admits solutions; let us note, moreover, that for $L \neq 0$ the critical proper acceleration of the string is always less than the maximal value m.

Now we can calculate the string size l in the Rindler frame; from (18) we obtain:

$$l = \int_{-\frac{L}{2}}^{\frac{L}{2}} dy \sqrt{1 + \left(1 + \frac{\beta^2}{m^2 L^2}\right) \sinh^2 \left(\frac{\beta}{L}y\right)} =$$

$$2\sqrt{\frac{L^2}{\beta^2} + \frac{1}{m^2}} \left[w F(v, \sqrt{1-w}) - E(v, \sqrt{1-w}) + \frac{\sinh \beta}{2\sqrt{w + \sinh^2 \frac{\beta}{2}}} \right] \qquad (22)$$

where

$$v = \arctan\left(\frac{1}{\sinh \frac{\beta}{2}}\right) \qquad\qquad w = \frac{m^2 L^2}{\beta^2 + m^2 L^2}$$

and F, E are the elliptic integrals of the first and second kind respectively. In the limit $m \to \infty$ we recover the result by Frolov and Sanchez [13]:

$$\tilde{l} = \frac{2L}{\beta} \sinh\left(\frac{\beta}{2}\right) \qquad (23)$$

which goes to zero if $L \to 0$. On the contrary, if a maximal acceleration exists, l remains finite even in the limit $L \to 0$. In fact we have:

$$l = \frac{2}{m} \sinh\left(\frac{\beta}{2}\right) \tanh\left(\frac{\beta}{2}\right) \qquad (24)$$

with $\beta = 2 arccosh(m/a)$. Eq. (24) shows moreover that l becomes zero only for $\beta = 0$, i.e. only for $a = m$. The string proper size in the Rindler frame is always greater than a minimal lenght, which is function of the proper acceleration a; this minimal length goes to zero only when a reaches its maximal value m. As in classical relativity only particles with zero mass can move at the maximal velocity c, so in our theory only point particles ($l = 0$) can move at maximal acceleration m.

References

[1] E.R. Caianiello: Nuovo Cimento, **59 B**, 350, (1980)

E.R. Caianiello: Lett. Nuovo Cimento, **32**, 65, (1981)

E.R. Caianiello, S. de Filippo, G. Marmo, G. Vilasi: Lett. Nuovo Cimento, **34**, 112, (1982)

E.R. Caianiello: Lett. Nuovo Cimento, **38**, 539, (1983)

E.R. Caianiello, G. Landi: Lett. Nuovo Cimento, **42**, 70, (1985)

[2] E.R. Caianiello, G. Vilasi: Lett. Nuovo Cimento, **30**, 469, (1981)

E.R. Caianiello, S. de Filippo, G. Vilasi: Lett. Nuovo Cimento, **33**, 555, (1982)

E.R. Caianiello, W. Guz: Lett. Nuovo Cimento, **43**, 1, (1985)

[3] E.R. Caianiello, G. Marmo, G. Scarpetta: Lett. Nuovo Cimento, **36**, 487, (1983)

E.R. Caianiello, G. Marmo, G. Scarpetta: Lett. Nuovo Cimento, **37**, 361, (1983)

E.R. Caianiello, G. Marmo, G. Scarpetta: Nuovo Cimento, **86A**, 337, (1985)

[4] E.R. Caianiello: Lett. Nuovo Cimento, **41**, 370, (1984)

[5] W.R. Wood, G. Papini, Y.Q. Cai, Nuovo Cimento, **104 B**, 361, (1989)

[6] B. Mashoon: Physics Letters , **122 A**, 67, (1987)

B. Mashoon: Physics Letters , **122 A**, 299, (1987)

B. Mashoon: Physics Letters , **126 A**, 393, (1988)

B. Mashoon: Physics Letters , **145 A**, 147, (1990)

[7] M. Toller: Nuovo Cimento, **102 B**, 261, (1988)

M. Toller: Int. J. Ther. Phys., **29**, (1990)

[8] H.E. Brandt: Lett. Nuovo Cimento, **38**, 522, (1983)

H.E. Brandt: Found. of Physcs Letters, **2**, 39, (1989)

[9] M. Carmeli: Lett. Nuovo Cimento, **41**, 551, (1984)

M. Carmeli: Found. Phys., **15**, 1263, (1985)

[10] C.W. Misner, K.S. Thorne and J.A. Wheeler: "Gravitation" (WH. Freeman and Company, San Francisco, 1973), chapt. 6

[11] N. Sanchez and G. Veneziano: Nucl. Phys., **333B**, 253, (1990)

M. Gasperini, N. Sanchez, G. Veneziano: Nucl. Phys. **36B** , 365, (1991)

M. Gasperini, N. Sanchez, G. Veneziano: Int. J. Mod. Phys. 6A, 3853, (1991)

[12] M. Gasperini: Phys. Lett., **258B**, 70, (1991)

     M. Gasperini: Gen. Rel. Grav., **24**, 219, (1992)

[13] V.P. Frolov and N. Sanchez: Nucl. Phys., **349B**, 815, (1991)

[14] G. Scarpetta: Lett. Nuovo Cimento, **41**, 51, (1984)

[15] A. Feoli: "String dynamics in Rindler space in a model with maximal acceleration". Preprint, Università di Salerno

[16] E.R. Caianiello, A. Feoli, M. Gasperini, G. Scarpetta, Intern. Journal of Theoretical Physics **29**, 131, (1990)

     A. Feoli and G. Scarpetta: *in* "Advances in Theoretical Physics" ed. E.R. Caianiello, (World Scientific, Singapore, 1991), p. 68

[17] E.R. Caianiello, M. Gasperini, G. Scarpetta: Nuovo Cimento, **195B**, 259, (1990)

[18] H.J. De Vega, N. Sanchez: Nucl. Phys. **299B**, 818, (1988)

[19] A. Vilenkin: Phys. Rep. **121**, 263, (1985)

[20] H.J. De Vega, N. Sanchez: Phys. Lett. **197B**, 320, (1987)

CONCEPTUAL ASPECTS OF STRING THEORY
AND THE PROBLEM OF MAXIMAL ACCELERATION

N. Sanchez

Observatoire de Paris, Section de Meudon, Demirm

92195 Meudon Principal Cedex, France.

Abstract

We discuss conceptual features of string theory in the presence of event horizons in the space-time: the need of a horizon regularization, the Hawking-Unruh effect, the problem of a limiting temperature, the instability of accelerated strings and the existence of a maximal acceleration. Conceptual features arising from the combination of General Relativity and relavistic quantum mechanics, (highest mass scale, breakdown of locality) and its implications for a quantum theory of gravity, are discussed.

1. Strings in Rindler Space and Horizon Regularization

The formulation of Quantum Field Theory in non-trivial (curved or flat) space-times has given new fundamental features with respect to the usual understanding of $Q.F.T.$ in trivial (Minkowski flat) space time, i. e.:

i) the possibility for a given field theory to have different alternative well defined Focks spaces (different "sectors" of the theory);

ii) the presence of "intrinsic" statistical features (temperature, entropy) arising from the non-trivial structure (geometry, topology) of the space-time and not from a superimposed statistical description of the quantum matter fields. Relevant examples are $Q.F.T.$ on the Rindler manifold and its analytic mapping extensions [1-2], black-holes and cosmological (de Sitter) space-times. Although Rindler space-time is flat, it possesses a space-time structure including an event horizon, similar to a black-hole manifold. The Rindler manifold is defined by

$$x^1 \pm x^0 = e^{\alpha(X^1 \pm X^o)}$$

$$x^i = X^i$$

$$i = 2, \ldots, D$$

where (x^1, x^0) and (X^1, X^0) are respectively inertial or Minkowskian (Kruskal-like) and accelerated or Rindler (Schwarzschild-like) coordinates. The constant α defines the proper acceleration of the Rindler observers (α is equal to the surface gravity in the black-hole case). The event horizon is at $x^1 \pm x^0 = 0$. This corresponds to $X^0 = \pm\infty$ and $X^1 = -\infty$.

Let us discuss some features of Rindler space which are important for our study of strings in this space. The Rindler tranformation maps the right-hand wedge $x^1 \geq |x^0|$ of Minkowski space onto the whole Rindler space $-\infty \leq X^1, X^0 \leq +\infty$ (the whole Minkowski space can be covered using four different Rindler patches). As is known, a point particle quantum field in Rindler space is in a thermal state with temperature $T = \alpha/2\pi$. In addition, ultraviolet divergences arise in the free energy and entropy of quantum fields from the existence of a horizon in the space-time. The same problem appears in the case of a four (or D-dimensional) black-hole [3]. This phenomenon is illustrated by considering (for simplicity) a free massive scalar field, with modes of positive frequency λ. The total number of wave modes with frequency less than λ. is given by [4]

$$\pi \mathcal{N}_\lambda = \int_{-H}^{a(\lambda)} dX^1 \int \prod_{i=1}^{D-2} \frac{dk^i}{2\pi} \sqrt{\lambda^2 - (m^2 + k^{i2})\alpha^2 e^{2\alpha X^1}} \, , \quad \left(a(\lambda) = \frac{1}{\alpha} \ln \frac{\lambda}{m\alpha} \right)$$

This has been evaluated in the WKB semiclassical approximation ($\lambda > \alpha$), which is enough to study the ultraviolet behaviour of the quantities interesting us. Here

120

H is a large cut-off ($H < \frac{1}{\alpha}$) on the negative Rindler coordinate X^1. The free energy (F) and entropy (S) at temperature T are given by

$$F = -\frac{\Gamma(D)}{(4\pi)^{D-\frac{1}{2}}} \frac{\alpha}{\pi^D(D-2)} \, e^{\alpha H(D-2)}$$

$$S = \frac{\Gamma(D)}{(4\pi)^{D-\frac{1}{2}}} \frac{2\pi D}{\pi^D(D-2)} \, e^{\alpha H(D-2)}$$

We explicitly see that F and S need the ultraviolet cut-off to be finite. This cut-off H shifts the horizon by replacing the light-cone $x^1 = |x^0|$ as a boundary of Rindler space-time by the hyperbola [4]

$$(x^1)^2 - (x^0)^2 = e^{-2\alpha H} .$$

This is equivalent to considering the following mapping defining the accelerated coordinates

$$x^1 - x^0 + \varepsilon = e^{\alpha(X^1-X^0)}$$
$$x^1 + x^0 + \varepsilon = e^{\alpha(X^1+X^0)}$$

where

$$\varepsilon \simeq e^{-\alpha H} , \quad -H \le X^1 \le +\infty.$$

The horizon is now at a finite distance $|X^1 \pm X^0| \sim (1/\alpha) \, \ln(1/\varepsilon)$. The longitudinal Rindler coordinate X^1 must be bounded below (it can not be $-\infty$ but $-\alpha \ln \alpha\varepsilon$). This regularization reflects the fact that a classical description of the geometry is not longer valid at distances of the order of the Panck length. Thus, $\varepsilon \sim \ell_{Planck}$.

In string theory, in order to quantize the string properly in Rindler's manifold, this horizon regularization is needed [4]. The need for such a cut-off appears already at the level of the definition of the positive frequency modes of the string with respect to Rindler's time.

2. The Hawking-Unruh effect in string theory

As it is known, reparametrization invariance of the world sheet coordinates allows to choose the world sheet metric in the conformal form and this still allows the reparametrizations

$$x_+ + \varepsilon = f(x'_+ + \ell)$$
$$x_- + \varepsilon = g(x'_- + \ell)$$

where

$$x_\pm = \sigma \pm \tau \quad , \quad x'_\pm = \sigma' \pm \tau' \ ,$$

and we have introduced the constants ε and ℓ whose meaning will be clear in the sequel. In this way, we can choose the light-cone gauge:

$$V \equiv X^0 + X^1 = p_+ x_+ \ ,$$

where the proportionality constant $p_+ > 0$ is the momentum of the center of mass of the string. An important step in field as well as string quantization is the definition of positive frequency states and its associated ground state. Canonical states for different coordinate systems are physically different (each timelike vector field leads to a separate indication of what constitutes a positive frequency). There are different ways (and there is thus an ambiguity) in choosing such a basis. This makes it possible for a given field or string theory to have different alternative well -defined Fock spaces (different "sectors" of the theory). Usually, the string is described in the Minkowski (inertial) frame where

$$\partial_{x_+} \partial_{x_-} X^A = 0 \quad , \quad A = 0, 1, \cdots, D - 1$$

and then, positive frequency modes are defined with respect to the inertial time X^0. In the light-cone gauge, X^0 is proportional to τ and therefore the modes $\phi_n \sim e^{\mp i n x_\pm}$ define the (inertial) particle states of the string.

Obviously, if we consider the (σ', τ') parametrization of the string world sheet defined above, i.e.,

$$\sigma' + \ell = \frac{1}{2}[G(\sigma + \tau + \varepsilon) + F(\sigma - \tau + \varepsilon)] , \quad F \equiv f^{-1}$$
$$\tau' = \frac{1}{2}[G(\sigma + \tau + \varepsilon) - F(\sigma - \tau + \varepsilon)] , \quad G \equiv g^{-1}$$

for which we have

$$\partial_{x'_+} \partial_{x'_-} X^A = 0 ,$$

positive frequency modes with respect to τ', namely $\Psi_n \sim e^{\mp i\lambda_n x'_\pm}$ do not define positive frequency modes with respect to X^0, i.e. are not inertial particle states of the string. However, they are positive frequency modes with respect to another inequivalent (accelerated) time $X^{0\prime}$ defined by [5]

$$X^1 - X^0 = p_- f(X'^1 - X'^0)$$
$$X^1 + X^0 = p_+ g(X'^1 + X'^0)$$

here, the mapping f, g is the same which acts on the world sheet variables. This corresponds to the description of the string in an accelerated reference frame $\{X'^0, X'^1, \cdots, X'^{D-1}\}$. In this description, we can always choose an accelerated, (in particular Rindler) light-cone gauge. Thus, the modes $\Psi_n \sim e^{\mp i\lambda_n x'_\pm}$ are accelerated particle modes of the string.

The manifold (σ', τ') is the convenient world sheet parametrization for a string in an accelerated space-time. This is true in either flat or curved space-time [although in the last case the free equation of motion, as well as the choice of the light-cone gauge, may only be valid asymptotically for $x'_\pm \to \pm\infty$].

An important point here is that of the boundary conditions which the transformations f in this context must satisfy: *Not all* conformal mappings are suitable in this context, but only those satisfying appropriate boundary conditions. As it is known, σ lies in a *finite* interval, and τ ranges from $-\infty$ to $+\infty$. In order to have τ' ranging from $-\infty$ to $+\infty$ and σ' lying in a *finite* interval, suitable boundary conditions on the mappings must be imposed. The condition on σ' is satisfied only if $\varepsilon = 0$. This yields

$$0 < \sigma' < L_\varepsilon$$

where

$$L_\varepsilon \;=\; F(\varepsilon + 2\pi) \;-\; F(\varepsilon)$$

$$\ell \;=\; F(\varepsilon) \;, \quad F = f^{-1}$$

The condition on τ' (and hence on X'^0) is not particular to strings. it is required to get a consistent quantization (of fields or strings) in accelerated manifolds and to have a complete ingoing (outgoing) basis [2]. On the other hand, it may occurs that the accelerated manifold cover only a bounded region (namely $X^1 > |X^0|$) of the original (inertial) one. Similarly, the manifold (σ', τ') may cover only a domain $(\sigma > |\tau|)$ of the inertial or global world sheet. All these considerations are satisfied by taking mappings such that

$$u_\pm \;=\; f(\pm\infty)$$

where u_+ , u_- are constants which can take independently, finite or infinite values, and $u_+ > u_-$. This means that the inverse mapping $F = f^{-1}$ has singularities at $X^1 - X^0 \;=\; u_\pm.$, i.e.,

$$F(u_\pm) \;=\; \pm\infty$$

Singularities of these mapping describe the asymptotic regions of the space-time. Critical points of f, i.e.,

$$f'(\pm\infty) \;=\; 0$$

describe event horizons at $X^1 - X^0 \;=\; u_-$ and $X^1 - X^0 = u_+$. In this case, u_- and u_+ are finite and the manifold covers the region $u_- < |X^1 \pm X^0| < u_+$ of the global space-time. If $u_\pm \;=\; \pm\infty$, there are no horizons. In this approach, the well-known Rindler's space corresponds to the mapping

$$f \;=\; e^{\alpha U} \quad (U \;=\; X^1 - X^0)$$

In this case, $u_- = 0$ and $u_+ = +\infty$; $X^1 - X^0 = u_- = 0$ is an event horizon and the manifold covers the right-hand wedge $|X^1 - X^0| > 0$ of Minkowski spacetime. (in order to cover the whole plane, four Rindler's patches are needed). For the string world sheet we have

$$x_\pm + \varepsilon = e^{\alpha(x'_\pm + \ell)}$$

and we see how the presence of the parameter ε is necessary in order to have a finite string period in the manifold (σ', τ'). For the Rindler's mapping we have

$$\ell = \frac{1}{\alpha} \log \varepsilon$$
$$L_\varepsilon = \frac{1}{\alpha} \log \left(\frac{2\pi}{\varepsilon} + 1 \right)$$

For $\varepsilon \to 0$, there is a stretching effect of the string due to the presence of an event horizon in the world sheet.

These mappings are not without consequences but change the ground state and, in general, the quantum states, except for mappings belonging to the $O(2,1)$ group, that is except for the Möbius or bilinear transformations. Under the mappings f, the states transform as

$$| > \to e^{i\hat{G}} |' >$$

$$\hat{G} = \sum_n \Theta_n (C_n\, C_n - C_n^+\, C_n^+)$$

here $a_n| > = 0$ defines the standard (inertial) Minkowskian ground state and $C_n|' > = 0$ defines the accelerated one. The operators a_n and C_n are related by a Bougolubov trasformation. [$\hat{G}$ is an operatorial representation for the Bogolubov transformation, with Bogoliubov coefficients $\cosh\theta_n$ and $\sinh\theta_n$].

The vacuum expectation value of the two dimensional (world sheet) energy-momentum tensor transforms as

$$\langle T_{\mu\nu} \rangle = \langle T'_{\mu\nu} \rangle + \Theta_{\mu\nu} + P_{\mu\nu}$$

$P_{\mu\nu}$ is any conserved traceless tensor taking into account the dependence of $T_{\mu\nu}$ on the quantum state. It represents the non-local part of $T_{\mu\nu}$. $\Theta_{\mu\nu}$ depends on the mapping and represents the local part. In the conformal gauge, the constraints

$$\langle T_{\mu\nu}\rangle \;=\; 0 \quad , \quad \langle T'_{\mu\nu}\rangle \;=\; 0$$

yield the equations

$$\frac{d^2}{dx'^2_{\pm}}\,\Psi_{\pm} \;-\; 12\pi\,\mathcal{U}_{\pm}(x'_{\pm})\Psi_{\pm} \;=\; 0$$

for the mappings f and g [5]. This is a zero-energy Schrödinger equation. Here $\mathcal{U}_{+}$ and $\mathcal{U}_{-}$ are arbitrary functions of the indicated variables. By giving the "potentials" $\mathcal{U}_{\pm}$, this equation allows to determine the "wave functions"

$$\Psi_{-} \;=\; \frac{1}{\sqrt{f'}} \quad , \quad \Psi_{+} \;=\; \frac{1}{\sqrt{g'}} \;,$$

that is the mappings. Notice that because $\mathcal{U}_{\pm}$ are arbitrary functions (compatible with the boundary conditions), these equations do not yield additional constraints on the mappings f, g but a way of connecting the mapping to a potential. The first term of the equation is the Schwarzian derivative of f:

$$\mathcal{D}[f] \;=\; \frac{f'''}{f'} \;-\; \frac{3}{2}\left(\frac{f''}{f'}\right)^2 \;,$$

which is invariant under the Möbius or bilinear transformations. Under these $O(2,1)$ transformations, f becomes a new function but $\mathcal{D}[f]$ is invariant determining the same ground state of the string. In particular, $\mathcal{U}_{+} \;=\; \mathcal{U}_{-} \;=\; O$ determines f, or g as

$$f \;=\; \frac{(\alpha x'_{-} + \beta)}{\gamma x'_{-} + \delta)} \quad , \quad (\alpha\delta - \beta\gamma) \;=\; 1$$

Interestingly, the ground state defined by this mapping can be considered as a reference or "minimal" state at zero temperature with respect to which other ground states corresponding to non-zero potencials $\mathcal{U}_{\pm}$ appear as excited or thermal ones. For constant potential, i.e. $\mathcal{U}_{+} = \mathcal{U}_{-} = constant > 0$, we have

$$\Psi_{\pm} = A\, e^{-kx'_{\pm}} \quad , \quad f = \frac{1}{A^2\alpha} e^{\alpha x'_{+}} - \varepsilon$$

This is the Rindler mapping. Here A is a normalizing constant (we choose $A^2 = \alpha^{-1}e^{-\alpha\ell}$) and K is the zero-energy transmission coefficient

$$K \equiv \frac{\alpha}{2} = \sqrt{12\pi\mathcal{U}_0} \quad , \quad (\mathcal{U}_{\pm} = const \equiv \mathcal{U}_0)$$

The surface gravity is the zero-energy transmission coefficient and the parameter ε arises naturally as an integration constant. For $\varepsilon \to 0$, this mapping defines an event horizon at $x'_{\pm} = \mathcal{O}(x_{\pm} = -\infty)$ and carries an intrinsic temperature $T_S = \frac{\alpha}{2\pi}$, as it can be seen by taking $\tau = i\tau(x_{\pm} = \sigma \pm i\tau)$ and then $0 < \tau < \frac{2\pi}{\alpha}$. The temperature is

$$T_S = \sqrt{12\mathcal{U}_0/\pi}$$

This temperature T_S characterizes the ground state spectrum of the string in Rindler space [4,5]. The expectation values of the Rindler number mode operator, as well as the square mass operator, in the ground state of the string follow a thermal distribution with temperature T_S. This is the same Hawking-Unruh value which appears in the field theoretical context. Here the frequencies of the Rindler modes are given by $\lambda_n = n\lambda_0$, that is, they differ in a large factor $\lambda_0 = 2\pi\alpha/(\ln 2\pi/\varepsilon + 1)$ from the inertial ones (n). Physically, this factor reflects the indefinite increasing of the string length when it approaches the event horizon. However, if one measures the frequency in dimensionless units (1,2,....) instead of multiples of λ_0, then, the temperature of the ground state is a (very large) pure number $T_0 = (T_S/\lambda_0) = \frac{1}{4\pi^2}\ln(2\pi/\varepsilon + 1)$.

We see that with appropriate boundary conditions, the holomorphic mappings of reparametrization invariance of string theory can be interpreted as a change of coordinate frame in the space-time in which strings are embedded. These mappings change the ground state in the quantized theory except for transformations belonging to the $O(2,1)$ group. This allows to discuss in a natural way the Hawking-Unruh effect in string theory. We also see that in order to discuss these effects, a change of global coordinate frame in the space-time must be accompanied by a similar change on the world sheet. The boundary conditions satisfied by these mappings are the same in string theory as well as in $Q.F.T.$, but for strings, the mappings carry an additional parameter ε (notice that $\ell = \ell(\varepsilon)$). This additional parameter plays a fundamental role in the string context. ε acts as a regulator or cut-off to avoid the presence of an event horizon in the world sheet and to get a finite string period. Its magnitude is of the order of the Planck length.

3. Strings near black holes

The results described above apply also to curved space-time. For the most important metrics in General Relativity, the presence of isometry groups allows the maximal analytic extension of the (D-dimensional) manifold to be performed through the extension of a relevant two-dimensional manifold containing the time axis and a suitable spatial coordinate. This maximal analytic extension is performed by the class of mappings we considered here, where X^1, X^0 are Kruskal (maximal) type coordinates and X'^1, X'^0 are of Schwarzschild type [4,5]. The major features of the string quantization in Rindler space presented here also hold for a string in a Schwarzschild space-time. An appropriate light-cone gauge can be introduced for left or right movers treated separately in the null Kruskal and Schwarzschild coordinates. The shifting of the horizon is equivalent to considering a shifting ε in the mapping

$$r_k \pm t_k + \varepsilon = e^{k(R^* \pm T)}$$

r_k, t_k are Kruskal coordinates, R^*, T are Schwarzschild's ones.

Here $\varepsilon \sim (M_{Planck}/M)^{(D-2)/(D-3)}$, M is the black-hole mass and K is the surface gravity $K \sim (M_{Planck}/M)^{1/(D-3)}$.

The expectation value of the string number operator of Schwarzschild string modes taken in the Kruskal ground state of the string gives a Planckian distribution at the temperature $T_S = K/2\pi$ or at the dimensionless temperature

$$T_0 = 1/4\pi^2 \ln(2\pi/\varepsilon) \gg 1$$

One can consider different higher dimensional black hole space-times, namely a 26-dimensional or a four dimensional black hole with extra 22 dimensions compactified in a torus. Intermediary situations can also be envisaged but it must be noticed that the qualitative properties of the string quantization will be the same since they depend upon the horizon structure in the two variables R, T (or X^0, X^1, for Rindler space). It can be noticed that the Hagedorn temperature (T_m) in this context is

$$T_m = 1/(\alpha')^{\frac{1}{2}} \simeq M_{Planck}$$

Then,

$$\frac{T_S}{T_m} \simeq \left(\frac{M_{Planck}}{M} \right)^{1/(D-3)}$$

and we have

$$T_S \lesssim T_m$$

since the basic requirement of the present semiclassical treatment is

$$M \gtrsim M_{Planck}$$

Here, we have only discussed the features corresponding to the Hawking temperature for strings near black-holes. Strong curvature effects and features like string mode excitation and scattering by black holes, are discussed in refs [7].

4. Accelerated strings

More recently [7], we have considered a simple model for *accelerated strings*. In the general case such a string must be affected by the action of some external force causing the acceleration. In order to make our study more concrete we consider that the string is open and that there are "heavy" particles at its ends (e.g. monopoles [8]); the force causing the acceleration being applied to these "heavy" particles. In such a simplified model all the information concerning the interaction of the string with the external forces is reduced to the specific choice of the boundary conditions. Then, one can consider a string whose ends are moving with uniform acceleration, that is they follow accelerated world lines (hyperbolae). Here the equations of motion and constraints are the usual ones but the boundary conditions are non-linear.

The main result of our study is the proof of the existence of a *critical value of the acceleration*. For accelerations smaller than the critical one, there exist equilibrium configurations of the accelerated string which represent its "rigid-body-like" motion. For accelerations larger than the critical one such a rigid equilibrium configuration is impossible; the *length* of the string begins to *grow* infinitely. First, we have considered a string moving as a rigid body without any excitation, and we have found that there exists a critical value of the acceleration a_c beyond which such a solution does not exist. Alternatively, this configuration can be interpreted as the static equilibrium configuration for a string in a homogeneous gravitational field, the inverse of the critical parameter (a_c^{-1}) giving the minimal distance from the string to the horizon. Then, we study linear small perturbations (string excitations) around this equilibrium or non-excited accelerated string solution. After linearization of the constraints and boundary conditions around this zero-order solution, a careful analysis of their solutions and the stability problem shows that:

(i) for $a < a_c$, the perturbations describe small oscillations, whereas that, for $a > a_c$ the solution contains an exponentially growing factor with time, that is, for $a > a_c$ the *equilibrium is unstable.*

(ii) For the value $a = a_c$, at which the instability develops, the growing factor in time is not exponential but a power of time. The notion of *limiting acceleration* and its *associated instability* naturally arises in the interpretation of our problem when the string is considered at rest in a homogeneous gravitational field: in this case we show that there exists a critical field strength, for which negative energy modes of the string perturbations arise. (This instability happens for the same value of the acceleration as we have found for the case of a uniformly accelerated string, which is exactly what corresponds to the equivalence principle). If for a given distance L between the (uniformly) accelerated string's ends, one tries to increase adiabatically the value of the acceleration, one finds out that the form of the accelerated equilibrium string will also be adiabatically changing until the moment when the string's acceleration reaches the critical value a_c. The string with this acceleration appears to be unstable. Hence, for higher values of the acceleration, the equilibrium configurations are already impossible.

The effect of the existence of a critical acceleration and of an accelerated string instability is of quite general nature. No extended system of size ℓ, say, can move as a rigid body with an acceleration a greater than ℓ^{-1}. The string is an elastic body and for a given distance L between the ends of the string, the string size in the direction of the acceleration depends not only on L but also on its tension and on the acceleration itself. That is why for the critical value of the acceleration, the size of the string in the direction of the acceleration, i.e.;

$$a^{-1}\left(1 - \cosh^{-1}\varsigma_0\right) \quad , \quad (\varsigma_0 = 1.2),$$

is less than the distance to the horizon a^{-1}. It is interesting to notice that the same situation occurs when the string is placed into the static gravitational field of a black hole, namely the equilibrium configurations of strings are possible until the distance between the string and the event horizon becomes less than some critical value [9].

What happens with the accelerated string after it loses its stability? At the beginning of this process, the length of the string in the direction of the acceleration becomes larger and larger until a part of the string crosses the horizon; after that, the string completely loses its rigidity. The causal signal propagating along the part of the string which is moving behind the horizon can reach the string ends. That is why this part of the string will be almost freely propagating and one may expect that the lenght of the string will be permanently growing.

This model allowed us to simplify the discussion because an external force inducing the acceleration was implied to the string ends while the string itself was not affected by any additional force. Now we make some general remarks based on the results obtained from this simplified model. Let us consider a closed string moving as a whole with some acceleration a. Of course this string must be affected by an external force. We may assume, for example, that the string possesses electric charge and that it is placed into a homogeneous electric field. An important point is that a uniformly accelerated string as well as any other system with internal degrees of freedom becomes a source of radiation, accompanied by the string's excitation process. The excitation of the accelerated system continues until the radiation of the modes (which decreases the energy of the system) becomes comparable to the radiation accompanied by the string's excitation. This effect is also valid for classical systems provided the system with internal degrees of freedom (a string in our case) at the beginning already is not at its lowest energy state. The quantum description of this process corresponds to the Unruh effect. The nature of the radiation depends of the nature of fields which interact with the system. In the case of the string, one of such fields must be the gravitational one. If the closed string is moving with acceleration a, or if it is at rest in a static gravitational field near the event horizon, then the equilibrium excitation of the string is characterized by the Unruh temperature $T_S = a/2\pi$. But the string heated up to this temperature must have a mass

$$M = 4\pi T_S \quad \text{and a size} \quad L = CT_S^{\frac{1}{2}} \alpha'^{\frac{3}{4}}.$$

On the other hand, the size L of the equilibrium string cannot exceed the distance a^{-1} to the horizon. This means that there must exist some *limiting acceleration*

$$a_c < \frac{2\pi^{\frac{1}{3}}}{C^2} \frac{1}{\alpha'^{\frac{1}{2}}} \quad , \quad C = 2\pi[(D-2)/6]^{\frac{1}{2}}.$$

Any attempt to accelerate a string beyond this value will result in an infinitely *growing mass* and *size* of the string. It is interesting to notice that the characteristic Unruh temperature connected with the limiting acceleration a_c is $T_c = (2\pi C)^{-\frac{2}{3}}(\alpha')^{-\frac{1}{2}}$ and it is comparable with the so-called Hagedorn temperature which arises as a *limiting temperature* in string thermodynamics. Some of these issues have been independently discussed in refs.[10].

String instabilities have been discovered and described in cosmological backgrounds of inflationary type [11-15]. They are characterized by a power type behaviour in time, (rather than an oscillatory time behaviour),and a growing proper size. More recently, this unstable behaviour (and other interesting new features) have been found for strings falling into space time singularities [16].

5. Conceptual Features

The results presented above have consequences in several directions. Let us discuss now, some of these consequences from a conceptual point of view.

The existence of a critical or maximal temperature $T_c \sim (\alpha')^{-\frac{1}{2}}$ in string theory can be considered as the indication of a *minimal length* beyond which the concept of *locality* breaks down. Assume we make two measurements at a very small distance Δx. Then,

$$\Delta p \sim \Delta E \sim \frac{1}{\Delta x}$$

where we set $\hbar = c = 1$. For sufficiently large ΔE, particles with masses $m \sim 1/\Delta x$ will be produced. The gravitational radius of such particles are of the order of

$$R_G \sim G_m \sim \frac{(\ell_{Planck})^2}{\Delta x}$$

where $\ell_{Planck} \sim 10^{-33}$ cm. Now, General Relativity allows measures at a distance Δx, provided

$$\Delta x > R_G \quad \Leftrightarrow \quad \Delta x > \frac{(\ell_{Planck})^2}{\Delta x}.$$

That is,

$$\Delta x > \ell_{Planck} \quad \text{or} \quad m < M_{Planck}$$

This means that no measurements can be made at distances smaller than the Planck lenght and that no particle can be heavier than M_{Planck}. This is a simple consequence of Relativistic Quantum Mechanics combined with General Relativity. In addition, the notion of locality and hence of space-time become meaningless at the Planck scale. Notice that the equality $m = M_{Planck}$ here, corresponds when the Compton length equals the Schwarzschild radius of a particle. Several consequences can be extracted from this discussion:

1) this allows to discuss from a conceptual point of view the renormalizability question of gravitation. The characteristic scale energy of quantum gravitation is M_{Planck}. But, as we have just seen, M_{Planck} is pecisely the *highest possible* particle mass. *Therefore, there cannot be any theory beyond it.*

If ultraviolet divergences appear in a quantum theory of gravitation, there is no way to interpret them as coming from a higher energy scale as it is usually done in Quantum Field Theory. That is, no physical understanding can be given to such ultraviolet infinities. The only logically consistent possibility would be to find a *finite* theory of gravitation. In other words, it is not possible to conceive a renormalizable Q.F.T. when the gravitation interaction is included. Let us explain in more detail this point.

Many attempts have been done to quantize gravity. The problem most often discussed in this connection is the one of renormalizability of Einstein theory (or its various generalizations) when quantized as a local quantum field theory. Actually, even deeper conceptual problems arise when ones tries to combine quantum concepts with General Relativity. *What is a renormalizable Q.F.T.?* This is a theory with some domain of validity characterized by energies E such that $E < \Lambda$. Here, the scale Λ is characteristic of the model under consideration: (e.g., $\Lambda = 1 Gev$ for QED, $100 Gev$ for the standard model or $10^{16} Gev$ for GUT, etc.). One always applies the QFT in question till finite energy (or zero distance) for virtual processes and finds usually ultra-violet infinities. These divergences reflect the fact that the model is unphysical for energies $\Lambda \ll E \ll \infty$. In a renormalizable QFT these infinities can be absorbed in a finite number (usually few) parameters like coupling constants and mass ratios, which are not predicted by the model in question. One would need a more general theory valid at energies beyond Λ in order to compute these renormalized parameters (presumably from others more fundamental). For example, M_W/M_Z is calculable in a Grand Unified Theory, whereas it must be fitted to its experimental value in the standard electro-weak model.

Now, what about quantization of gravity? The relevant energy scale is now the Planck mass $(M_{Planck} \sim 10^{19} Gev.)$. At this mass, the Schwarzschild radius (R_G) equals the Compton wavelength (λ_c) of a particle. Then, if we imagine particles heavier than M_{Planck}, their size λ_c will be smaller than (R_G). In other words, to localize them in a region of size λ_c will be in conflict with what we know about the Schwarzchild radius from general relativity. Such heavy objects $(M_{Planck} \sim 10^{-5} g)$ can not behave (if they really exist) as usual point particles do in relativistic QFT. This means that M_{Planck} gives the order of magnitude for the heaviest point particles. There cannot be point particles beyond M_{Planck} in a relativistic QFT as soon as gravity is included. This shows that we can not conceive a renormalizable QFT including gravity since there can not exist a theory at energies higher than M_{Planck} whose ignorance is responsible for the infinities of quantum gravity. (If ultraviolet divergences appear in a quantum theory of gravity, there is no way to interpret them as coming from a higher energy scale as it is usually done in QFT). Hence, a consistent theory including gravity must be *finite*. All dimensionless physical quantities must be computable in it. These conceptual arguments are consistent with all failed attempts to construct renormalizable field theories of quantum gravity.

2) Another consequence of these arguments is the following: Since a quantum theory of gravitation would describe the highest possible mass scale, such model must also include all other interactions in order to be consistent and true. That is, one may ignore higher energy phenomena in a low energy theory, but the opposite is not true. To give an example, a theoretical prediction for graviton-graviton scattering at energies of the order of M_{Planck} must include *all* particles produced in a real experiment. That is, in practice, all existing particles in nature, since

gravity couples to *all matter.*

These simple arguments, based on the renormalization group lead us to an important conclusion: a consistent quantum theory of gravitation must be a *theory of everything (TOE).* So rich a theory should be very complicated to find and to solve. In particular, it needs the understanding of the present desert beween 1 and 10^{16} Gev. There is an additional dimensional argument about the inference that a Quantum Theory of Gravity implies a *TOE.* There exist only three fundamental physical magnitudes: length, time and energy and hence three fundamental dimensional constants: c, $\hbar$ and G. All other parameters being dimensionless, they must be calculable in a unified quantum theory including gravity, and therefore, a theory like this must be a *TOE.* From the purely theoretical side the only serious candidate at present for a *TOE* is string theory. Unfortunately, most of the research work done on strings consider the strings in Minkowski space-time. All string models exhibit particle spectra formed by tower of massive particles going up to infinite mass and hence passing by M_{Planck}. If these states are to be considered as point particles, one arrives, for masses *larger than* M_{Planck}, to the *clash* between general relativity and quantum mechanics described above. A solution of this paradox could be that the particle spectrum of string models is at energies $E \gg M_{Planck}$ very different from what we know today on the basis of perturbation theory in flat (10 or 26 dimensional) space-time. Our results about strings in strong gravitational fields support this suggestion.

Since the most relevant new physics provided by strings concerns quantization of gravity, we must, at least, understand string quantization in a curved space-time. Actually, one wuold like to extract the space-time and the particle spectrum from the solution of string theory, but we are still far from doing that explicitly. Practically, all what we know about strings comes from their study in flat critical (10 or 26) dimensional space-time. It must be noticed that expanding in perturbation around the Minkowski metric is not better since the non-trival features appear in the strong curvature regimes, in the presence of horizon and of the intrinsic singularities. Curved space-times, besides their evident relevance in classical gravitation are also important at energies of the order of the Planck scale. At such energy scales, the picture of particles propagating in flat space-times is no longer valid and one must take into account the curved geometry created by the particles themselves. In other words, gravitational interactions are at least as important as the rest and can not anymore being neglected as it is usually the case in particle physics.

As a first step in the understanding of quantum gravitational phenomena in the framework of string theory, we started in 1987 a programme of string quantization in curved space-times. A summary of the developements and results till now in that programme is given in ref. [17].

References

[1] See for example, N.D. Birrell and P.C.W. Davies, *Quantum Fields in Curved Space*, Cambridge University Press, Cambridge, England, 1982 and references therein.

[2] N. Sánchez and Phys. Rev. 24D, 2100 (1981).

N. Sánchez and B.F. Whiting, Phys. Rev. 34D, 1056 (1986).

[3] G. 't Hooft, Nucl. Phys. 256B, 727 (1985).

[4] H.J. de Vega and N. Sánchez, Nucl. Phys. 299B, 818 (1988).

[5] N. Sánchez, Phys. Lett. 195B, 160 (1987).

[6] H.J. de Vega and N. Sánchez, Nucl. Phys. 309B, 552 and 309B, 577 (1988)

[7] V.P. Frolov and N. Sánchez, Nucl. Phys. 349B, 815 (1991).

[8] A Vilenkin, Phys. Rep. 121, 263 (1985).

[9] V.P. Frolov, V.D. Skarzhinsky, A.I. Zelnikov and A.O. Heinrich, Phys. Lett. 224B, 255 (1989).

[10] R. Parentini and R. Potting, Phys. Rev. Lett. 63, 945 (1989).

M. Bowick and S. Giddings, Nucl. Phys. 325B, 631 (1989).

[11] H. J. de Vega and N. Sánchez, Phys. Lett. 197B, 320 (1987).

[12] N. Sánchez and G. Veneziano, Nucl, Phys. 333B, 253 (1990).

[13] M. Gasperini, N. Sánchez and G. Veneziano, CERN-TH5893/90, DFFTT30/90 and MEUDON 90091 preprint (to appear in IJMPA).

[14] M. Gasperini, N. Sánchez and G. Veneziano, CERN-TH 6010/91, DFTT 06/91 and MEUDON-DEMIRM 91004 preprint.

[15] M. Gasperini, DFTT-38/90 preprint (to appear in Phys. Lett. B).

[16] H.J. de Vega and N. Sánchez, LPTHE 90-48 preprint.

[17] N. Sánchez, *String Theory and the Quantization of Gravity*, to appear in *Gravitation and Modern Cosmology. A Great Problem: The Cosmological Constant*, Volume in Honor of P.G. Bergmann's 75th Birthday.

HYPERBOLIC GEOMETRY
AND LANDAU QUANTIZATION

V. Barone[a], V. Penna[b] and P. Sodano[a]

[a] *Dipartimento di Fisica and Sezione INFN,*
Università di Perugia, I–06100, Perugia, Italy

[b] *Dipartimento di Fisica and Unità INFM,*
Politecnico di Torino, I–10129, Torino, Italy

Abstract

The quantum mechanics of a particle moving under the action of a constant magnetic field in a hyperbolic space is studied in some detail. We present an algebraic treatment of the problem and a careful investigation of the Hilbert space for the discrete part of the spectrum. Coherent states and Lindblad–Nagel states are explicitly obtained and some developments are briefly outlined.

1. Introduction

The study of quantum mechanical problems on a two–dimensional space of constant negative curvature (hyperbolic surface, also called pseudosphere) has been recently prompted by the widespread interest in the relationship between classical and quantum mechanics of chaotic systems. In order to clarify the question as to how the classical stochastic behavior manifests itself at the quantum level, some authors[1,2] have undertaken the study of the free motion of a particle constrained to move on a pseudosphere. This extremely simple system exhibits two intriguing features that make it a precious theoretical tool. First of all its classical motion is chaotic and second, due to the non–compactness of the hyperboloid, it possesses a quantum energy spectrum with both a discrete and a continuous part, thus allowing one to analyze quantum chaoticity in presence of either bound or scattering states.

In a different framework, a closely related problem has been thoroughly investigated in a series of papers by Alhassid, Gürsey and Iachello[3]. With the aim of presenting a group–theoretical approach to the S–matrix for some one–dimensional potentials, they were led to a study of the $SU(1,1)$ algebra of the constants of motion for a free particle on the pseudosphere.

The effect of a constant magnetic field on the system here discussed has been explored by Comtet[4], whose main goals were a test of the semiclassical approximation (which turns out to give exact results) and the construction of the resolvent.

A more detailed inspection of the Hilbert space for this problem has been carried out in a paper by Oshima[5], where a complete set of eigenstates of the Hamiltonian, which was recognized as a Maass laplacian with a weight given by the magnetic field strength, was obtained by analytical methods, with no consideration of the group representation aspects of the underlying algebra.

The purpose of the present paper is to study the quantum mechanics of a particle moving on a hyperbolic plane acted on by a constant magnetic field, focusing our attention on the group representation analysis of the Hilbert space of the system. By relying on the techniques developed by Lindblad and Nagel (LN)[6], and the results of a previous work[7], we explicitly find the generalized eigenstates for the noncompact $SU(1,1)$ generators which we write as continuous superpositions of coherent states. Along this route the diagonalization of the compact generators is performed, providing the standard basis for both the discrete series $D_\ell^\pm$ (related to the two possible orientations of the background magnetic field).

We believe that the interest of our discussion and further investigations of the

140

problem can be at least twofold.

From a physical point of view, it seems worthwhile studying the effect of a non ordinary geometry (in particular that of a curved noncompact manifold) on the Landau quantization, which appears to be quite sensitive to all geometrical and topological properties of the space.

From a mathematical point of view the representation theory of $SU(1,1)$ presents some points not yet fully understood, and a richness which has been so far only partially explored. On the other hand, a wide variety of physical situations (e.g. nonlinear quantum optics models[8,9], dissipative Hamiltonian systems[10], Thermo Field Dynamics[11], XY models on the hyperboloid[12]) involve an $SU(1,1)$ algebraic structure.

2. Classical Mechanics

As a starting point we briefly review the classical dynamics of a particle constrained to move on a pseudosphere under the action of a constant magnetic field.

The two–dimensional ambient space of this particle is given by the upper sheet of a double–sheeted hyperbolic surface parametrized as

$$x_1 = R \, \mathrm{sh}\,\tau \, \cos\varphi$$

$$x_2 = R \, \mathrm{sh}\,\tau \, \sin\varphi \qquad (2.1)$$

$$x_3 = R \, \mathrm{ch}\,\tau \ ;$$

where the coordinates x_1, x_2, x_3 in the embedding $3d$ space satisfy $x_3^2 - x_1^2 - x_2^2 = R^2$.

A different parametrization of the pseudosphere corresponds to the so–called Poincaré disk with radius R, and is described by the complex coordinate $\xi = \xi_1 + i\xi_2$, which is related to the x_i's by

$$x_1 = (2R^2/\Delta)\,|\xi|\cos\varphi$$

$$x_2 = (2R^2/\Delta)\,|\xi|\sin\varphi \qquad (2.2)$$

$$x_3 = (R/\Delta)(R^2 + |\xi|^2) \ ;$$

where $\Delta = R^2 - |\xi|^2$, and to (τ, φ) by

$$\xi = R\, e^{i\varphi}\, \text{th}\frac{\tau}{2}. \tag{2.3}$$

In order to allow an easy comparison with other treatments in the literature, we quote the formula relating ξ to the Poincaré half–plane complex coordinate $z = x + iy$ $(y > 0)$

$$\xi = R\,\frac{i - z}{i + z}. \tag{2.4}$$

The metric of the pseudosphere in terms of the disk coordinates is

$$ds^2 = \left(\frac{4R^4}{\Delta^2}\right) d\xi\, d\bar{\xi} \; ; \tag{2.5}$$

while the volume 2–form is

$$\omega = \left(\frac{8iR^4}{\Delta^2}\right) d\xi \wedge d\bar{\xi}. \tag{2.6}$$

A constant magnetic field on the hyperboloid is the 2–form proportional to ω

$$\mathcal{B} = d\mathcal{A} = \left(\frac{2iBR^4}{\Delta^2}\right) d\xi \wedge d\bar{\xi} \; ; \tag{2.7}$$

where B is the intensity of the field (in the following for convenience we shall use the quantity $b \equiv BR^2$). The vector potential $\mathcal{A} = \mathcal{A}_\xi d\xi + \mathcal{A}_{\bar{\xi}} d\xi$ solving eq. (2.7) has the general form

$$\mathcal{A}_\xi = \frac{-ibR}{\Delta}\left(\frac{R + \bar{\xi}}{R + \xi}\right) + \partial_\xi \Lambda$$

$$\mathcal{A}_{\bar{\xi}} = \bar{\mathcal{A}}_\xi \; ; \tag{2.8}$$

where the Landau gauge and the symmetric gauge correspond to the choices $\Lambda = 0$ and $\Lambda = i\,b\,\ln\left[(R + \xi)/(R + \bar{\xi})\right]$ respectively.

We have now the necessary ingredients to write down the Lagrangian of the particle, and it reads (we set the mass and the charge of the particle equal to 1)

$$\mathcal{L} = \frac{4R^4}{\Delta^2}|\dot{\xi}^2| - (\dot{\xi}\mathcal{A}_\xi + \dot{\bar{\xi}}\mathcal{A}_{\bar{\xi}}). \tag{2.9}$$

Even though the subsequent discussion will be kept at a general level without specifying the gauge, for the sake of clarity we give the explicit form of $\mathcal{L}$ in the symmetric gauge:

$$\mathcal{L} = \frac{4R^4}{\Delta^2}|\dot{\xi}^2| + \frac{ib}{\Delta}(\dot{\xi}\bar{\xi} - \dot{\bar{\xi}}\xi) \; ; \tag{2.10}$$

and we notice that in the limit $R \to \infty$ the ordinary flat space Lagrangian is recovered.

From eq. (2.9) we can derive the canonical momenta

$$p_\xi \equiv \frac{\partial \mathcal{L}}{\partial \dot{\xi}} = \frac{4R^4 \dot{\xi}}{\Delta^2} - \mathcal{A}_{\bar{\xi}} \tag{2.11a}$$

$$p_{\bar{\xi}} \equiv \frac{\partial \mathcal{L}}{\partial \dot{\bar{\xi}}} = \bar{p}_\xi \; , \tag{2.11b}$$

with the usual Poisson brackets $\{\xi, p_{\bar{\xi}}\} = \{\bar{\xi}, p_\xi\} = 1$, and construct the Hamiltonian

$$\mathcal{H} = \frac{\Delta^2}{4R^4}\left|p_\xi + \mathcal{A}_{\bar{\xi}}\right|^2 . \tag{2.12}$$

The constants of motion of the system are

$$J_1 = \frac{1}{\hbar}\left\{\left[\frac{1}{iR}(R^2 + \bar{\xi}^2)(p_\xi + \partial_{\bar{\xi}}\Lambda) + c.c.\right] + \frac{2b\Delta}{|R + \xi|^2}\right\} \tag{2.13a}$$

$$J_2 = \frac{1}{\hbar}\left[\frac{1}{R}(\bar{\xi}^2 - R^2)(p_\xi + \partial_{\bar{\xi}}\Lambda) + c.c.\right] \tag{2.13b}$$

$$J_3 = \frac{1}{\hbar}\left\{\left[-2i\bar{\xi}(p_\xi + \partial_{\bar{\xi}}\Lambda) + c.c.\right] - \frac{2b\Delta}{|R + \xi|^2}\right\} \; , \tag{2.13c}$$

and obey the $SU(1,1)$ algebra

$$\{J_1, J_2\} = -J_3/\hbar$$

$$\{J_2, J_3\} = J_1/\hbar \tag{2.14}$$

$$\{J_3, J_1\} = J_2/\hbar \; ;$$

where the Planck constant has been introduced in order to have the ordinary $SU(1,1)$ commutators after quantizing the system.

We can recast the Hamiltonian in the following form (where it is related to the $SU(1,1)$ Casimir operator $J_0 \equiv J_3^2 - J_1^2 - J_2^2$):

$$\mathcal{H} = \frac{1}{4R^2}(b^2 - \hbar^2 J_0). \tag{2.15}$$

In the Poincaré half plane realization (see, e.g. Ref.[4]) the equation of the trajectories takes the rather simple form

$$\left[x - \frac{J_2}{J_1 + J_3}\right]^2 + \left[y + \frac{b}{\hbar(J_1 + J_3)}\right]^2 = \frac{b^2 - \hbar^2 J_0}{\hbar^2(J_1 + J_3)^2} \qquad (2.16)$$

and shows that the classical paths can be classified as follows:

i) closed orbits for $0 < J_0 < b^2/\hbar^2$;

ii) open paths for $J_0 < 0$.

As we shall see, in the quantum case these two classes correspond in terms of $SU(1,1)$ unitary irreducible representations (UIR's) to the discrete series $D_\ell^{\pm}$, and to the continuous series C_ℓ^{δ} and E_ℓ respectively.

3. Group Representation Analysis of the Hilbert Space

Quantizing the system by the replacement $\{\ \} \to i\hbar\,[\ ]$ leads to the $SU(1,1)$ commutators for the symmetry operators $\hat{J}_i$:

$$\begin{aligned}
\left[\hat{J}_1, \hat{J}_2\right] &= -i\hat{J}_3 \\
\left[\hat{J}_2, \hat{J}_3\right] &= i\hat{J}_1 \\
\left[\hat{J}_3, \hat{J}_1\right] &= i\hat{J}_2
\end{aligned} \qquad (3.1)$$

or equivalently, after defining $\hat{J}_\pm = \hat{J}_1 \pm i\hat{J}_2$,

$$\begin{aligned}
\left[\hat{J}_+, \hat{J}_-\right] &= -2\hat{J}_3 \\
\left[\hat{J}_3, \hat{J}_\pm\right] &= \pm\hat{J}_\pm.
\end{aligned} \qquad (3.2)$$

The quantum canonical momenta p_ξ and $p_{\bar{\xi}}$ on the Poincaré disk are obtained by DeWitt's prescription[13]

$$\begin{aligned}
p_\xi &= -i\hbar\left(\partial_{\bar{\xi}} - \partial_{\bar{\xi}}\ln\Delta\right) \\
p_{\bar{\xi}} &= -i\hbar\left(\partial_\xi - \partial_\xi\ln\Delta\right)\ ;
\end{aligned} \qquad (3.3)$$

and obey the canonical commutators $[\xi, p_{\bar{\xi}}] = [\bar{\xi}, p_\xi] = i\hbar$.

In terms of $(\xi, \bar{\xi})$ and $(p_\xi, p_{\bar{\xi}})$ the $SU(1,1)$ generators $\hat{J}_\pm, \hat{J}_3$ in the ordered form are

$$\begin{aligned}
\hat{J}_+ &= \frac{1}{R\hbar}\left(-iR^2 Q + i\xi^2 Q^\dagger + \hbar\xi + \frac{Rb\Delta}{|R + \xi|^2}\right) \\
&= \frac{1}{R\hbar}\left(-iR^2 q + i\xi^2 q^\dagger + \frac{Rb\Delta}{|R + \xi|^2}\right)
\end{aligned} \qquad (3.4a)$$

144

$$\hat{J}_- = \frac{1}{R\hbar}\left(iR^2Q^\dagger - i\bar{\xi}^2 Q - \hbar\bar{\xi} + \frac{Rb\Delta}{|R+\xi|^2}\right)$$
$$= \frac{1}{R\hbar}\left(iR^2 q^\dagger - i\bar{\xi}^2 q + \frac{Rb\Delta}{|R+\xi|^2}\right) \tag{3.4b}$$

$$\hat{J}_3 = \frac{1}{\hbar}\left(-i\bar{\xi}Q + i\xi Q^\dagger - \frac{b\Delta}{|R+\xi|^2}\right)$$
$$= \frac{1}{\hbar}\left(-i\bar{\xi}q + i\xi q^\dagger - \frac{b\Delta}{|R+\xi|^2}\right) \; ; \tag{3.4c}$$

where we have defined $Q \equiv p_\xi + \partial_{\bar{\xi}}\Lambda$, $Q^\dagger \equiv p_{\bar{\xi}} + \partial_\xi\Lambda$ and $q \equiv -i\hbar\partial_{\bar{\xi}} + \partial_{\bar{\xi}}\Lambda$, $q^\dagger \equiv -i\hbar\partial_\xi + \partial_\xi\Lambda$. The ordering adopted in eqs. (3.4) allows one to recover the usual action of the raising and lowering operators $\hat{J}_\pm$ on the $\hat{J}_3$ eigenstates.

The Casimir operator $J_0 \equiv J_3^2 - \frac{1}{2}(J_+ J_- + J_- J_+)$ turns out to be

$$J_0 = -\frac{\Delta^2}{R^2\hbar^2}\left\{ qq^\dagger + \frac{iRb}{\Delta|R+\xi|^2}\left[(\xi+R)^2 q^\dagger - (\bar{\xi}+R)^2 q\right]\right\} \; . \tag{3.5}$$

The algebra $su(1,1)$ can be represented in the standard basis $\{|\ell,m\rangle\}$, which is used in the secular equations for $\hat{J}_3$ and $\hat{J}_0$,

$$\hat{J}_3|\ell,m\rangle = m|\ell,m\rangle \tag{3.6}$$

$$\hat{J}_0|\ell,m\rangle = \ell(\ell+1)|\ell,m\rangle, \tag{3.7}$$

while the action of $\hat{J}_\pm$ is given by

$$\hat{J}_+|\ell,m\rangle = \sqrt{[(m+\ell+1)(m-\ell)]}|\ell,m+1\rangle \tag{3.8a}$$

$$\hat{J}_-|\ell,m\rangle = \sqrt{[(m+\ell)(m-\ell-1)]}|\ell,m-1\rangle. \tag{3.8b}$$

The UIR's are labeled by the index ℓ and can be grouped into the following three series[14,6]:

$i)$ the discrete principal series $D_\ell^\pm$ with $\ell = -q/2$ and $m = \pm(-\ell+N-1)$, where $q, N \in \mathbf{Z}^+$;

$ii)$ the continuous principal series C_ℓ^δ with $\ell = ip - 1/2$ and $m = n + \delta/2$, where $p \in \mathbf{R}^+$, $\quad n \in \mathbf{Z}$, $\quad \delta = 0,1$;

$iii)$ the supplementary series E_ℓ with $-1/2 < \ell < 0$ and $m \in \mathbf{Z}$.

Furthermore the non–compact operators $\hat{J}_2$, $\hat{J}_1 + \hat{J}_3$ can be diagonalized utilizing the standard basis and possess continuos eigenvalues.

The generic eigenstates Φ_m of $\hat{J}_3$ and $\hat{J}_0$ are easily constructed, and for $b > 0$ are explicitly given by

$$\Phi_m(\xi, \bar{\xi}) = \mathcal{N} e^{-\frac{i}{\hbar}\Lambda} R^m \Delta^{b/\hbar} \left(\frac{R + \bar{\xi}}{R + \xi}\right)^{b/\hbar} \bar{\xi}^{-m-b/\hbar} \; ; \tag{3.9}$$

where $\mathcal{N}$ is a normalization factor to be fixed. Making use of eq. (3.9) the $\hat{J}_0$ secular equation reads

$$\hat{J}_0 \, \Phi_m = \frac{b}{\hbar}\left(\frac{b}{\hbar} - 1\right) \Phi_m \; ; \tag{3.10}$$

and compared to eq. (3.7) allows one to set

$$\ell = -b/\hbar = -(BR^2)/\hbar \tag{3.11}$$

and recover eqs. (3.8). The normalizability condition on Φ_m combined with the identification (3.11) requires

$$\begin{cases} \ell < -\frac{1}{2} \\ m < 1 + \ell \; ; \end{cases} \tag{3.12}$$

and shows that only the discrete series $D_\ell^{\,-}$ is represented by the states Φ_m.

Recalling that ℓ for the discrete series is integer or half–integer, eq. (3.11) appears as a quantization condition for the magnetic flux, derived by means of purely algebraic methods and not by topological arguments, as for the case of compact manifolds.

The normalized eigenstates turn out to be

$$\langle \xi | \ell, m \rangle_- =$$
$$\frac{(-1)^m}{4\pi^{1/2}} \sqrt{\frac{\Gamma(-\ell - m)}{\Gamma(1 + \ell - m)\Gamma(-1 - 2\ell)}} \; e^{-\frac{i}{\hbar}\Lambda} R^{m+\ell-1} \Delta^{-\ell} \left(\frac{R + \bar{\xi}}{R + \xi}\right)^{-\ell} \bar{\xi}^{-m+\ell} \; . \tag{3.13}$$

When b is taken to be negative the $\hat{J}_0$ and $\hat{J}_3$ eigenfunctions become

$$\langle \xi | \ell, m \rangle_+ =$$
$$\frac{1}{4\pi^{1/2}} \sqrt{\frac{\Gamma(-\ell + m)}{\Gamma(1 + \ell + m)\Gamma(-1 - 2\ell)}} \; e^{-\frac{i}{\hbar}\Lambda} R^{-m+\ell-1} \Delta^{-\ell} \left(\frac{R + \bar{\xi}}{R + \xi}\right)^{\ell} \xi^{m+\ell} \; ; \tag{3.14}$$

and represent the discrete series $D_\ell{}^+$; the normalizability condition now leading to (in this case $\ell = b/\hbar$)

$$\begin{cases} \ell < -\frac{1}{2} \\ m > -\ell - 1 \ . \end{cases} \tag{3.15}$$

The discrete part of the energy spectrum can be read from the quantum version of eq. (2.15)

$$E_\ell = \frac{1}{4R^2}\left(b^2 - \hbar^2\ell(\ell+1)\right) = \frac{\hbar|b|}{4R^2} \ ; \tag{3.16}$$

and reflects in a natural way the fact that the classical paths fulfilling condition *i)* of Sec. 2 are closed orbits.

One can notice an intriguing feature of the eigenstates (3.9), namely their behavior under a shift of the $\hat{J}_3$ eigenvalue m, $m \to m + \alpha$. Eqs. (3.6)-(3.8) now become

$$\hat{J}_3|\ell, m + \alpha\rangle = (m + \alpha)\,|\ell, m + \alpha\rangle \tag{3.17}$$

$$\hat{J}_+|\ell, m + \alpha\rangle = \sqrt{[(m + \alpha + \ell + 1)(m + \alpha - \ell)]}\,|\ell, m + \alpha + 1\rangle \tag{3.18}$$

$$\hat{J}_-|\ell, m + \alpha\rangle = \sqrt{[(m + \alpha + \ell)(m + \alpha - \ell - 1)]}\,|\ell, m + \alpha - 1\rangle. \tag{3.19}$$

$$\hat{J}_0|\ell, m + \alpha\rangle = \ell(\ell + 1)\,|\ell, m + \alpha\rangle, \tag{3.20}$$

showing that the quantum number ℓ is unaffected.

Nevertheless the single–valuedness of the eigenfunctions is not preserved as long as α is a generic real number. Requiring that we have ordinary (fermionic and bosonic) statistics, i.e. $\exp\left(2\pi i \hat{J}_3\right) = \pm\mathbf{1}$, constrains α to be integer or half–integer. The general choice $\alpha \in \mathbf{R}$ would lead to wavefunctions which acquire a non–trivial phase under a 2π rotation, thus exhibiting anyon–like statistics[15]. We should remark that the shift of m considered above is not equivalent to introducing $SU(1,1)$ projective representations, which allow $m \in \mathbf{R}$ provided ℓ is a (negative) real number.

At this point it is interesting to compare our algebraic treatment with the results of Oshima[5], who directly diagonalizes $\hat{J}_3$ and $\hat{J}_0$ (the latter being recognized as a Maass laplacian[16]) without imposing the eigenfunctions to obey the standard equations of raising and lowering operators. Satisfying these equations is a necessary step towards the construction of coherent states and LN states; and hence is a condition that we strictly observe throughout the present paper.

On the other hand, Oshima's less restrictive procedure allows one to obtain generalized representations for both the discrete and the continuous series, given explicitly by

$$e^{im\varphi} P^m_{|\ell|,b/\hbar}(u) \; ;$$

where $u = (R^2 + |\xi|^2)/(R^2 - |\xi|^2)$, $\varphi = \frac{1}{2i}\ln(\xi/\bar{\xi})$ and $P^m_{|\ell|,b/\hbar}$ are generalized associated Legendre functions.

Notice that in Oshima's wavefunctions, $|\ell|$ is not identified with $b/\hbar$, but rather one has $\ell = -|\frac{b}{\hbar}| + v$ with $v \in \mathbf{Z}$, $0 \le v < |\frac{b}{\hbar}| - \frac{1}{2}$. Setting $v = 0$ in the eigenfunctions of Ref.[5], we recover our shifted eigenfunctions $\langle \xi | \ell, m + \alpha \rangle_\pm$ with the choice $\alpha = -|\frac{b}{\hbar}|$.

As a final remark, we point out that in the flat space limit $(R \to \infty)$ a straightforward calculation shows that the eigenstates (3.13) and (3.14) lead to the familiar Landau level wavefunctions[17]. By taking $|\ell| \to \infty$ and $k \equiv (m - |\ell|)$ finite one gets in the case of D_ℓ^+ and for the symmetric gauge

$$\Psi_k(\xi) = \frac{1}{4(\pi k!)^{1/2}} \left(\frac{2|B|}{\hbar} \right)^{\frac{k+1}{2}} \xi^k e^{-\frac{|B|}{\hbar}|\xi|^2} \; , \tag{3.21}$$

and an analogous result for D_ℓ^-. In the same gauge the $SU(1,1)$ generators assume a rather transparent form:

$$\hat{J}_1 = \frac{1}{2\hbar R}[(R^2 + \xi_1^2 - \xi_2^2)p_2 - 2\xi_1\xi_2 p_1] - \frac{b}{\hbar R}\xi_1$$

$$\hat{J}_2 = -\frac{1}{2\hbar R}[(R^2 + \xi_2^2 - \xi_1^2)p_1 - 2\xi_1\xi_2 p_2] - \frac{b}{\hbar R}\xi_2 \tag{3.22}$$

$$\hat{J}_3 = \frac{1}{\hbar}(\xi_1 p_2 - \xi_2 p_1) - \frac{b}{\hbar} \; ;$$

where $p_1 = p_\xi + p_{\bar{\xi}}$ and $p_2 = -i(p_\xi - p_{\bar{\xi}})$. $\hat{J}_1$ and $\hat{J}_2$ play the role of magnetic translators on the hyperboloid. In fact, in the limit $R \to \infty$ the operators $\hat{W}_1 \equiv -(\hbar/R)\hat{J}_2$ and $\hat{W}_2 \equiv (\hbar/R)\hat{J}_1$ reduce to the usual magnetic translators on the plane obeying the commutation rule[17]

$$[\hat{W}_1, \hat{W}_2] = i\hbar B \; .$$

It is interesting to notice that whereas in the flat case, setting $B = 0$ changes the symmetry of the problem (from the magnetic group to the euclidean group $E(2)$), on the pseudosphere the algebra does not change when the magnetic field vanishes.

4. Coherent States and Lindblad–Nagel States

In this section we explicitly calculate the coherent states for the series $D_\ell^\pm$ and then construct the eigenstates of the $SU(1,1)$ noncompact operators $\hat{J}_2$ and $\hat{J}_1 + \hat{J}_3$, following the LN prescription[6].

By definition the generalized $SU(1,1)$ coherent states for D_ℓ^+ are given by[18]

$$|\mu\rangle_+ \equiv e^{\mu \hat{J}_+ - \bar{\mu}\hat{J}_-}|\ell, |\ell|\rangle_+ \;, \mu \in \mathbf{C} \; ; \tag{4.1}$$

$|\ell, |\ell|\rangle$ being the highest weight vector. Reparametrizing the label μ in the following way:

$$\mu = \frac{1}{2}e^{i\theta} \ln\frac{1 + |\varsigma|^2}{1 - |\varsigma|^2} \;, \quad (\varsigma = e^{i\theta} \,\mathrm{th}\,|\mu|) \; ;$$

where ς is now the complex coordinate of the unit disk $|\varsigma| < 1$, and using Baker–Campbell–Hausdorff formulas for $SU(1,1)$, one can express the coherent state in the form

$$|\varsigma\rangle_+ = (1 - |\varsigma|^2)^{-\ell} \sum_{m=|\ell|}^{\infty} C(m,\ell)\,\varsigma^{m-|\ell|}|\ell, m\rangle_+ \; ; \tag{4.2}$$

with

$$C(m,\ell) = \sqrt{\frac{\Gamma(m - \ell)}{(m + \ell)!\,\Gamma(-2\ell)}} \; .$$

From eqs. (3.14) and (4.2) it is an easy task to obtain the wavefunction representation of $|\varsigma\rangle_+$ in the ambient space:

$$\langle\xi|\varsigma\rangle_+ = \frac{\sqrt{2|\ell| - 1}}{4\pi^{1/2}R}\,e^{-\frac{i}{\hbar}\Lambda}\Delta^{-\ell}\left(\frac{R + \bar{\xi}}{R + \xi}\right)^\ell \frac{(1 - |\varsigma|^2)^{-\ell}}{(R - \xi\varsigma)^{-2\ell}} \; . \tag{4.3}$$

The corresponding results for D_ℓ^- are

$$|\varsigma\rangle_- = (1 - |\varsigma|^2)^\ell \sum_{m=-|\ell|}^{-\infty} C(-m,\ell)\,(-\bar{\varsigma})^{-m-|\ell|}|\ell, m\rangle_- \tag{4.4}$$

$$\langle \xi | \varsigma \rangle_- = \frac{\sqrt{2|\ell|-1}}{4\pi^{1/2}R}(-1)^{\ell}e^{-\frac{i}{\hbar}\Lambda}\Delta^{-\ell}\left(\frac{R+\bar{\xi}}{R+\xi}\right)^{-\ell}\frac{(1-|\varsigma|^2)^{\ell}}{(R-\bar{\xi}\varsigma)^{-2\ell}} \ . \tag{4.5}$$

Let us now proceed to diagonalize $\hat{J}_2$ (for the necessary mathematical background we refer the reader to the original Lindblad–Nagel paper [6]).

The $\hat{J}_2$ eigenstates are labeled by a real number $\lambda \in (-\infty, +\infty)$ and, as we have shown in a previous paper[7], can be written in a very simple form as continuous superpositions of $SU(1,1)$ coherent states

$$|\ell, \lambda\rangle_+ = S(\ell, \lambda)\sqrt{\Gamma(2|\ell|)}\int_{-\infty}^{+\infty} ds \ e^{-2i\lambda s}|\varsigma(s)\rangle_+ \ ; \tag{4.6}$$

with $\varsigma(s) = \text{th}\, s$ and $S(\ell, \lambda) = \Gamma(\frac{|\ell|+i\lambda}{2})\Gamma(\frac{|\ell|+1-i\lambda}{2})$. The configuration space realization of this state is

$$\begin{aligned}
\langle \xi | \ell, \lambda\rangle_+ =\ &A(\ell, \lambda)\, e^{-\frac{i}{\hbar}\Lambda}\Delta^{-\ell}\left(\frac{R+\bar{\xi}}{R+\xi}\right)^{\ell}\left(\frac{R-\xi}{R}\right)^{2\ell} \\
&\times\ F\left(2|\ell|, i\lambda + 2|\ell|, 2|\ell|; \frac{2\xi}{R-\xi}\right)
\end{aligned} \tag{4.7}$$

$$A(\ell, \lambda) = 2^{1-2\ell}\frac{|\Gamma(i\lambda + |\ell|)|^2}{\sqrt{\Gamma(2|\ell|)}}\frac{\sqrt{2|\ell|-1}}{\pi^{1/2}R}S(\ell, \lambda) \ ;$$

where $F(a, b, c; z)$ is the hypergeometric function. The diagonalization of the other non–compact operator $\hat{J}_1 + \hat{J}_3$ exhibits some novel features. The eigenstates (now labeled by a positive real number η) can still be expressed by an integral over coherent states

$$|\ell, \eta\rangle_+ = N(\ell, \eta)\sqrt{\Gamma(2|\ell|)}\int_{\gamma} dt \ e^{2\eta t}(2t-1)^{\ell}|\varsigma(t)\rangle_+ \ ; \tag{4.8}$$

with $\varsigma(t) = (1-t)/t$. The contour γ is now encircling the cut $(-\infty, 0)$ in the t complex plane in order to guarantee a well behaved expansion of $|\ell, \eta\rangle$ in the standard basis $\{|\ell, m\rangle\}$, as prescribed by LN. All along this contour we have $|\varsigma| > 1$, which implies that γ lies outside the unit disk $|\varsigma| < 1$ where the coherent states were originally defined. However, an explicit computation of the integral (4.8) in the configuration space basis provides a wavefunction which is manifestly free of pathologies and is given by

$$\langle \xi | \ell, \eta \rangle_{+} =$$

$$N(\ell,\eta)\sqrt{\Gamma(2|\ell|)}\,\frac{i\pi^{1/2}\sqrt{2|\ell|-1}}{R(2\eta)^{2\ell+1}}\,e^{-\frac{i}{\hbar}\Lambda}\Delta^{-\ell}\left(\frac{R+\bar{\xi}}{R+\xi}\right)^{\ell}(R+\xi)^{2\ell}\exp\left(-\frac{2\eta\xi}{R+\xi}\right)\,,$$

$$N(\ell,\eta) = \frac{-i}{\sqrt{2\pi^2}}\,|2\eta|^{\ell+1/2}\,e^{-\eta}\,.$$

$$(4.9)$$

The set of coherent states mapped into the complement of the disk $|\varsigma| < 1$ (which corresponds to the half plane $\mathrm{Re}\,t < 1/2$) can be regarded as the analytic continuation of the originary set.

5. Further Developments

In the present work we have not touched upon some open issues (mainly of mathematical origin) which are certainly worth being explored in the future.

First of all, our discussion was confined to the discrete series of $SU(1,1)$. The reason for this arises from the well–known and still unsolved problem of deriving explicitly (that is in a wavefunction representation) the standard basis for the continuous series C_ℓ^δ and the supplementary series E_ℓ. Although Oshima in Ref.[5] seems to have succeeded in representing C_ℓ^δ, it is still unclear whether the solutions he obtains for the eigenvalue equations of $\hat{J}_0$ and $\hat{J}_3$ form a standard basis, that is a set of states obeying eqs. (3.8) for the $\hat{J}_\pm$ action.

The problem of consistently representing C_ℓ^δ has also remarkable physical implications, since one of the peculiarities of the Landau levels on the hyperboloid is the existence of a continuous spectrum and scattering states, which are missing in the flat space. Constructing the S–matrix for such a purely geometrical and nonpotential scattering might be an interesting task.

A whole body of knowledge still to be acquired concerns the relationship between the physical problem we are dealing with and the mathematical theory of the Maass operators[16,19,20]. Such operators increase or decrease the magnetic flux by one unit, and in view of eq. (3.11), make a shift from one representation to another, thus having perhaps a deep connection with the intertwining operators of $SU(1,1)^{[21]}$. The fact that the Hamiltonian of our system can be written as a product of a raising and a lowering Maass operator has been recognized[4] but not exploited yet in all its implications.

We are currently working to address the questions raised above.

Acknowledgments

We would like to thank M. Rasetti for enlightening discussions during the course of this work.

We hope that Prof. E.R. Caianiello will enjoy reading this paper which deals with the relationship between quantization and geometry.

References

[1] M. C. Gutzwiller: Physica $\underline{7D}$, (1983), 341.

[2] N. L. Balazs and A. Voros: Phys. Rep. $\underline{143}$, (1986), 109.

[3] Y. Alhassid, F. Gürsey and F. Iachello: Ann. Phys. (NY) $\underline{148}$, (1983), 346; Ann. Phys. (NY) $\underline{167}$, (1986), 181.

[4] A. Comtet: J. Math. Phys. $\underline{26}$, (1985), 185; Ann. Phys. (NY) $\underline{173}$, (1987), 185.

[5] K. Oshima, Progr. Theor. Phys. $\underline{81}$, (1989), 286.

[6] G. Lindblad, B. Nagel: Ann. Inst. H. Poincaré $\underline{13}$, (1970), 27. Throughout our work we adopt the mathematical notation of this paper. For recent developments see also: V. Penna: Mod. Phys. Lett. $\underline{B5}$, (1991), 1947 and Ref. 7.

[7] V. Barone, V. Penna and P. Sodano: Phys. Lett. $\underline{A161}$, (1991), 41.

[8] E. Celeghini, M. Rasetti, M. Tarlini and G. Vitiello: Mod. Phys. Lett. $\underline{B3}$, (1989), 1213.

[9] *Squeezed and Nonclassical Light*, P. Tombesi and E. R. Pike eds., Plenum, New York, (1989).

[10] H. Dekker: Phys. Rep. $\underline{80}$, (1981), 1; S. K. Bose, U. B. Dubey and N. Varma, Fortschr. Phys. $\underline{37}$, (1989), 10.

[11] H. Umezawa and Y. Yamanaka: Adv. Phys $\underline{37}$, (1988), 531; Phys. Lett. $\underline{A155}$, (1991), 75.

[12] C. G. Callan and F. Wilczek: Nucl. Phys $\underline{B340}$, (1990), 366; R. Link: Mod. Phys. Lett. $\underline{B5}$, (1991), 1655.

[13] B. S. DeWitt: Rev. Mod. Phys. $\underline{29}$, (1957), 377.

[14] V. Bargmann: Ann. Math. $\underline{48}$, (1947), 568.

[15] See for instance: S. Forte: Rev. Mod. Phys. $\underline{64}$, (1992), 193.

[16] J. Fay: J. reine angew. Math. $\underline{293}$, (1977), 143.

[17] See for instance: G. Morandi, *Quantum Hall Effect*, Bibliopolis, Naples, (1988); S. Fubini: Int. J. Mod. Phys. $\underline{A5}$, (1990), 3533.

[18] A. M. Perelomov: *Generalized Coherent States and Their Applications*, Springer, Berlin, (1986).

[19] H. Maass: Math. Ann. $\underline{121}$, (1949), 141; Math. Ann. $\underline{125}$, (1952), 235.

[20] W. Roelcke: Math. Ann. $\underline{167}$, (1966), 292; Math. Ann. $\underline{168}$, (1967), 261.

[21] E. M. Stein and A. W. Knapp: Ann. Math. $\underline{93}$, (1971), 489.

Thermal Fluctuation of Microscopic Origin

H. Umezawa
The Theoretical Physics Institute
University of Alberta
Edmonton, Alberta T6G 2J1
CANADA

October 15, 1991

Abstract

It is shown that in thermal quantum field theories, quasi particles move through a bundle of thermal vacua which belong to the set
of zero hat-energy degenerate states. Distribution of these intermediate thermal vacua controls temporal thermal behavior. This can
be seen in the time-representation (t-representation) rather than in
the energy Fourier representation. This formalism explicitly shows
how thermal fluctuations behave. This approach is applied to equilibrium TFD, time-dependent TFD and time-space dependent TFD. The
fluctuations in equilibrium situation may control the thermal behavior of high energy particle collisions. The space-time dependent TFD
is useful in derivation of heat conduction theory. Renormalization of
self-energy diagrams leads to the Boltzmann equation which will play
a vital role in relating TFD to thermodynamics.

1 Introduction

I was born around the time of discovery of quantum mechanics. Within less than a decade the quantum field theory started. However, most of physicists at that time regarded quantum field theory as a simple extension of quantum mechanics, although there were some anomalies suggesting a drastic difference between quantum mechanics (QM) and quantum field theory (QFT). For example, the infra-red catastrophe in perturbative calculation of Bremsstrahlung cross section suggested that observed electron contains, in addition to the bare electrons, infinite number of soft photons. This suggested that use of observed dressed electron instead of bare electron picture is significant. This view was carried over to the renormalization theory.

My physics research has come through this dramatic change in understanding of QFT. My research started at the beginning of the post-war development in physics and I was one of those who took it as their research target to formulate QFT in a systematic form.

In those days we were told that we have two distinctive levels in nature: macroscopic world described by the classical physics and microscopic world ruled by quantum physics. The latter consisted of quantum mechanics and quantum field theory. The transitions between these different levels were not made by smooth derivation but by revolutionary changes. This was the physics view in those days.

The physics view made a dramatic change in last four decades. The fundamental entities in modern physics are quantum fields. We now know that in quantum field theories a Hamiltonian has many solutions. Electron gas in metals appears sometimes in normal conducting states, while in other times in superconducting states. Furthermore, a quantum field model has solutions which contain parameters which do not exist in the original Lagrangian. For example, solutions with solitons in sine-Gordon model have soliton velocities which are not seen in the Lagrangian. These new parameters can exhibit new symmetries which are not seen from the Lagrangian. For example two soliton solutions of sine-Gordon models exhibit double Lorentz symmetry, though original Lagrangian has single Lorentz symmetry. This is the reason for appearance of more conserving quantities than the Nöther current given by original Lagrangian (emergent symmetries).

Modern quantum field theory is capable of describing both phenomena; the spontaneous breakdown of symmetries and the emergent symmetries.

This large variety of solutions originate from large variety in choice of state vector spaces: superconducting electron Fock space built on vacuum with Cooper pair condensate is inequivalent to normal conducting electron Fock space which is built on empty Fock space. A variety in choice of Fock spaces is the reason for richness in choice of solutions. Each Fock space is built on its own vacuum. Since most of vacua are not empty but contain certain forms of particle condensate, they exhibit a variety of properties. The properties associated with vacua are of macroscopic nature. Thus, the old view in physics has changed; *modern quantum field theories accommodate, not only quantum field particles, but also objects with quantum mechanical and classical degrees of freedom.*

I strongly felt this transition in view of physics around the end of nineteen fifties. This motivated me to look for a new research environment of fresh atmosphere and of wide disciplines. Since I had known Prof. Caianiello who had built such an institute in Naples, I moved to his institute. Looking back my life now, I see that I made a right decision. I owe this to Eduardo Caianiello. Therefore, I like to present a brief sketch of some of my present view on physics in this conference for celebrating the seventieth birthday of Eduardo.

As was pointed above the concept of vacuum in modern quantum field theory has a rich and complex material properties. This is very similar to the old concept of ether. The concept of ether was abandoned by its confliction with the relativity, because the rest system of ether defines a particular coordinate. The concept of vacuum with particle condensate does not meet this difficulty, because the Lorentz invariance is recovered by the zero energy Goldstone modes. This illustrates a rich power of QFT which neither classical physics nor quantum mechanics possess.

2   Thermal Quantum Field Theories

I have been interested in the question how large the domain of nature are accommodated by quantum field theory. So far I have not seen any border line. One of such effort is to see how quantum field theory can accommodate thermal physics. This development created many forms of thermal quantum field theories such as the c*-algebra approach of Haag and others, the path-ordering method, the thermo field dynamics and others.

These theories for equilibrium situations have been well established. Still, the problem of thermal fluctuations has not been well understood. Consider, for example, high energy particle reactions. There have been many speculation that there may appear strong thermal effects during reactions. Since most of reactions seem to occur in equilibrium initial states, such thermal effects may be fluctuation effects. According to quantum physics, many kinds of states show up during a short time phenomenon. Therefore, it is possible that these intermediate states may include many thermal states. An aim of this talk is to present a method for separating out this fluctuation. In the next section we consider equilibrium thermo field dynamics (TFD). In later sections we will show how this leads to Boltzmann equation in nonequilibrium TFD.

3   The t-Representation in Equilibrium TFD

The thermal quantum field theories for equilibrium situations are usually formulated in terms of the Fourier representation. However, to study nature of thermal fluctuation during particle reactions, we wish to follow behavior of a quasi particle in time rather than in the Fourier representation. Therefore, we make use of the time representation (t-representation).

To make our story short, let me describe my arguments in terms of quantum field oscillator operators $a_k(t)$. In TFD every degree of freedom is doubled. Thus $a_k(t)$ has its partner $\tilde{a}_k(t)$. Any nontilde operator is related to its tilde partner through the so called tilde conjugation rules, which have been widely known among physicists working on thermal quantum field theories. With half an hour available for this talk I should omit explanations of things known in literatures. The thermal doublet notation $a_k(t)^\mu$ is commonly used. This is $a_k^1 = a_k$, $a_k^2 = \tilde{a}_k^\dagger$, $\bar{a}_k^1 = a^\dagger$, $\bar{a}_k^2 = -\tilde{a}_k$. Here, for simplicity, we consider only bosons.

Consider a system, the dynamics of which is described by Hamiltonian H. This is the Hamiltonian for nontilde operators. One of fundamental aspects of TFD is that the Hamiltonian $\hat{H}$ for both nontilde and tilde operators has the form

$$\hat{H} = H - \tilde{H}, \tag{3.1}$$

where $\tilde{H}$ is obtained from H by the tilde conjugation rules. Thus, knowledge of H is sufficient for construction of $\hat{H}$. The eigenvalues of $\hat{H}$ are called hat-energy.

In TFD thermal average of any observable A is obtained by its vacuum expectation value:

$$\langle A \rangle = \langle 0|A|0 \rangle. \tag{3.2}$$

All of the thermal vacua, $|0\rangle$, are eigenstates of $\hat{H}$ and are required to be invariant under tilde conjugation. Then, it has been shown that the thermal vacua are states of zero hat-energy:

$$\hat{H}|0\rangle = 0. \tag{3.3}$$

Since $\tilde{H}$ in $\hat{H}$ carries negative sign, these hat-energy degenerate thermal vacua form a continuous set $[|0(\theta)\rangle]$ with a continuous parameter θ. The equilibrium states form a subset of this. In this subset θ is the temperature.

Thus, the modes moving through members of the set of thermal vacua are zero hat-energy mode. This situation is same as the appearance of Goldstone modes associated with spontaneous breakdown of symmetries. Indeed, TFD can be formulated as spontaneous breakdown of thermal SU(1,1)-symmetry.

Note that these thermal vacua are not degenerate in dynamical energies which are given by expectation values of H. The difference of the dynamical energy between any two different vacua is the heat energy.

Thermal degree of freedom is the freedom of moving through these thermal vacua. Thus, temporal development of any thermal process goes through the vacua in this set. In other words, This set is the space through which thermal processes move around. In this sense the thermal vacuum set plays a role analogous to the phase space in classical physics. This is important when we try to derive thermodynamics from TFD.

In the interaction representation with free quasi particle picture, thermal effects are treated by the thermal Bogoliubov transformation

$$a_k(\theta)^\mu = B_k^{-1}(\theta)^{\mu\nu}\xi_k^\nu. \tag{3.4}$$

We can formulate the Feynman diagram methods in this framework of TFD. Introducing the operator $u(t)$ by $|t\rangle = u(t)|-\infty\rangle$ for states $|t\rangle$ in the interaction representation and defining $S = u(t = \infty)$, we can show that thermal equilibrium states form a subset of the above vacuum set with the property $S|0\rangle = |0\rangle$, $\langle 0|S = \langle 0|$ (equilibrium vacuum stability).

158

As it is well known, the unperturbed propagator (Feynman line) with momentum $\vec{k}$ in TFD is a two by two matrix given by

$$\Delta_k(t - t') = B^{-1}[n_k]\frac{1}{k_0 - \omega_k + i\epsilon_F\tau_3}B[n_k], \qquad (3.5)$$

where ϵ_F is the positive infinitesimal and τ_3 is

$$\tau_3 = \begin{bmatrix} 1 & 0 \\ 0 & -1 \end{bmatrix}. \qquad (3.6)$$

The matrix $B[n_k]$ is the Bogoliubov matrix

$$B_k = \begin{bmatrix} 1 + \sigma n_k & -n_k \\ -\sigma & 1 \end{bmatrix}. \qquad (3.7)$$

(A note for specialists: we chose $\alpha = 1$ and $s_k = \ln[1 + \sigma n_k]^{1/2}$). In equilibrium situations n_k is given by the Boltzmann distribution with the temperature β^{-1}:

$$n_k = \frac{1}{e^{\beta\omega_k} - 1}. \qquad (3.8)$$

Writing a one body propagator for the Heisenberg oscillator operators $a^\mu_{Hk}(t)$ and $\bar{a}^\mu_{Hk}(t)$ (suffix H means Heisenberg operators) as

$$D_k(t - t')^{\mu\nu} = \int\frac{dk_0}{2\pi}e^{-ik_0(t-t')}D_k(k_0)^{\mu\nu}, \qquad (3.9)$$

we have the spectral representation

$$D_k(k_0)^{\mu\nu} = \int_{-\infty}^{\infty} d\kappa \left[B^{-1}[n(\kappa)]\frac{\rho(\kappa, \vec{k})}{k_0 - \kappa + i\epsilon\tau_3}B[n(\kappa)]\right]^{\mu\nu}, \qquad (3.10)$$

where $n(\kappa)$ is given by the Boltzmann distribution with the temperature β^{-1},

$$n(\kappa) = \frac{1}{e^{\beta\kappa} - 1}. \qquad (3.11)$$

The function ρ is the spectral function which usually has a peak at $\kappa = \omega_k$ and a width κ_k. The latter is the dissipative constant due to thermal instability of quasi particles. The equal-time oscillator commutation relation for $a_{Hk}^{\mu}(t)$ and $\bar{a}_{Hk}^{\mu}(t)$ gives

$$\int d\kappa\, \rho(\kappa, \vec{k}) = 1. \tag{3.12}$$

To find temporal behavior of the propagator, we should perform the k_0-integration. We then find

$$D_k(t - t')^{\mu\nu} = B^{-1}[N_k(t - t')]^{\mu\mu'}$$
$$\times \int d\kappa\, e^{-i\kappa(t-t')}\rho(\kappa, \vec{k})\begin{bmatrix} -i\theta(t - t') & 0 \\ 0 & i\theta(t' - t) \end{bmatrix}^{\mu'\nu'} B[N_k(t - t')]^{\nu'\nu}. \tag{3.13}$$

Here

$$N_k(t - t') = \frac{\int d\kappa\, e^{-i\kappa(t-t')}\rho(\kappa, \vec{k})n(\kappa)}{\int d\kappa\, e^{-i\kappa(t-t')}\rho(\kappa, \vec{k})} \tag{3.14}$$
$$= N_{R,k} + \nu_k(t - t') \tag{3.15}$$

with

$$N_{R,k} \equiv N_k(0) = \int d\kappa\, \rho(\kappa, \vec{k})n(\kappa) \tag{3.16}$$

and $\nu(t - t')$ has the property

$$\nu(t - t') = 0 \text{ for } t = t'. \tag{3.17}$$

The equation (3.16) shows that the observable particle number fluctuates around the Boltzmann number with fluctuation of order of width of the spectral function ρ. The above result shows a remarkable result that *even in equilibrium situation each quasi particle wave propagation feels the time-dependent fluctuation $\nu(t - t')$*. We have calculated $\nu(t - t')$ in terms of the spectral function, but we omit here the result.

In a similar way two particle states may feel time dependent thermal fluctuations in the t-representation. When the fluctuation is very large, quarks may become free for a short time during extremely high energy particle collisions. Study of nature of thermal fluctuation demands many model calculations.

4 Time-Space Dependent TFD

With the t-representation formalism, extension of TFD to space-time dependent situations becomes easy. Let us summarize results for one body propagator.

We begin with situations dependent on time only. Then the one body propagator corrected with interaction is:

$$
D_k(t,t')^{\mu\nu} = B^{-1}[N_{2,k}(t,t')]^{\mu\mu'}
$$
$$
\times \begin{bmatrix} -i\theta(t-t')g(t,t':\vec{k})^{11} & 0 \\ 0 & i\theta(t'-t)g(t,t':\vec{k})^{22} \end{bmatrix}^{\mu'\nu'} B[N_{1,k}(t,t')]^{\nu'\nu},
$$

$$(4.18)$$

where $N_{i,k}(t,t')$ have the form

$$
N_{1,k}(t,t') = N_k(t) + \nu(t,t'), \tag{4.19}
$$
$$
N_{2,k}(t,t') = N_k(t') + \nu^*(t',t). \tag{4.20}
$$

This again shows that particle number corrected by interaction contains the fluctuation $\nu(t,t')$.

Note that in t-representation any one body propagator has the form B^{-1} (diagonal propagator) B. In this way the thermal effects are accumulated mostly by each vertices and no thermal mixing occurs in the propagation. This is the remarkable feature of the t-representation.

The self-energy diagrams have a similar structure. Furthermore, the (11)-component is proportional to the step function $\theta(t-t')$ and the (22)-component to $\theta(t'-t)$. Let us compare this with the structure of self-energy term in equilibrium cases. In equilibrium situation the self-energy Σ is a function of $t-t'$. The $(t-t')$-dependence appears through step functions. The on-shell self-energy is obtained by using the Fourier transform with $k_0 = \omega_k$. In the case under consideration too, we define the on-shell part by applying the Fourier transformation to the step functions, $\theta(t-t')$ and $\theta(t'-t)$, and by replacing k_0 with ω_k. This determines the one body propagator for a quasi particle with energy ω_k, from which follows the Hamiltonian of the quasi particle. When we require that this Hamiltonian is diagonal in terms of the operators ξ_k^μ and $\bar{\xi}_k^\mu$, we find a kinetic equation for $n_k(t)$. The one loop approximation without the vertex corrections this equation reduces to the backward-time Boltzmann equation. This result is due to the fact that I chose the coincidence time t_0 of the Heisenberg and interaction representation as $t_0 = -\infty$. It is expected that this choice may cause some trouble, because with starting at minus infinite time any state at present may be an equilibrium one. The reasonable choice is $t_0 = +\infty$. Indeed, this choice gives the right Boltzmann equation (For specialists; the chronological Feynman diagram with this choice needs $\alpha = 0$. This is remarkable!!)

To extend this argument to space dependent situations is rather straightforward. At first glance, we might suppose that spatial dependence in thermal situation could be introduced by making the Bogoliubov matrix B_k in $a_k = B_k^{-1}\xi$ dependent of space. This would lead to $\vec{x}$-dependent a_k. However, this approach does not work, because the commutation relation

$$[a_k(t)^\mu, \bar{a}_l(t)^\nu] = \delta_{\mu\nu}\delta(\vec{k} - \vec{l}) \tag{4.21}$$

forbids a_k from depending on $\vec{x}$.

However, the formula

$$B^{-1}(n)B(N) = 1 - \sigma(n - N)T_0 \tag{4.22}$$

shows us how to introduce space dependent Bogoliubov transformations. Here $B(n)$ means the Bogoliubov matrix with the number parameter n. This and the following formulae hold true for both boson and fermion fields; $\sigma = 1(-1)$ for boson (fermion). Now note the identity

$$\int d^3q \int d^3x \int d^3y \, e^{i[-(\vec{k}-\vec{q})\cdot\vec{x}+(\vec{l}-\vec{q})\cdot\vec{y}]}$$
$$\times \left(n[t, \vec{x} : \frac{\vec{k} + \vec{q}}{2}] - n[t, \vec{y} : \frac{\vec{l} + \vec{q}}{2}] \right) = 0. \tag{4.23}$$

This together with (4.22) leads to the formula

$$\frac{1}{(2\pi)^6} \int d^3q \int d^3x \int d^3y \, e^{i[-(\vec{k}-\vec{q})\cdot\vec{x}+(\vec{l}-\vec{q})\cdot\vec{y}]}$$
$$\times \left(B^{-1}(n[t, \vec{x} : \frac{\vec{k} + \vec{q}}{2}])B(n[t, \vec{y} : \frac{\vec{l} + \vec{q}}{2}]) \right)^{\mu\nu} = \delta_{\mu\nu}\delta(\vec{k} - \vec{l}).$$
$$\tag{4.24}$$

This shows that the Bogoliubov transformations

$$a_k(t)^\mu = \frac{1}{(2\pi)^3} \int d^3q \int d^3x \, e^{-i(\vec{k}-\vec{q})\cdot\vec{x}}$$
$$\times \; B^{-1}(n[t,\vec{x}:\frac{\vec{k}+\vec{q}}{2}])^{\mu\nu}\xi_q(t)^\nu, \qquad (4.25)$$

$$\bar{a}_k(t)^\mu = \frac{1}{(2\pi)^3} \int d^3q \int d^3x \, e^{i(\vec{k}-\vec{q})\cdot\vec{x}}\bar{\xi}_q(t)^\nu$$
$$\times \; B(n[t,\vec{x}:\frac{\vec{k}+\vec{q}}{2}])^{\nu\mu}, \qquad (4.26)$$

satisfy the commutation relation (4.21). This Bogoliubov transformation is the basis of time-space dependent TFD. The analysis of the self-energy diagrams and derivation of Boltzmann equation follows the same steps as those in time-dependent cases. Detailed Accounts will appear in a series of recent preprints of the University of Alberta authored by myself together with T. Arimitsu, Y. Yamanaka and K. Nakamura.

5   A concluding remark

Presence of the concept of vacuum in quantum field theory sharply distinguishes quantum field theory from quantum mechanics and classical physics. At the time of Boltzmann one had the classical theory which does not have the concept of the representation spaces. In TFD, which is an extension of quantum field theory, there is a set of infinite number of hat-energy degenerate thermal vacua. Any thermal process moves through a bundle of thermal vacua. It is expected that distribution of this bundle of intermediate thermal vacua may provide an information entropy, which may be the origin of second law of thermodynamics. In the usual quantum field theory a lower unbounded Hamiltonian is prohibited because the lower unbounded nature leads to instability. Therefore, it is remarkable that in TFD the lower unbounded $\hat{H}$ leads to the second law of thermodynamics (increasing entropy in time), which saves TFD systems from instability. We (Arimitsu, Nakamura, Yamanaka and myself) are actively pursuing this possibility.

Quantum Langevin Equation
for
A Non-Linear Damped Oscillator *

T. Arimitsu, M. Ban[†] and T. Saito

Institute of Physics, University of Tsukuba

Ibaraki 305, Japan

[†]Advanced Research Laboratory, Hitachi, Ltd.

Hatoyama, Saitama 350-03, Japan

May 15, 1991

Abstract

The *quantum stochastic Liouville equation*, within Non-Equilibrium Thermo Field Dynamics, is applied to a quantum system of a non-linear damped oscillator in order to obtain the correlation of the random force operators for the quantum Langevin equation of the system. The case of Gaussian and white process in the quantum system is considered.

---

*Dedicated to Prof. Eduardo R. Caianiello on the occasion of his seventieth birthday.

1 Introduction

The *quantum stochastic Liouville equation*[*], formulated by one of the authors [4, 5] within the formalism of Non-Equilibrium Thermo Field Dynamics (NETFD) [6]-[10], is applied to a quantum system of a non-linear damped oscillator in order to obtain the correlation of the random force operators for the quantum Langevin equation of the system. The correlation is determined so that the dynamics of the system described by the Langevin equation is equivalent to that described by the Schrödinger equation within NETFD, which is nothing but the Fokker-Planck equation of the system [4, 5]. The case of *Gaussian* and *white* process in the quantum system is considered, which may be one of the most interesting cases for the fundamental understanding of the method both for physicists and mathematicians. Extension to other processes is straightforward. It may be interesting to note here that the approach given in this paper will provide us with a good tool for the construction of quantum Ito formula [11, 12].

An operator A within the formalism of NETFD has its partner operator $\tilde{A}$. The tilde conjugation $\sim$ is defined by $(A_1 A_2)^\sim = \tilde{A}_1 \tilde{A}_2$, $(c_1 A_1 + c_2 A_2)^\sim = c_1^* \tilde{A}_1 + c_2^* \tilde{A}_2$, $(\tilde{A})^\sim = A$, $(A^\dagger)^\sim = \tilde{A}^\dagger$, with arbitrary bosonic operators A_i and c-numbers c_i. The tilde and non-tilde operators are related with each other through the relations

$$\langle 1|\tilde{A} = \langle 1|A^\dagger. \tag{1}$$

The thermal vacuums $\langle 1|$ and $|0\rangle$ are tilde invariant:

$$\langle 1|^\sim = \langle 1|, \qquad |0\rangle^\sim = |0\rangle. \tag{2}$$

The expectation value of an operator A is given by $\langle 1|A|0\rangle$, and the inner product of the thermal vacuums are normalized as $\langle 1|0\rangle = 1$.

2 Langevin Equation for a Non-Linear Damped Oscillator

We are going to investigate the system of a non-linear damped oscillator which is specified by the Hamiltonian

$$H = \omega a^\dagger a + \frac{1}{2} g a^\dagger a^\dagger a a, \tag{3}$$

---

and is agitated by the Gaussian and white random force operator $F(t)$. The random force operator may depend on the annihilation and creation operator, a and $a^\dagger$, of the non-linear oscillator, which satisfy the canonical commutation relation $[\, a\,,\, a^\dagger\,] = 1$. The Langevin equation for the system can be written down as

$$D_t a(t) = i\left[H - \tilde{H}\,, a(t)\right] + F[t, a(t)]$$
$$= -i\omega a(t) - iga^\dagger(t)a(t)a(t) + F[t, a(t)]\,, \tag{4}$$

where $a(t)$ is an operator in the Heisenberg representation, and $D_t = d_t + \kappa$ with $d_t = d/dt$. The Langevin equation for $\tilde{a}^\dagger(t)$ is obtained by taking the tilde and Hermite conjugation of (4). The form of the random force is assumed to be given by

$$F(t, a) = f(t) + 2k(t)a, \tag{5}$$

where the characteristics of the random force operators $f(t)$ and $k(t)$, which may depend on the operators $a, a^\dagger$ and their tilde conjugate, will be determined in the following in consistent with the Fokker-Planck equataion of the system as already stated above.

3 Fokker-Planck Equation

The Fokker-Planck equation of the system is given by

$$\partial_t |0(t)\rangle = -i\hat{H}|0(t)\rangle, \tag{6}$$

with

$$\hat{H} = H - \tilde{H} + i\hat{\Pi}\,, \tag{7}$$

where $|0(t)\rangle$ is the thermal vacuum state, and $\partial_t = \partial/\partial t$.

Assuming, for simplicity, the linear dissipative coupling between the system and reservoir, we can derive $\hat{\Pi}$ in the form:

$$\hat{\Pi} = \left(a - \tilde{a}^\dagger\right)\left\{e^{\tilde{a}^\dagger \tilde{a}g\partial/\partial\omega}\tilde{a}\kappa_0(\omega)\left[1 + \bar{n}_0(\omega)\right] - a^\dagger e^{a^\dagger ag\partial/\partial\omega}\kappa_0(\omega)\bar{n}_0(\omega)\right\} + \text{t.c.}$$
$$+i\left(a - \tilde{a}^\dagger\right)\left[-e^{\tilde{a}^\dagger \tilde{a}g\partial/\partial\omega}\tilde{a}\phi''_{-+}(\omega) + a^\dagger e^{a^\dagger ag\partial/\partial\omega}\phi''_{+-}(\omega)\right] + \text{t.c.}, \tag{8}$$

where t.c. indicates to take a tilde conjugation, and

$$\bar{n}_0(\omega) = \left(e^{\omega/T} - 1\right)^{-1}, \qquad (k_B = 1) \tag{9}$$
$$\kappa_0(\omega) = \Re e\,\left[\phi_{-+}(\omega) - \phi_{+-}(\omega)\right], \tag{10}$$

with

$$\phi_{-+}(\omega) = \phi'_{-+}(\omega) + i\phi''_{-+}(\omega) = \lambda^2 \int_0^\infty dt\, e^{i\omega t} \langle R(t)R^\dagger \rangle_R, \tag{11}$$

$$\phi_{+-}(\omega) = \phi'_{+-}(\omega) + i\phi''_{+-}(\omega) = \lambda^2 \int_0^\infty dt\, e^{i\omega t} \langle R^\dagger R(t) \rangle_R. \tag{12}$$

Here R and $R^\dagger$ are operators in the reservoir, and $\langle \cdots \rangle_R$ indicates the canonical average with temperature T. In deriving (8), we have taken the long-time limit and the weak coupling limit with respect to the coupling constant λ between the system and reservoir. In the following in this paper, we will neglect the imaginary part in (8) as it is nothing to do with the energy relaxation of the system. The influence of the imaginary part will be investigated in a separate paper. Note that κ in (4) is given by

$$\kappa(a^\dagger a) = e^{a^\dagger a g \partial / \partial \omega} \kappa_0(\omega) = \kappa_0(\omega + g a^\dagger a). \tag{13}$$

The damping generator $\hat{\Pi}$ contains the full information about its dependence on the non-linear coupling within the system, which corresponds to the non-conventional treatment within the damping theory [13]-[21]. Therefore, the damping generator $\hat{\Pi}$ garantees the true final equilibrium state with reservoir temperature T, which is characterized by the thermal state condition:

$$a\,|0(\infty)\rangle = \tilde{a}^\dagger \exp\left[-\left(\omega + g\tilde{a}^\dagger\tilde{a}\right)/T\right] |0(\infty)\rangle, \tag{14}$$

$$a^\dagger\,|0(\infty)\rangle = \exp\left[\left(\omega + g\tilde{a}^\dagger\tilde{a}\right)/T\right] \tilde{a}\,|0(\infty)\rangle. \tag{15}$$

It means that

$$\hat{\Pi}\,|0(\infty)\rangle = 0, \tag{16}$$

and that

$$\langle 1|a^\dagger a|0(\infty)\rangle = \langle 1|\exp\left[-\left(\omega + g a^\dagger a\right)/T\right] aa^\dagger|0(\infty)\rangle. \tag{17}$$

In the proof of (16), the thermal state conditions (14) and (15) were used in order to rewrite the terms in (8). For example,

$$\begin{aligned}
&\left\{e^{\tilde{a}^\dagger\tilde{a}g\partial/\partial\omega}\kappa_0(\omega)\left[1 + \bar{n}_0(\omega)\right]\right\}\tilde{a}|0(\infty)\rangle \\
&= e^{\tilde{a}^\dagger\tilde{a}g\partial/\partial\omega}\kappa_0(\omega)\left[1 + \bar{n}_0(\omega)\right]e^{-\tilde{a}^\dagger\tilde{a}g\partial/\partial\omega}\tilde{a}|0(\infty)\rangle \\
&= a^\dagger e^{-ga^\dagger a/T}e^{\tilde{a}^\dagger\tilde{a}g\partial/\partial\omega}\kappa_0(\omega)\left[1 + \bar{n}_0(\omega)\right]e^{-\tilde{a}^\dagger\tilde{a}g\partial/\partial\omega}e^{-\omega/T}|0(\infty)\rangle \\
&= a^\dagger\left[e^{a^\dagger a g\partial/\partial\omega}\kappa_0(\omega)\bar{n}_0(\omega)\right]|0(\infty)\rangle,
\end{aligned} \tag{18}$$

where we used $\tilde{a}^\dagger \tilde{a}|0(\infty)\rangle = a^\dagger a|0(\infty)\rangle$, which is also derived by the thermal state conditions (14) and (15). If one puts $g = 0$ in the expression (8), it reduces to the damping generator within the conventional treatment of the damping theory where the influence of non-linear coupling within the system upon its relaxation is completely neglected, i.e., $\langle 1|a^\dagger a|0(\infty)\rangle = \bar{n}_0(\omega)$. Then, $\hat{\Pi}$ is the same as that of a linear harmonic oscillator [6]-[8].

4 Quantum Stochastic Liouville Equation

The quantum stochastic Liouville equation is given by [4, 5]

$$\partial_t |0_f(t)\rangle = -i\hat{H}_f(t)|0_f(t)\rangle, \tag{19}$$

with

$$\hat{H}_f(t) = \frac{1}{2}\left(a^\dagger - \tilde{a}\right) id_t \left(a + \tilde{a}^\dagger\right) - \text{t.c..} \tag{20}$$

The operator $d_t a$ and $d_t \tilde{a}^\dagger$ for the flow in the Schrödinger representation in (20) is specified by the equations whose formal structures are the same as the quantum Langevin equations in (4), i.e.,

$$D_t a = -i\omega a - iga^\dagger aa + F(t, a), \tag{21}$$

and the one obtained by taking the tilde and Hermite conjugation of (21).

5 Random Average of Stochastic Liouville Equation

The time-evolution generator $\hat{H}$ in the Fokker-Planck equation (6) is related to the time-evolution generator $\hat{\Pi}_f(t)$ in the stochastic Liouville equation (19) by [4, 5]

$$-i\hat{H} = \lim_{\Delta t \to 0} \frac{1}{\Delta t} \int_t^{t+\Delta t} dt_1 \hat{K}(t_1), \tag{22}$$

where

$$\hat{K}(t) = \sum_{n=1}^{\infty} \hat{K}_n(t), \tag{23}$$

with

$$\hat{K}_n(t) = (-i)^n \int_0^t dt_1 \int_0^{t_1} dt_2 \cdots \int_0^{t_{n-2}} dt_{n-1} \langle \hat{H}_f(t)\hat{H}_f(t_1)\cdots \hat{H}_f(t_{n-1})\rangle_{o.c.}. \tag{24}$$

Here the symbol $\langle \cdots \rangle_{o.c.}$ indicates the *ordered cumulants* [22] defined, for example, by

$$\langle \hat{H}_f(t)\rangle_{o.c.} = \langle \hat{H}_f(t)\rangle, \tag{25}$$

$$\langle \hat{H}_f(t)\hat{H}_f(t_1)\rangle_{o.c.} = \langle \hat{H}_f(t)\hat{H}_f(t_1)\rangle - \langle \hat{H}_f(t)\rangle\langle \hat{H}_f(t_1)\rangle, \tag{26}$$

168

$$\langle \hat{H}_f(t)\hat{H}_f(t_1)\hat{H}_f(t_2)\rangle_{o.c.}$$

$$= \langle \hat{H}_f(t)\hat{H}_f(t_1)\hat{H}_f(t_2)\rangle - \langle \hat{H}_f(t)\hat{H}_f(t_1)\rangle\langle \hat{H}_f(t_2)\rangle$$

$$- \langle \hat{H}_f(t)\hat{H}_f(t_2)\rangle\langle \hat{H}_f(t_1)\rangle - \langle \hat{H}_f(t)\rangle\langle \hat{H}_f(t_1)\hat{H}_f(t_2)\rangle$$

$$+ \langle \hat{H}_f(t)\rangle\langle \hat{H}_f(t_1)\rangle\langle \hat{H}_f(t_2)\rangle + \langle \hat{H}_f(t)\rangle\langle \hat{H}_f(t_2)\rangle\langle \hat{H}_f(t_1)\rangle, \tag{27}$$

where the symbol $\langle \cdots \rangle$ is the random average reffering to the Langevin force operator in (20). The thermal vacuum $|0(t)\rangle$ of the Fokker-Planck equation (6) is given by the random average of the thermal vacuum $|0_f(t)\rangle$ of the stochastic Liouville equation (19), i.e.,

$$|0(t)\rangle = |\langle 0_f(t)\rangle\rangle. \tag{28}$$

6 Correlation of Random Force Operators

The stochastic time-evolution generator (20) becomes

$$\hat{H}_f(t) = H - \tilde{H} + \frac{1}{2}g\left[\left(aa^\dagger a^\dagger a + a^\dagger \tilde{a}^\dagger \tilde{a}^\dagger \tilde{a} + a\tilde{a}^\dagger \tilde{a}\tilde{a}\right) - \text{t.c.}\right]$$

$$-i\frac{1}{2}\left[\left(a^\dagger - \tilde{a}\right)\left(\kappa a + \tilde{a}^\dagger \tilde{\kappa}\right) + \text{t.c.}\right] + \frac{1}{2}i\left\{\left(a^\dagger - \tilde{a}\right)[F(t) + \tilde{F}^\dagger(t)] + \text{t.c.}\right\}. \tag{29}$$

Substituting (29) into (20), we have

$$\hat{H} = \lim_{\Delta t \to 0} \frac{1}{\Delta t}\int_t^{t+\Delta t} dt_1 \left\{\langle \hat{H}_f(t_1)\rangle - i\int_0^{t_1} dt_2 \langle \hat{H}_f(t_1)\hat{H}_f(t_2)\rangle_{o.c.}\right\}$$

$$= H - \tilde{H} + i\hat{\Pi}, \tag{30}$$

with H and $\hat{\Pi}$ being given by (3) and (8), respectively, *when* the random force operators have the following correlations:

$$\langle f(t)\rangle = 0, \quad \langle f^\dagger(t)\rangle = 0, \tag{31}$$

$$\langle \mathcal{F}(t)\, a\, \mathcal{F}^\dagger(s)\rangle = e^{a^\dagger a g \partial/\partial \omega} a\kappa_0(\omega)\left[1 + 2\bar{n}_0(\omega)\right]\delta(t-s), \tag{32}$$

$$\langle \mathcal{F}^\dagger(t)a^\dagger \mathcal{F}(s)\rangle = a^\dagger e^{a^\dagger a g \partial/\partial \omega}\kappa_0(\omega)\left[1 + 2\bar{n}_0(\omega)\right]\delta(t-s), \tag{33}$$

and

$$\langle k(t)\rangle = 0, \qquad \langle k^\dagger(t)\rangle = 0, \tag{34}$$

$$\langle k(t)k^\dagger(s)\rangle = \langle k^\dagger(t)k(s)\rangle = \frac{1}{2}ig|t-s|\partial_t\delta(t-s). \tag{35}$$

Here we have introduced a random force operator $\mathcal{F}(t)$ by

$$\mathcal{F}(t) = \left[f(t) + \tilde{f}^\dagger(t) \right] /2, \tag{36}$$

which satisfies $\tilde{\mathcal{F}}^\dagger(t) = \mathcal{F}(t)$. Note that the random forces $f(t)$ and $k(t)$ are mutually independent.

The equation of motion for $n(t) = \langle a^\dagger a \rangle_t$ can be obtained by using the Fokker-Planck equation (6) in the form

$$d_t n(t) = -2 \langle e^{a^\dagger a g \partial/\partial\omega} \left[a^\dagger a - \bar{n}_0(\omega) \right] \rangle_t \kappa_0(\omega)$$
$$+2 \langle e^{a^\dagger a g \partial/\partial\omega} a^\dagger a \rangle_t \left(1 - e^{-g\partial/\partial\omega} \right) \kappa_0(\omega) \left[1 + \bar{n}_0(\omega) \right], \tag{37}$$

where we have introduced an abbreviation for the vacuum expectation: $\langle \cdots \rangle_t = \langle 1 | \cdots | 0(t) \rangle$.

7 Heisenberg Equation of Motion

The Heisenberg equation of motion is given by

$$d_t a(t) = i[\hat{H} , a(t)]$$
$$= -i\omega a(t) - iga^\dagger(t)a(t)a(t)$$
$$- \left\{ 1 + \left[a^\dagger(t) - \tilde{a}(t) \right] a(t) \left(1 - e^{-g\partial/\partial\omega} \right) \right\} a(t) e^{-g\partial/\partial\omega} \kappa$$
$$- \left\{ a(t) \left(1 + e^{-g\partial/\partial\omega} \right) - \tilde{a}^\dagger(t) \left(1 + \exp \left[\left[\tilde{a}^\dagger(t)\tilde{a}(t) - a^\dagger(t)a(t) \right] g\partial/\partial\omega \right] \right) \right.$$
$$\left. + \left[\left(a(t) - \tilde{a}^\dagger(t) \right) a^\dagger(t) + \left(a^\dagger(t) - \tilde{a}(t) \right) a(t) e^{-g\partial/\partial\omega} \right] a(t) \left(1 - e^{-g\partial/\partial\omega} \right) \right\} \kappa\bar{n}, \tag{38}$$
$$d_t \tilde{a}^\dagger(t) = i[\hat{H} , \tilde{a}^\dagger(t)]$$
$$= -i\omega \tilde{a}^\dagger(t) - ig\tilde{a}^\dagger(t)\tilde{a}^\dagger(t)\tilde{a}(t)$$
$$- \left\{ 1 + \left[a^\dagger(t) - \tilde{a}(t) \right] \tilde{a}^\dagger(t) \left(1 - e^{g\partial/\partial\omega} \right) \right\} \tilde{a}^\dagger(t) \tilde{\kappa}$$
$$+ \left\{ \tilde{a}^\dagger(t) \left(1 + e^{-g\partial/\partial\omega} \right) - a(t) \left(1 + \exp \left[\left[a^\dagger(t)a(t) - \tilde{a}^\dagger(t)\tilde{a}(t) \right] g\partial/\partial\omega \right] \right) e^{-g\partial/\partial\omega} \right.$$
$$\left. + \left[\left(a(t) - \tilde{a}^\dagger(t) \right) \tilde{a}(t) e^{-g\partial/\partial\omega} + \left(a^\dagger(t) - \tilde{a}(t) \right) \tilde{a}^\dagger(t) \right] \tilde{a}^\dagger(t) \left(1 - e^{g\partial/\partial\omega} \right) \right\} \tilde{\kappa} (1 + \bar{n})^\sim, \tag{39}$$

where κ and $\bar{n}$ are defined by (13) and $\bar{n}(a^\dagger a) = e^{a^\dagger a g\partial/\partial\omega} \bar{n}_0(\omega) = \bar{n}_0(\omega + ga^\dagger a)$, respectively. The existence of the Heisenberg equation of motion for dissipative annihilation and creation operators, like (38) and (39), enabled us to construct a quantum field theory of disspative fields [24]-[31]. Note that the Heisenberg equation describes the dynamics of a coarse grained system, i.e., a random averaged system in the sense given by (22). We can derive the equation of motion (37) using (38) and (39), and taking the thermal vacuum expectation $\langle 1 | \cdots | 0 \rangle$ with the property (1).

8 Discussion

Although we used the same notations for the annihilation and creation operator in the Langevin equation (4), for those in the Fokker-Planck equation (6) and for those in the Heisenberg equation of motion (38), (39), one should be very carefull about their mutual difference. The annihilation and creation operator in the Langevin equation is a microscopic ones, since they are stochastic operators. On the other hand, those in the Fokker-Planck equation and in the Heisenberg equation are coarse grained operators, since they are random averaged in the process (22) to obtain the time-evolution generator $\hat{H}$ in the Fokker-Planck equation from the generator $\hat{H}_f(t)$ in the stochastic Liouville equation. We would like to emphasize that the size of the *thermal space* [6, 7], the representation space of NETFD, permits the above variety of the levels of coarse graining. This also permits us to introduce the *white* stochastic process for non-linear quantum systems, as was shown in this paper.

The relation between the Fokker-Planck equation (6) and the Langevin equation (4) through the stochastic Liouville equation (19), developed in this paper, will provide us with a good tool for the construction of the quantum Ito formula [11, 12]. The existence of both the Fokker-Planck equation and the quantum stochastic equation (the Langevin equation) may be essential in the construction of the stochastic integration and the stochasic differentiation in quantum systems as was in classical systems. In the formulation of the quantum Ito stochastic equation which is presently been developed by mathematicians [11, 12], it seems that they do not have a corresponding quantum Fokker-Planck equation.

The success of the inclusion of the Langevin and the stochastic Liouville equation approach has considerably enlarged the applicability of the formalism of NETFD, since many attractive problems in far-from-equilibrium states are specified in terms of stochastic equations, e.g., a Langevin equation, a stochastic TDGL equation and so on. The stochastic Liouville equation enabled us to derive the time evolution generator $\hat{H}$ which describes the dynamics of the system compatible with the Langevin equation. The generator $\hat{H}$ makes it possible to construct a quantum field theory for these problems [24]-[31].

In writing down the stochastic equation, we should specify a set of *gross variables* which are dependent on macroscopic space and time coordinates. The choice of the set of gross variables may be crucial in the choice of the representation space in the disspative quantum field theory. Each representation space, which is labeled by the macroscopic space and time, are connected by the help of a set of equations of motion for the gross variables [32]. This concept was never introduced into the usual quantum field theory, and is very attractive in conjunction with the *dynamical symmetry breaking* [27, 32] which is a common feature of the dynamical evolutions in far-from-equilibrium states (see [20, 21] for the applications).

References

[1] R. Kubo, J. Phys. Soc. Japan **9** (1954) 935.

[2] P. W. Anderson, J. Phys. Soc. Japan **9** (1954) 316.

[3] R. Kubo, M. Toda and N. Hashitsume, *Statistical Physics II* (Springer, Berlin 1985).

[4] T. Arimitsu, *Thermal Field Theories*, eds. H. Ezawa, T. Arimitsu and Y. Hashimoto (North-Holland 1991) pp. 207–221.

[5] T. Arimitsu, Phys. Lett. A **153** (1991) 163.

[6] T. Arimitsu and H. Umezawa, Prog. Theor. Phys. **74** (1985) 429.

[7] T. Arimitsu and H. Umezawa, Prog. Theor. Phys. **77** (1987) 32.

[8] T. Arimitsu and H. Umezawa, Prog. Theor. Phys. **77** (1987) 53.

[9] T. Arimitsu, M. Guida and H. Umezawa, Europhys. Lett. **3** (1987) 277.

[10] T. Arimitsu, M. Guida and H. Umezawa, Physica **A148** (1988) 1.

[11] R. L. Hudson and K. R. Parthasarathy, Commun. Math. Phys. **93** (1984) 301.

[12] K. R. Parthasarathy, Rev. Math. Phys. **1** (1989) 89.

[13] T. Arimitsu, Y. Takahashi and F. Shibata, physica **100A** (1980) 507.

[14] T. Arimitsu, Physica **104A** (1980) 126.

[15] T. Arimitsu, J. Phys. Soc. Japan **51** (1982) 1054.

[16] T. Arimitsu, J. Phys. Soc. Japan **51** (1982) 1720.

[17] T. Arimitsu and T. Tomonaga, J. Phys. Soc. Japan **51** (1982) 3102.

[18] M. Ban and T. Arimitsu, Physica **129A** (1985) 455.

[19] F. Haake, H. Risken, C. Savage and D. Walls, Phys. Rev. A **34** (1986) 3969.

[20] T. Tominaga, M. Ban, T. Arimitsu, J. Pradko and H. Umezawa, Physica **A149** (1988) 26.

[21] T. Tominaga, T. Arimitsu, J. Pradko and H. Umezawa, Physica **A150** (1988) 97.

[22] F. Shibata and T. Arimitsu, J. Phys. Soc. Japan **49** (1980) 891.

[23] M. Lax, Phys. Rev. **145** (1966) 110.

[24] T. Arimitsu and H. Umezawa, J. Phys. Soc. Japan **55** (1986) 1475.

[25] T. Arimitsu, H. Umezawa and Y. Yamanaka, J. Math. Phys. **28** (1987) 2741.

[26] T. Arimitsu, J. Pradko and H. Umezawa, Physica **A135** (1986) 487.

[27] T. Arimitsu, Y. Sudo and H. Umezawa, Physica **A146** (1987) 433.

[28] T. Arimitsu and H. Umezawa, in: *Advances on Phase Transitions and Disordered Phenomena*, eds. G. Busiello, L. De Cesare, F. Mancini and M. Marinaro (World Scientific, Singapore 1987) pp. 483-504.

[29] H. Umezawa and T. Arimitsu, in: *Foundation of Quantum Mechanics — In the Light of New Technology*, eds. M. Namiki, Y. Ohnuki, Y. Murayama and S. Nomura (Physical Society of Japan, Tokyo 1987) pp. 79-90.

[30] T. Arimitsu, H. Umezawa, Y. Yamanaka and P. Papastamatiou, Physica **A148** (1988) 27.

[31] T. Arimitsu, Physica **A148** (1988) 427.

[32] T. Arimitsu, Physica **A158** (1989) 317.

Structure in Particle Interactions:
The Static-Ultralocal Model and its Descendants.

R.J. Rivers,

The Blackett Laboratory, Imperial College, London SW7 2BZ.

The Static-Ultralocal model proposed by Eduardo Caianiello and others in the 1970s has become a useful tool in the hunt for noncanonical field theories.

Prologue.

My appreciation of Eduardo goes back to the early 1970s, when he enlightened and entertained us for a year at Imperial College. Those times were strange for particle physics. A period of stagnation, that had degenerated into metaphysical wrangling, was coming to an end. The ideological battles between the holists (behind the bootstrap banner of Chew) and the purportedly reductionist field theorists (sensing the ubiquity of the gauge principle) had not yet led to the rout of the former but it was apparent which way the fight was going.

S-matrix camp-followers like myself, who had a field-theory background, were reluctantly changing sides. Reluctantly, because whatever the merits of the field theories in question, the methods of solution were feeble. [Remember that these were the days before the ready availability of large computers to make numerical simulations feasible]. The major successes of field theory had been in Quantum Electrodynamics, for which perturbation theory in the fine structure constant was ideal. However, it was not apparent that perturbative methods had any relevance for the dynamical realisation of the Goldstone mechanism, or the confinement of the newly found partons.

While Eduardo was at Imperial College he gave an elegant series of lectures that later became his book on combinatorics [1]. In these lectures he tried hard to dissuade us from thinking along conventional perturbative lines. Indeed, I remember him saying, in his gently provocative way, that the trouble with Feynman was that he had left us Feynman diagrams. Having only recently returned to the fold I found this somewhat shocking, but it was an apt comment. Let me remind you of some of the strengths and weaknesses of the Feynman series.

Consider the simplest theory, that of a single scalar field ϕ experiencing a quartic $g\phi^4/4!$ interaction. What interests us are the Green functions

$$G_m(x_1, ..., x_m) \; = \; < 0|T(\phi(x_1)\phi(x_2).....\phi(x_m)|0 >, \tag{1}$$

the space-time correlation functions of the theory from which all physical observables can be derived, represented diagrammatically as

$$G_m(x_1\, x_2 ... x_m) = \tag{2}$$

The white circle represents the interaction experienced by the field. We can think of it as a ball of string with many ends, in which the string is multiply knotted, each knot being weighted with a factor of g. To see its structure we tug on one line, say one ending at x. Then it either comes off in our hands, as in the first term in (3), or it pulls a four-string knot out of the interior of the circle. The equations representing this, the Nakano-Dyson-Schwinger (NDS) equations, embody *both* Heisenberg's equations *and* the canonical equal-time commutation relations $[\phi(\mathbf{x}), \pi(\mathbf{y})] = i\delta(\mathbf{x} - \mathbf{y})$.

$$\tag{3}$$

As it stands the ball need not be connected, and it is useful to rewrite the equations in terms of *connected* balls (shaded in (4)). Then, for example, if we extract two ends from the ball, and pull, it either unknots or a knot is pulled clear, either totally (as in the second term) or not (as in the third).

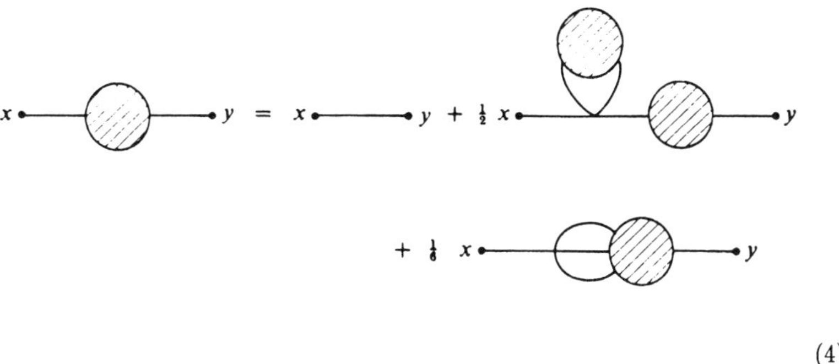

$$(4)$$

The presence of the inhomogeneous term (the unwound ball) enables us to solve iteratively as a series in the number of knots (i.e. in powers of g). This is the Feynman series. For example, for elastic scattering we have

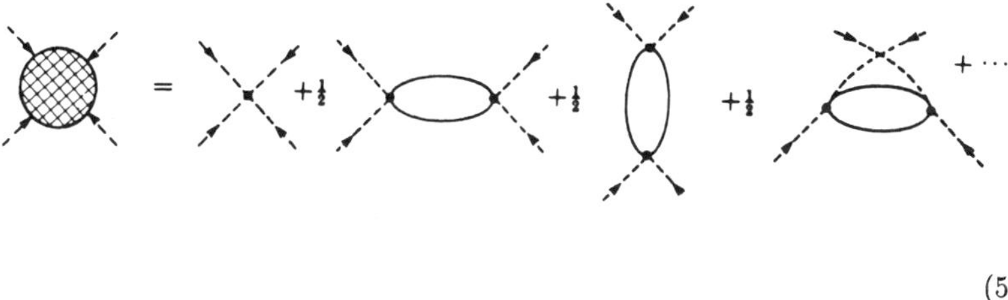

$$(5)$$

where the cross-hatching denotes the four-string ball that is one- string irreducible. [That is, it will not come into two by making a single cut].

The terms shown are only of order g, g^2. In general, the more intermediate lines the smaller the distances that are being probed. The Feynman series is an expansion in which higher and higher terms (that are harder and harder to calculate) correspond to shorter and shorter distance events. If g is small we have a potentially sensible calculational scene. For QED the results are stunningly good.

The Static-Ultralocal Model.

The main weakness of the approach is that many things can occur at small distances that will not be easily accessible. In particular, there is the possibility of charge shielding. In $d > 4$ and, in some circumstances, in $d = 4$ dimensions, the $g\phi^4$ theory is a *free* theory [2]. That is, despite the existence of nontrivial Feynman diagrams there is *total screening* of the interaction.

We understand this as follows. For $d > 4$ dimensions the diagrams are uncontrollably singular in the UV. The theory is not renormalisable order by order in g and the Feynman series cannot be used in principle. The only way to tame such a theory is to make it trivial. The UV fever has killed the patient. For $d = 4$ dimensions the UV singularities are just manageable and can be compensated by (infinite) counterterms. The theory is renormalisable. However, individual diagrams of order n in g exist (see (6)) of magnitude $O(n!)$ and of the *same* sign.

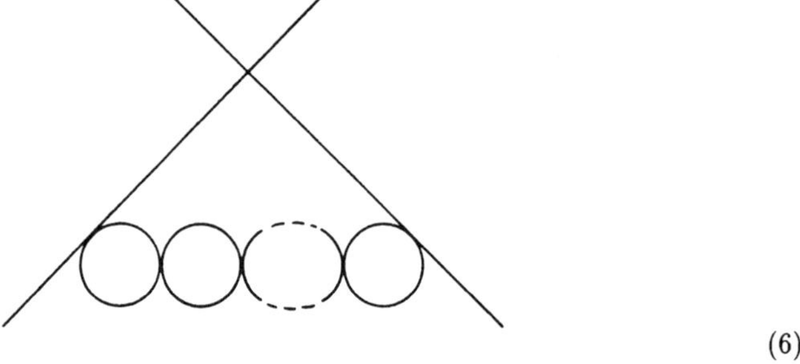

$$\tag{6}$$

The series (which Eduardo taught us to be asymptotic at best) is *not* Borel summable and the first few terms are fundamentally misleading, even if they have low-energy utility.

It is not obvious how to set up alternative *analytic* approaches that make sense *a priori* in large dimension (which, from this viewpoint includes $d = 4$). Rather than unroll the ball from the outside, counting knots as we do so, Eduardo and his co-workers made the surprising suggestion that, in the first instance, *we compress it to a point* [3, 4]. The link between two knots is the propagator $\Delta_F(x)$ of the ϕ-field, satisfying

$$(\partial^2 + m^2)\Delta_F(x) = -\delta(x). \tag{7}$$

Suppose, as a first step, we replaced $\Delta_F(x)$ by $\Delta(x)$, where

$$m^2 \Delta(x) = -\delta(x) \tag{8}$$

neglects the kinetic D'Alembertian. Then each line shrinks to a point, collapsing the ball. For this reason the model was termed the Static-Ultralocal (SUL) model. However, we can now solve the NDS equations exactly without resorting to perturbative methods. With lattice regularisation (of which more later), the result is highly non-trivial, corresponding in one sense to a theory in *zero* space-time dimensions. [In particular, it mimics the analytic structure (in g) of the full nontrivial theory in $d < 4$ dimensions remarkably well].

In order to tease out this tight knot we restore the field derivatives $D^{-1}(x,y) = \partial^2\delta(x-y)$ perturbatively. There is an immediate technical problem in that the terms in the series are also very singular in the UV (even more so than in the Feynman series). However, without the artifice of counterterms, we can attempt to control them directly. As an intermediate step the theory is made finite by putting it on a lattice of cell size a in d dimensions, requiring us to go Euclidean. The kinetic inverse propagator $D^{-1}(x,y)$ requires some care. In one dimension it can be written as

$$D^{-1} = a^{-3}(\delta_{i,j+1} - 2\delta_{i,j} + \delta_{i,j-1}) \qquad (9)$$

but in more than one dimension a more elaborate notation is needed.

It is now straightforward, in principle, to develop a diagrammatic expansion for n-point functions in powers of D^{-1}. Specifically, the generating functional

$$Z[j] = \int \mathcal{D}\phi\, exp\left(-\int dx \mathcal{L}(\phi) - j\phi\right), \qquad (10)$$

with Euclidean Lagrangian density

$$\mathcal{L} = \frac{1}{2}(\nabla\phi)^2 + \frac{1}{2}m^2\phi^2 + \frac{g}{4!}\phi^4 \qquad (11)$$

can be written as

$$Z[j] = \left(exp\int\left(\frac{\delta}{\delta j}D^{-1}\frac{\delta}{\delta j}\right)\right) Z_{SUL}[j] \qquad (12)$$

where

$$Z_{SUL}[j] = \int \mathcal{D}\phi\, exp -\int dx[\frac{1}{2}m^2\phi^2 + \frac{1}{4!}g\phi^4 - j\phi] \qquad (13)$$

is the SUL generating functional, with fields uncorrelated at adjacent spacetime points. Since, on scaling ϕ to $g^{-1/4}\phi$, (11) becomes

$$\mathcal{L} = \frac{1}{2}g^{-1/2}(\nabla\phi)^2 + \frac{1}{2}g^{-1/2}m^2\phi^2 + \frac{1}{4!}\phi^4 \qquad (14)$$

we are performing a strong-coupling expansion in powers of $g^{-1/2}$.

Surprisingly, in the light of the preceding comments, the SUL, with formal path integral realisation (13), is *trivial* in the continuum limit of vanishing a. That is, just as for the full theory in high dimension, *no trace of the interaction survives*. This starting-point, in which we have triviality in the first instance, is arguably more realistic for renormalisable (but not superrenormalisable) and non-renormalisable theories than the Feynman series, which is predicated on nontriviality. In the context of lattice regularisation, non-trivial results arise, when they do, because a very singular operation is being applied to an almost trivial quantity.

Two properties of this expansion are noteworthy.

i) Vertices occur with all numbers of legs. A typical diagram contributing to the four-point function is given in (15), in which the *crossed* lines denote D^{-1}.

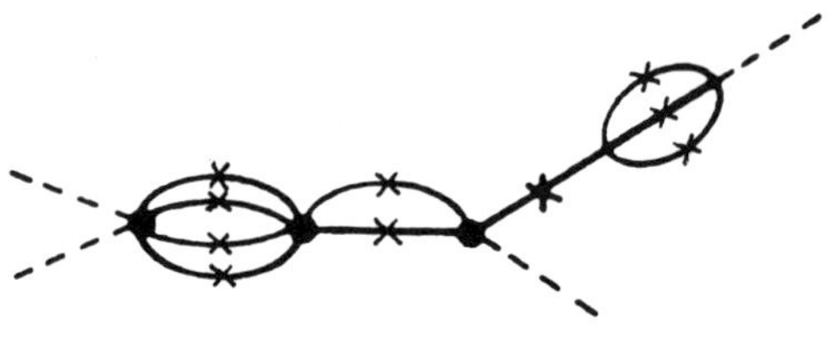

$$(15)$$

ii) From the above, individual diagrams diverge as powers of a^{-1} in the continuum limit. Because of this difficulty, much of the work by the Salerno group was concerned with many-body systems (e.g. spin systems, liquids) for which a continuum limit could be avoided (see [4]). However, as we mentioned earlier, this divergence need not be a problem. Suppose we were given the partial series for a function f,

$$f(a) = a^{-1} - a^{-2} + a^{-3} - a^{-4} + O(a^{-5}), \qquad (16)$$

and asked to identify its continuum limit $f(0)$. It requires little imagination to anticipate that the series is an attempt to recreate

$$f(a) = (1 + a)^{-1} \qquad (17)$$

with limit $f(0) = 1$.

The breakthrough in the use of expansions about the SUL came with the development, by Carl Bender and collaborators [5] of approximant schemes for handling short series of divergent terms, more general than (16), whose reliability could be determined [6]. In particular, it became possible to study the role of space-time dimension in determining whether a canonical theory was trivial or not. The details of this combinatoric *tour de force* will not be discussed here, beyond saying that it works.

Scale-Invariant Quantisation.

If the SUL itself showed interactions the full theory would be more likely to be non-trivial. However, in order to obtain a non-trivial SUL in the lattice continuum limit it is necessary to change the quantisation procedure. *A priori* this seems unsatisfactory, if not impermissable.

The structure of the ball shown in (2) and (3) looks so natural that we have to remind ourselves that it is a consequence of the *canonical* equal-time commutation relations. For very different motives than the examination of triviality, John Klauder had made a case for an enquiry into *affine* quantisation [7]. The affine equal-time commutation relations for the ϕ field are

$$[\phi(\mathbf{x}), K(\mathbf{y})] = \imath\phi(\mathbf{x})\delta(\mathbf{x} - \mathbf{y}) \tag{18}$$

where $K(\mathbf{x})$, formally equal to $\frac{1}{2}(\phi(\mathbf{x})\pi(\mathbf{x}) + \pi(\mathbf{x})\phi(\mathbf{x}))$, is the generator of scale-transformations for the ϕ field. The relations (18) are perfectly acceptable quantum-mechanically. They were motivated by the suggestion, in quantum gravity, that field (metric) scalings be unitarily implementable, but not field (metric) translations. In this way the signature of the metric would be unchanged by quantum fluctuations. However, it was appreciated [8] that a byproduct of such quantisation could be the nontriviality of otherwise trivial theories and, as such, worthy of further study.

[A closely related approach was discussed at Eduardo's 60th birthday meeting by Prof. Fubini and his collaborators [11], who considered the effect of *conformally-invariant* measures. Again motivated by gravity, a change of measure arises naturally in their two-time formulation of stochastic QFT. However, the models proposed there are 'effective' phenomenological models, requiring a different approach [9] from the one that I shall now present.]

The diagrammatic equations [7, 10] for the *connected* Green functions, that are a consequence of (18), are given in (19).

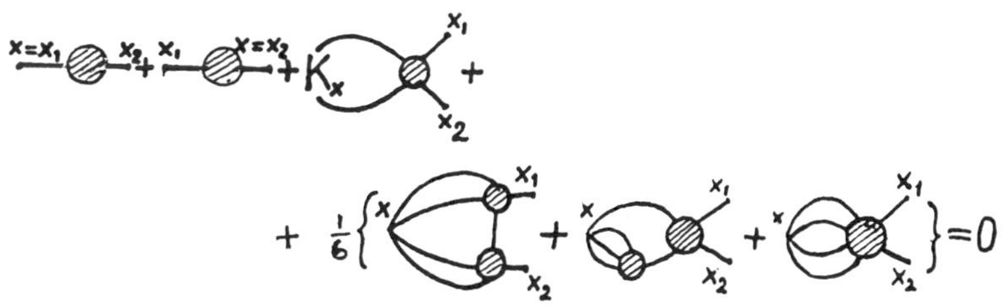

$$\tag{19}$$

Two points should be noted. Firstly, there is an ambiguity in how the legs are brought together at the point x, before operating with $K_x = \partial_x^2 + m^2$. Secondly, there is no inhomogeneous term to kickstart an iterative series expansion in g. Feynman diagrams are totally inappropriate [12]. However, as before, we can crush the ball to a point, and develop the series expansion in D^{-1} about this static ultralocal limit (which Klauder termed the Independent-Valued-Model (IVM)). Rather than (13), the formal generating functional for the IVM is now

$$Z_{IVM}[j] = \int \mathcal{D}'\phi \, exp - \int dx [\frac{1}{2}m^2\phi^2 + \frac{1}{4!}g\phi^4 - j\phi] \tag{20}$$

where, for all it positive $\Lambda(x)$,

$$\mathcal{D}'(\Lambda\phi) = \mathcal{D}'\phi. \tag{21}$$

This is to be contrasted with the translationally-invariant canonical measure $\mathcal{D}\phi$, for which $\mathcal{D}(\phi + \Lambda) = \mathcal{D}\phi$. In a lattice regularisation [7], $\mathcal{D}'\phi$ can be written as

$$\mathcal{D}'\phi = \mathcal{D}\phi \prod_x |\phi(x)|^{-1+2f} \tag{22}$$

where, in d dimensions, $f = O(a^{2n})$, $2n = d$. Only for this value of n is the IVM nontrivial in the continuum limit (see [13]). (In the language of spin-systems this is because the measure is so singular as to violate the Lebowitz inequality [7], an essential ingredient in demonstrating triviality [2]). If we were to take $2n < d$ we would find a *trivial* continuum limit (the continuum SUL), whereas taking $2n > d$ is too singular for the model to be defined. Further, we have a nontrivial model even when $g = 0$ (the *pseudofree* IVM).

Thus, if the measure is carefully chosen (the reflection of the ambiguity in (19), we expect that even a *free* scalar theory, quantised affinely, can be non-trivial in $d = 4$ (or more) dimensions. To see this we need to develop a derivative expansion about the pseudofree IVM as was done about the Salerno SUL. So little is understood about affinely quantised theories that the demonstration of the nontriviality of such a simple theory can provide a touchstone to their utility.

In fact, the Salerno paper of 1976 [4] addresses, but does not solve; just this case (among others). This is a problem that I have been considering recently with Carl Bender and James Wong [14] as part of a programme for evaluating noncanonical theories. So, as a belated birthday present I am happy to offer a qualified answer to the question as to whether nontrivial affinely quantised theories exist (especially in $d = 4$ dimensions).

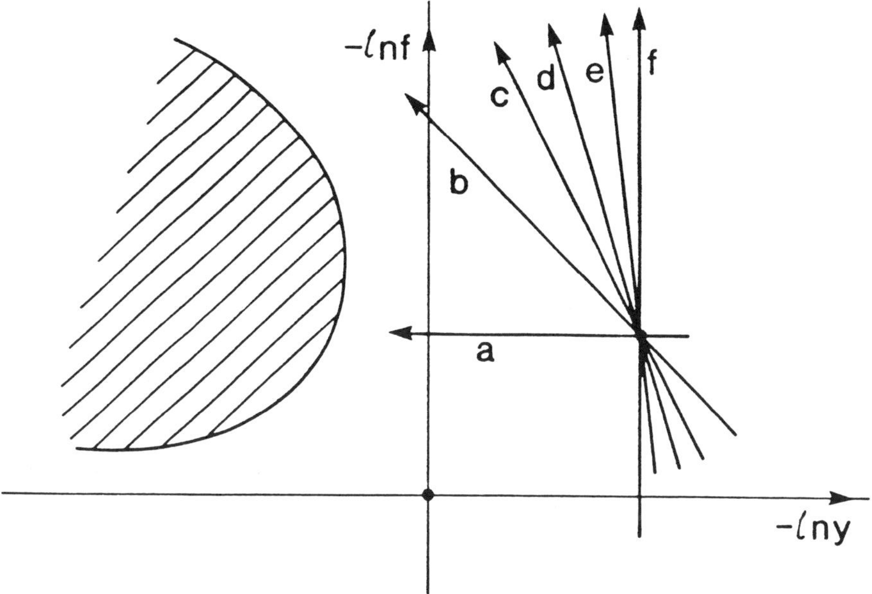

Fig.1

Once the field derivatives have been introduced there is no reason to require that the exponent f of the measure $D'\phi$ vanish in the same way as for the IVM. As before, we take

$$f = O(a^{2n}) \tag{23}$$

where $n > 0$ is to be determined numerically.

We have calculated the derivative expansion about the pseudofree IVM to 5th order in D^{-1}. Using the approximants of [5] for 5- term series the behaviour of the theory is shown in Fig.1. for different values of n. (For convenience f is plotted against $y = (a^2 M^2)^{-1}$ where M is the renormalised mass of the field quanta). Each value of n gives a straight line in the plane. Lines b, c, d, e, lead to *triviality* for all d whereas line f gives *singular* behaviour (compatible with our earlier results for $n = 0$ [15]). The shaded area is forbidden in $d = 4$ dimensions. This suggests that, as for the IVM, a path can be taken between the Scylla and Charybdis of triviality and nonexistence that leads to a nontrivial nonsingular affine theory.

However, it has to be said that the strength of the method is in making affine theories plausible, rather than tractable. An operator-based approach is, ultimately, the only way to avoid the fine-tuning implicit in the path-integral but as yet I do not know how to procede in that way.

References

[1] E. Caianiello, *Combinatorics and Renormalisation in Quantum Field Theory* (Benjamin Inc., New York, 1973).

[2] J. Frolich, Nucl. Phys. **B200**, 281 (1982).

[3] E.R. Caianiello & G. Scarpetta, Nuov. Cim. **22A**, 668 (1974); Lett. Nuov. Cim. **11**, 283 (1974).
E.R. Caianiello, M. Marinaro & G. Scarpetta, in *Particles and Fields*, ed. A.N. Kamal (New York, 1978).

[4] E.R. Caianiello, M. Marinaro & G. Scarpetta, Nuov. Cim. **44B**, 299 (1978).

[5] C. M. Bender, F. Cooper, G.S. Guralnik & D.H. Sharp, Phys. Rev.**D19**, 1865 (1979).
C.M. Bender, F. Cooper, G.S. Guralink, R. Roskies & D.H. Sharp, Phys. Rev. **D23**, 2976 (1981); **D23**, 2999 (1981); **D23**, 3010 (1981).

[6] R.J. Rivers, Phys. Rev. **D20**, 3425 (1979).

[7] J.R. Klauder, Acta Phys. Austr. **41**, 237 (1975).
J.R. Klauder, Ann. Phys. **117**, 19 (1979).

[8] J.R. Klauder, Phys. Rev. **D24**, 2599 (1981).

[9] R.J. Rivers & C.C. Wong, J. Phys.**A24**, 3859 (1991).

[10] M. Nouri-Moghadam & T. Yoshimura, Phys. Rev. **D17**, 3284 (1978).

[11] V. De Alfaro, S. Fubini & G. Furlan, *Theoretical Physics Meeting*, Pub. dell'Universita degli Studi di Salerno,7, 37 (1984): Nuov. Cim. **A74**, 365 (1983).

[12] J.M. Ebbutt & R.J. Rivers, J. Phys. **A15**, 2947 (1982).

[13] R. Kotecky & D. Preiss, Phys. Rev. **D18**, 2187 (1979).

[14] C.M. Bender, R.J. Rivers & C.C. Wong, Phys. Rev. **D**, to be published.

[15] C.M. Bender, R.J. Rivers & C.C. Wong, Phys. Rev. **D41**, 455 (1990).

MATRIX MODELS IN STATISTICAL MECHANICS
AND IN QUANTUM FIELD THEORY IN THE LARGE ORDER LIMIT.

Giovanni M.Cicuta

Dipartimento di Fisica, Universitá di Bari
and INFN, Sezione di Bari
Via Amendola 173, 70126 Bari, Italy

1. Introduction.

In the past few years considerable efforts were devoted to investigate statistical mechanics models where the degrees of freedom are square matrices. The action (or the Boltzmann weight) is chosen to be invariant under a large group of matrix transformations (global or local ones). One wishes to compute "observables", invariant under the symmetry group of transformations. A new calculus for the "observables" emerges when the limit of infinite size of the matrices is performed.

Important results were obtained that seem to give insights to the theory of random surfaces, the behaviour of statistical mechanics models in presence of quantum gravity in two dimensions, a systematic non-perturbative approach to quantum field theory, unexpected suggestions for the theory of strings, and a variety of mathematical results.

It is difficult at this stage to summarize the main results or the relevant literature. Various reviews and lecture notes may help [1−5].

The aim of this short contribution is to comment on few points of the present knowledge which seem worth further investigation (of course the choice and content of the comments are very subjective).

In Sect.2 an outline of some basic steps in the development of the topic is given, to provide the reference for the comments briefly exposed in the Sect.3-5 that follow.

2. An outline of some basic steps.

The interest for quantum field theory in the large order limit of matrix models begins with the works by G.t'Hooft [6] and G.Veneziano [7] on the topological expansion. The non abelian theory of the Lorentz-vector field $A_\mu(x)$, invariant under the local (gauge) group SU(N) is studied in the $N \to \infty$ limit, with a proper (unique) rescaling for the gluon coupling constants. The infinite number of Feynman graphs contributing to any correlation function $G_{2p}(x_1,..x_{2p}) = < \Omega|TA_{\mu_1}(x_1)...A_{\mu_{2p}}(x_{2p})|\Omega >$ fall into distinct classes characterized by increasing powers $(\frac{1}{N^2})^k$, the power k being related to the genus of the orientable surface on which the Feynman graph may be properly embedded. This result stems from the group factor associated to each Feynman graph (a convolution of f_{ijk}) and it holds true also for the globally SU(N) invariant model of Lorentz-scalar fields with euclidean action

$$\mathcal{A} = \int d^d x \, tr[\frac{1}{2}(\partial_\mu A(x))^2 + g_2 A^2(x) + \frac{g_4}{N} A^4(x)] \tag{1}$$

where the space-time dimension d is arbitrary, $A(x)$ is a Hermitian matrix of order N (here a convolution of symmetric tensors d_{ijk} replaces the previous convolution of f_{ijk}). It does not seem possible to evaluate analytically the sum of the leading contributions in the $N \to \infty$ limit, that is the series associated to the Feynman planar graphs. Only for the small dimensions d=0 and d=1 the matrix model of type eq.(1) were "solved" by the two methods of frequent use today : saddle point analysis and orthogonal polynomials [8].

The generating functional Z of vacuum Feynman graphs, for the model of eq.(1), with d=0, after an integration over angular variables, is written in terms of the eigenvalues λ_i of the matrix A

$$Z(g_2, g_4) = \int \prod_{i=1}^{N} d\lambda_i \, e^{-V(\lambda_i)} \Delta(\lambda_i)^2 \tag{2}$$

$$\Delta(\lambda_i) = \prod_{1 \leq i < j \leq N} (\lambda_i - \lambda_j) \, , \, V(\lambda_i) = \sum_{i=1}^{N} [g_2 \lambda_i^2 + \frac{g_4}{N} \lambda_i^4]$$

Here the two methods depart. The saddle point analysis proceeds by exponentiating the Jacobian

$$\Delta(\lambda_i) = e^{\sum_{1 \leq i < j \leq N} log|\lambda_i - \lambda_j|} \tag{3}$$

by replacing, in the $N \to \infty$ limit, the sums with integrals :

$$\lambda_i \to \sqrt{N} \lambda(\frac{i}{N}), \qquad \sum_i (\lambda_i)^k \to N \int_0^1 dx [\sqrt{N} \lambda(x)]^k$$

$$2 \sum_{1 \leq i < j \leq N} log|\lambda_i - \lambda_j| \to N^2 \int_0^1 \int_0^1 dx \, dy \, log|\lambda(x) - \lambda(y)| \tag{4}$$

The saddle point corresponds to a singular integral equation for the function $\lambda(x)$. By introducing the non-negative density of eigenvalues $u(\lambda) = \frac{dx}{d\lambda} \geq 0$, $\int_L d\lambda \, u(\lambda) = 1$, L being the (unknown) support of the eigenvalues, the singular integral equation is

$$\frac{1}{2} \frac{dV(\lambda)}{d\lambda} = g_2 \lambda + 2g_4 \lambda^3 = P \int_L d\lambda' \frac{u(\lambda')}{\lambda - \lambda'} \tag{6}$$

Or, with the orthogonal polynomials technique, one uses the invariance of the Jacobian $\Delta(\lambda_i)$, being a Vandermonde determinant, with the replacement $\lambda^n \to P_n(\lambda)$, $\Delta(\lambda_i) = \det[(\lambda_i)^{j-1}] = \det[P_{(j-1)}(\lambda_i)]$ where $P_n(\lambda)$ are monic polynomials ($P_n(\lambda) = \lambda^n +$ lower order polynomial) which are chosen orthogonal with respect to the measure

$$\int_{-\infty}^{+\infty} d\lambda \, e^{-V(\lambda)} P_n(\lambda) P_m(\lambda) = h_n \delta_{nm}$$

The orthogonal polynomials $P_n(\lambda)$ satisfy a recurrence relation

$$\lambda P_n(\lambda) = P_{n+1}(\lambda) + S_n P_n(\lambda) + R_n P_{n-1}(\lambda), \text{ with } R_n = \frac{h_n}{h_{n-1}} \tag{7}$$

If $V(\lambda)$ has only even powers of λ (as in this example) and a solution without spontaneous symmetry breaking is expected, $S_n = 0$, and a the recurrence relation is found

$$n = R_n[2g_2 + 4\frac{g_4}{N}(R_{n-1} + R_n + R_{n+1})] \tag{8}$$

186

sometime called pre-string equation.

A smooth ansatz for R_n as $N \to \infty$ reproduces the results of the planar theory, whereas a different, smooth ansatz, in the double-scaling limit $N \to \infty$ and $g_i \to g_{i,crit}$, resums contributions from any topology.

3. Multiple solutions: saddle point versus orthogonal polynomials.

It might seem that both eq.(6) and eq.(8) have a unique solution. Singular integral equations of type (6) are easily inverted, if the support L is known (one may use the Poincare'-Bertrand formula). In a generic model, however (except for the pure gaussian model of random matrices), several choices for the support L are possible that lead to multiple solutions of the model in the planar limit. The support may be one segment or the union of more segments, and, even if $V(\lambda)$ is an even function of λ, L may be symmetrical with respect to the origin or not (thus obtaining a spontaneous symmetry breaking solution) [9−11]. Each solution has a domain of existence in the space of parameters g_i of the model. If, in a given region in such space, more than one solution exist, the solution of the model (in the planar limit) corresponds to the solution of the eq.(6) that provides the lowest free energy. Then the boundaries, in the space of parameters, for the existence of solutions not necessarely coincide with phase transition surfaces.[12,13]. However, in order to obtain a non-linear equation in the double-scaling limit, the "string equation" , it is necessary to approach the boundary of existence of a given solution (obviously possible from one side only), so, for that purpose, the former boundaries are relevant.

It was shown, mainly by Molinari [14], that a period-two ansatz for R_n in eq.(8), that is R_{2n} and R_{2n+1} approach two different smooth functions, reproduces, in the planar limit, the results of the symmetric, two-segment support, solution of eq.(6).

Generalizing the Molinari ansatz to solutions with support on more than two segments seems problematic : the conjecture [13] that the proper ansatz for the continuum limit of R_n should have k-periodicity, to reproduce a k-segment saddle point solution, seems to fail [15].

Ever more disturbing and relevant is the chaotic behaviour [16] exhibited by R_n, : as it generally happens with non-linear recurrence relations, R_n are very unstable with respect to the initial conditions (small n values) and exhibit a chaotic behaviour for large values of n. This seems to prevent a smooth ansatz to reproduce the saddle point results in the planar limit and it casts doubts on the validity of the smooth ansatz used in the double-scaling limit.

To summarize this comment, it seems obvious that each of the two methods here considered has advantages and weaknesses : the saddle point analysis easily produces the multiple solutions and the phase diagram of the model in the planar limit, but it is forbiddingly impractical to analyze, in a systematic way, the topological expansion. The orthogonal polynomial technique is the basic way to perform the double-scaling limit but, so far, it is still an art to guess **proper** ansatzs. Certainly, an understanding of the non-linear recurrence relations for R_n (pre-string equation) to know the regions in the space of the parameters g_i where regular/periodic/chaotic behaviours arise, would be very welcome.

4. Systematic evaluation in arbitrary dimension and rectangular matrices.

Models of square matrices are generalized to models with rectangular matrices. Consider M a rectangular matrix with N_1 rows and N_2 columns with complex entries.

$$A = \int d^d x \, tr[\partial_\mu M^\dagger(x)\partial_\mu M(x) + g_2 M^\dagger(x)M(x) + \frac{g_4}{N} M^\dagger(x)M(x)M^\dagger(x)M(x)] \tag{9}$$

Let $L = \frac{N_1}{N_2}$ fixed as $N_1, N_2 \to \infty$. Again Feynman graphs are arranged into the topological expansion. The sum of the infinite Feynman graphs of fixed genus contributing to the free energy or to a correlation function in the model, may further be expanded in a Taylor series in the parameter L. Well defined classes of Feynman graphs contribute to a given term in the topological expansion and to a given power in L, they are the classes identified in the various orders of the expansion of a **n-vector** model in the limit $n \to \infty$. The analysis only relies on the group factors of each Feynman graph, then it is (formally) valid in any dimension of space-time.

The method provides a systematic way to evaluate in approximate form the sum of the planar graphs in any theory with large global symmetry group (it is not known how to extend it to non abelian gauge symmetry). For those models where exact planar results for the free energy or the propagator are known, like the models in d=0 or d=1, it was checked [17,18] that the first few orders in the Taylor expansion in L provide an approximation accurate in the whole range $-\infty < g_2 < \infty, g_4 > 0$, for L=1 (that is for the planar theory).

In d=0, models with rectangular matrices are analyzed by the technique of orthogonal polynomials, and one obtains a system of two recurrence relations (pre-string equations). By a suitable (L dependent) smooth ansatz, a triple-scaling limit may be performed, that may reproduce the critical behaviour of continuum models interpolating between branched polymers and random surfaces [19].

Perhaps it may help to recall the well known description of diluted polymers (self-avoiding chains) derived by de Gennes as the $n \to 0$ limit of a **n-vector** model [20]. In the same spirit, a model of self-avoiding random surfaces (with non fixed genus) was derived by a $n \to 0$ limit of a model of square $n \times n$ matrices [21].

Given the large amount of knowledge available for n-vector models in statistical mechanics and in quantum field theory, both in the $n \to 0$ and in the $n \to \infty$ limit, it seems that investigation of models with rectangular matrices, which interpolate between the vector results ($L = 0$ or ∞) and the square matrix results ($L = 1$) might be fruitful.

5. Statistical Models and Colouring Problems.

It is well known that a number of matrix models (in d = 0) describe interesting statistical models of matter on a fluctuating lattice (matter in d = 2, in presence of quantum gravity) : the Ising model [22] , also with external magnetic field [23], the self-avoiding walks [24], the $O(n)$ model [25], the Lee-Yang edge singularity and the dimer coverings [26],...

Most of the interest focused at the critical values in each model, where an exciting agreement was found between critical coefficients and the KPZ formula from conformal quantum field theory [27].

Since the matrix models are analytically solved for generic values of the couplings (not just the critical values), it seems that a more general analysis of the solutions may be worthwhile. The features that may emerge may lack universality, but may uncover physical phenomena that occur on statistical models in presence of gravity and have no correspondence on a rigid lattice.

It would also be very desirable to enlarge the class of the two- (or multi-) matrix models, that may analytically be solved in the planar limit.

After the solution [28,29], in the planar limit, in d = 0, of the two-matrix model with "action" $\mathcal{A} = V(A) + V(B) + c\,tr[AB]$, the analysis was extended [30] to models involving a set of random matrices $\{M^{(i)}\}$:

$$Z = \int \prod_i dM^{(i)} e^{-\mathcal{A}(M^{(i)})}$$

$$\mathcal{A} = \sum_i V(M^{(i)}) + \sum_{i<j} c_{ij}\, tr[M^{(i)}M^{(j)}] \tag{10}$$

This integrand may be represented by a graph : the matrix $M^{(i)}$ being represented by a vertex i and the non-vanishing coefficient c_{ij} by the edge (i,j). It is known how to perform the sequence of angular integrations (first step in the large N analysis) only if the graph contains no cycle (it is the Mehta tree).

I wish to suggest that the solution of a simple looking matrix model may lead to a straightforward proof of the famous 4 colour theorem [31] [32].

All planar maps with regular vertices of valence three are generated by the large N limit of the one matrix model, $V(A) = tr[A^2 + \frac{g}{\sqrt{N}}A^3]$, with d = 0. Its solution in the planar limit is in Ref. [8]. It is well known, in the mathematical literature, that a proper edge coloring, with 3 colours, of the edges of any planar trivalent map, corresponds to a proper coloring, with 4 colours, of the regions of the same map [33]. An evaluation of the number of proper edge colorings, with 3 colours, may be obtained by the large N limit of the free energy of the simple looking three matrix model, in d $= 0$, with $V(A,B,C) = tr[A^2+B^2+C^2+c_1ABC]$, which, after integration over one matrix, is the two matrix model $V(A,B) = tr[A^2 + B^2 + c_2ABAB]$.

So far nobody succeeded in obtaining a representation of the partition function Z in terms of the eigenvalues $\{\lambda_i, \mu_i\}$ for this two matrix model. If this (technical ?) difficulty could be solved [34], one may hope to derive an algebraic proof of the 4 colour theorem by comparing the free energy of this model with that of the one matrix with cubic coupling (after subtraction of tadpoles and double edges).

Acknowledgements

I wish to thank Drs. L.Molinari and V.Periwal for useful discussions on this subject and the Scientific Committee of the International Conference in honour of Prof. Eduardo R. Caianiello for the very interesting Conference.

References.

[1] V.A.Kazakov, Lecture at Les Houches, Session XLIX, edited by E.Brezin et J.Zinn-Justin, North-Holland 1990 and Lecture at Cargese 1990, LPTENS 90/30, to appear in Random Surfaces and Quantum Gravity, ed. by O.Alvarez et al.

[2] Two dimensional quantum gravity and random surfaces, ed. by D.Gross, T.Piran, S.Weinberg, World Scientific, to appear 1992.

[3] F.David, Lecture at Trieste 1990, S.Ph-T/90-127.

[4] Y.Lozano, J.L.Manes, Introduction to Nonperturbative 2D Quantum Gravity, UB-ECM-PF 3/91, Barcelona 1991.

[5] T.Tada, Ph.D.Thesis at Univ.of Tokio, Komaba, 1990.

[6] G.t'Hooft, Nucl.Phys. B72 (1974) 461 and B75 (1974) 461.

[7] G.Veneziano, see review talk at the 1978 Tokio meeting.

[8] E.Brezin, C.Itzykson, G.Parisi, J.B.Zuber, Comm.Math.Phys.59 (1978) 35. D.Bessis, Comm.Math.Phys. 69 (1979) 147. D.Bessis,C.Itzykson, J.B.Zuber, Adv.Appl.Math.1 (1980) 109.

[9] Y.Shimamune, Phys.Lett. 108B (1982) 407. I.Ya.Arefieva, 3rd Int. Symposium Dubna 1984.

[10] Similar multiple solutions, in unitary matrix models, were found by J.-Jurkiewicz and K.Zalewski, Nucl.Phys. B220 (1983) 167 and T.L.Chen, C-I.Tan, Z.T.Zheng, Phys.Lett. 123B (1983).

[11] Unaware of Ref.[9],[10], the multiple solutions in hermitian models in dimension d=0 and d=1 were discussed in G.M.Cicuta, L. Molinari, E.Montaldi, Mod.Phys.Lett. A1 (1986) 125 and Jour.of Phys. A20 (1987) L67.

[12] J. Jurkiewicz, Phys.Lett. 245B (1990) 178. G.M. Cicuta, L. Molinari, E.-Montaldi, Jour.of Phys.A23 (1990) L421.

[13] K.Demeterfi, N.Deo, S.Jain, C-I. Tan, Phys.Rev. D42 (1990) 4105.

[14] L.Molinari, J.of Phys.A21 (1988) 1. G.M.Cicuta, L.Molinari, E.Montaldi, Nucl.Phys. B300 (1988) 197. L.Molinari, E.Montaldi, J.of Phys. A23 (1990) 4995.

[15] G.Bhanot, G.Mandal, O.Narayan, Phys.Lett. B251 (1990) 388. O. Lechtenfeld, R.Ray, A.Ray, Jour.Mod.Phys. A6 (1991) 4491. M.Sasaki, H.Suzuki, Phys.Rev. D43 (1991) 4015.

[16] O.Lechtenfeld, Princeton preprint IASSNS-HEP 91/2 (1991) to appear in Jour.Mod.Phys.A. J.Jurkiewicz, Copenhagen preprint NBI-HE-91-06 (1991). D.Senechal, LAVAL-PHY-22/91 (1991) to appear in Jour.Mod.Phys.

[17] A. Barbieri, G.M. Cicuta, E. Montaldi, Nuovo Cim. 84A (1984) 173. C.M. Canali, G.M. Cicuta, L. Molinari, E. Montaldi, Nucl.Phys. B265 (1986) 485. G.M. Cicuta, L. Molinari, E. Montaldi, F. Riva, J.Math.Phys.28 (1987) 1716.

[18] A similar expansion is described in A.A.Slavnov, Lecture at Schladming 1983 and in F.Green, S.Samuel, Nucl.Phys. B190 (1981) 113.

[19] A.Anderson, R.C.Myers, V.Periwal, Phys.Lett. 254B (1991) 89 and Nucl.Phys. B360 (1991) 463. R.C.Myers, V.Periwal, NSF-ITP-91-01, Princeton 1991.

[20] P.G.de Gennes, Scaling Concepts in Polymer Physics, Cornell Univ.Press, Ithaca, N.Y. 1979.

[21] A.Maritan, A.L.Stella, Nucl.Phys. B280 (1987) 561. P.D.Gujrati, Phys. Rev. B39 (1989) 2494.

[22] V.A.Kazakov, Phys.Lett. A119 (1986) 140.

[23] D.V.Boulatov, V.A.Kazakov, Phys.Lett. B186 (1987) 379.

[24] B.Duplantier, I.Kostov, Phys.Rev.Lett. 61 (1988) 1433.

[25] I.Kostov, Mod.Phys.Lett. A4 (1989) 217.

[26] M.Staudacher, Nucl.Phys. B336 (1990) 349.

[27] V.G.Knizhnik, A.M.Polyakov, A.B.Zamolodchikov, Mod.Phys.Lett. A3 (1988) 819.

[28] C.Itzykson, J.B.Zuber, J.Math.Phys. 21 (1980) 411.

[29] M.L.Mehta, Comm.Math.Phys. 79 (1981) 327.

[30] S.Chadha, G.Mahoux, M.L.Mehta, J.Phys.A14 (1981) 579.

[31] T.L.Saaty, P.C.Kainen, The Four Color Problem, McGraw-Hill, New York 1977.

[32] An early attempt to evaluate the number of proper colorings with methods

of quantum field theory (without using random matrices) is in N.Nakanishi, Comm.Math.Phys. 32 (1973) 167.

[33] See, for instance, B.Bollobas, Graph Theory, Springer-Verlag, New York, 1979, exerc.27, pag.101, or the Tait theorem in S.Fiorini, R.J. Wilson, Edge-colourings of graphs, Pitman, London 1977, pag.26, or Ref.[31], pag.103.

[34] This, unsolved, two matrix model has been considered by S.Das, A.Dhar, A.M.Sengupta, S.R.Wadia, Mod.Phys.Lett. A5 (1990) 1041 as a model of random surfaces with curvature terms. It was there approximated by a one matrix model with powers $(tr A^2)^k$, which was easily solved. This solution was reconsidered in G.M.Cicuta, E.Montaldi, Mod.Phys.Lett. A5 (1990) 1927.

The Structure of the Standard Model of Particle Interactions

by

R.E. Marshak

Department of Physics

Virginia Polytechnic Institute and State University

Blacksburg, VA 24061 USA

I have known Eduardo Caianiello for almost a half-century and I have always admired his vibrancy, his versatility and his intellectual sophistication in science and culture. My modest contribution on the structure of the standard model of particle interactions in physics is intended to honor Prof. Cainaniello's wide-ranging accomplishments by taking a broader look—as he might do it—at the plateau of excellence that has been reached in particle physics after a half-century of vigorous world-wide activity in both experiment and theory. During the past fifteen years, much effort has gone into trying to move from the plateau of the standard model to a much higher plateau or, more accurately, to a "steeple of excellence" which is called TOE (theory of everything, including gravity). If the recent frenetic fifteen-year period of speculation is successful, or even partially successful—which is problematic at this time—it is likely that there will be a major "face-lifting" in the structure of particle physics. Meanwhile, I am content to try to describe the essential ingredients of the structure of particle physics at "plateau time" and leave it to the speaker thirty years hence—on the occasion Prof. Cainaniello's 100th birthday celebration—to survey the new structures that have evolved.

It is a truism that the structure of mathematics is more variegated, more creative and more universal than the structure of physics because the structure of mathematics need satisfy only the laws of logic (themselves products of the human mind) whereas the structure of physics must be not only in accord with the laws of logic but also consonant with the phenomenological strictures of the four-dimensional world in which we live. Fortunately, for particle physics, the scope of the subject matter is so broad (the universe) and the questions so fundamental (the basic forces in nature and the elementary constituents of matter) that the most sophisticated branches of mathematics can be used to define its structure. This is one reason why the structure of particle physics is relatively stable until there is good and sufficient reason—usually from the experimental side—for a major modification. The twentieth century has also been good to particle physics with the formulation of the special theory of relativity and quantum mechanics (the general theory of relativity—Einsteinian gravity—has been very important for astronomy and cosmology and its full impact on particle physics must still be realized). Moreover, during the past half century, particle physics has been the beneficiary of the extensive construction of high energy accelerators and large telescopes whose discoveries have made it possible for the structure of the present day theory of strong and electroweak interactions to be clarified and refined more rapidly than would otherwise have been the case.

What then is the structure of the present-day so-called standard model of particle interactions that has been so remarkably successful? First and foremost, has been the acceptance of relativistic quantum field theory (and the renormalization method required to solve the quantum field equations) as the vehicle for understanding the strong, electromagnetic and weak interactions among the quarks and leptons of three generations. This acceptance did not come easy—renormalization is not a pretty technique—and it took some time for the interplay of experiment and theory to compel the realization that the QED of 1950—as a Lorentz-invariant, gauge-invariant and renormalizable quantum field theory—is, after all, the paradigm for a theory of strong and weak interactions as well. This realization was slow in coming (the standard model was not mature until the early 1970s) because the strong and weak interactions, in different ways, did not appear to have some of the key properties characterizing the structure of QED [global Abelian internal symmetry (electric charge), long-range interaction (Coulomb force), parity conservation]. The early attempts—during the 1950s and 1960s—to determine the structure of the strong and weak interactions led to a number of interesting developments [S matrix theory, Regge pole theory, dual models] that deepened our understanding of relativistic quantum field theory but the main thrust that brought the strong and weak interactions under the single rubric of relativistic quantum field theory came from a different direction. I propose to explain how the structure of the present-day standard model was arrived at under four headings: (1) global symmetry structure; (2) gauge symmetry structure; (3) anomaly structure; and (4) topological structure.

(1) <u>Global symmetry structure of the standard model</u>: under this heading, I distinguish between the space-time symmetries and global internal symmetries. At the dawn of the "modern particle physics" era—which I like to date with the

termination of World War II (ca. 1945)—continuous space-time symmetries (translations, spatial rotations, "Lorentz boosts") were well-known and the properties of the Poincaré group fully explored. There was absolutely no experimental evidence for deviations from the conservation laws (energy-momentum conservation, angular momentum conservation, Lorentz invariance, two helicities for massless particles with $J > 0$) predicted by the Poincaré group and these conservation laws were taken as absolute in quantum field theory. The discrete symmetries (space inversion P, time inversion T and charge conjugation C) were another matter and the discovery of maximal P and C violation in weak interactions (in 1956), and of weak CP violation (in 1964) opened a new chapter in the formulation of quantum field theories and brought further refinement in our understanding of the conditions and status of the CPT theorem and the relation between spin and statistics. [It led, for example, to the recognition of the possibility of fractional spin and fractional statistics for fewer than four space-time dimensions and to the exploitation of these ideas in condensed matter physics (polyacetylene, fractional quantum Hall effect, etc.).]

The developments connected with global internal symmetries in quantum field theory after 1945 were equally innovative and far-reaching. The global Abelian internal symmetries of baryon charge invariance and lepton charge invariance, in analogy to electric charge invariance, were known but the discovery of increasingly higher-dimensional (non-Abelian) global internal symmetries in the strong and weak interactions was a novel and gratifying experience and had much to do with the determination of the structure of the standard model. Starting with the $SU(2)_I$ isospin group for the pion-nucleon interaction, there was a progression of enlargements—reflecting the growth in experimental knowledge of strange particles and so on—through the isospin-hypercharge group $SU(2)_I \times U(1)_Y$ (Y is the "strong" hypercharge) through the $SU(3)_F$ flavor group and the quark flavor model of hadrons, until finally the global $SU(6)$ quark flavor-spin group, mixing $SU(3)_F$ and the $SU(2)$ spin group, produced such a phenomenologically impressive classification of the low-lying (in mass) multiplets of mesons and baryons that one was compelled to introduce a new internal symmetry for the quark (global "color") in order to satisfy the conditions of the Pauli principle.

While the global internal symmetry structure was being sharply defined for the quarks and their strong interactions, progress was also being made in defining more sharply the global internal symmetry structure for both the quarks and the leptons participating in the weak interactions. Here the chirality-invariant

(V-A) theory of the weak interaction of 1957 showed the way. The requirement of chirality invariance (a global internal symmetry) led uniquely to the maximal parity-violating V-A theory and pinpointed the vector (V) and axial vector (A) charged weak currents as having a very special status [chirality invariance applied to massive Dirac fields in the V-A theory was inspired by the exact chirality invariance of the Weyl equation for a massless neutrino.] The basic ideas of the V-A theory and the quark flavor model of hadrons were brought together into the global chiral quark flavor symmetry group that led to the algebra of vector (V) and axial vector (A) currents—current algebra—and its quasi-dynamical successes in the strong and weak interactions. In addition, when the PCAC (partially conserved axial vector current) hypothesis was joined to current algebra, the concept of the spontaneous symmetry breaking of a global internal symmetry (and its corollaries; the Goldstone theorem and the generation of massless pseudoscalar Nambu-Goldstone boson particles) entered the structure of modern particle physics.

(2) <u>Gauge symmetry structure of the standard model</u>: by the mid-1960s, the stage was set for the big jump from global internal symmetries to gauge symmetries in the quantum field theories of the strong and weak interactions. On the negative side, the S matrix—Regge pole program was running into snags, attempts to "relativize" the non-relativistic $SU(6)$ quark flavor-spin model came afoul of the "no-go" theorem (that it was impossible to combine the Poincaré group and any global internal symmetry into a compact Lie group), and the limitation of the V-A theory at high energies had become apparent. On the positive side, the quark flavor model of hadrons was in good shape [the best days were still to come with the deep inelastic lepton-hadron scattering experiments establishing quarks as "physical entities" rather than mere "mathematical fictions"], the mixed $SU(6)$ quark flavor-spin model model had given strong indications of a new quark degree of freedom (global "color"), the successes of current algebra with PCAC (giving respectability to the spontaneous breaking of global internal symmetries) were in full swing and a second generation of quarks and leptons (revealing a basic quark-lepton symmetry) loomed on the horizon.

The time was ripe to rediscover the Yang-Mills 1954 paper—as well as Gell-Mann-Low's paper of the same year on the renormalization group—and to recognize that the application of the gauge principle to non-Abelian global internal symmetries could bring success. With hindsight, the magical power of gauging a quantum field theory—to covert "symmetry into dynamics"—should not have come at such a great surprise. After all, the highly successful gauged non-chiral Abelian quantum field theory of electromagnetism (QED) had been in existence since ca. 1950. And yet it took two decades before the seminal ideas of QED were combined with some new ideas to transform the global, non-chiral, non-Abelian "color" symmetry of the strong interactions into QCD and to convert the global chirality-invariant, non-Abelian quark-lepton flavor symmetry of the weak and electromagnetic interactions into QFD.

The new ideas that made possible the successful formulation of QCD and QFD as gauge theories encompassed the following: (1) instead of gauging one global internal symmetry—that of electric charge to yield a $U(1)$ gauge group, as is done in QED—the global non-Abelian $SU(3)$ color group was gauged in a non-chiral fashion to yield the $SU(3)$ color gauge theory of the strong interaction (QCD). As for the weak interaction, the chiral quark flavors and lepton quark flavors of each generation were gauged as lefthanded doublets and righthanded singlets under the minimal group chiral $SU(2)_L \times U(1)_Y$ (Y here is the weak hypercharge) that can accommodate the chirality-invariant (maximally parity-violating) weak interaction and (parity-conserving) electromagnetic interaction; (2) the short-range character of the strong interaction among quarks in apparent contradiction to the masslessness of the gluon gauge fields—was explained by exploiting the asymptotic freedom and confinement properties of unbroken non-Abelian $SU(3)$ color in combination with the scale introduced through the renormalization of QCD. In the case of the extremely short-range character of the weak interaction, the Higgs mechanism to generate massive gauge bosons by spontaneously breaking a gauge group was utilized. The spontaneous breaking of the chiral $SU(2)_L \times U(1)_Y$ group down to the non-chiral $U(1)_{EM}$ provides the link between the weak and electromagnetic interactions; and (3) the recognition that the renormalizability of a quantum field theory can be achieved not only for Abelian gauge theories but also for non-Abelian gauge theories, both broken and spontaneously broken. The basic point of applying the gauge principle to appropriate non-Abelian global internal symmetries—and taking advantage of spontaneous symmetry breaking if necessary—was the key to the success of the dynamical, renormalizable and powerfully predictive gauge theory of the standard model.

It should be noted that the success of the gauge symmetry structure of the standard model owes much to several fortuitous circumstances: (1) the adroit choice of the "particles" that are subjected to the gauge-invariant strong and electroweak interactions, namely the quarks and leptons of each of the three generations. This "adroitness" is possible because of a built-in quark-lepton symmetry for each generation that is not yet fully understood and whose full explanation may require a solution of the "fermion generation problem" (why there are just three generations of quarks and leptons, identical in their quantum numbers but differing wildly in their masses). Curiously enough, the gauge symmetry structure of the standard model does not depend on the existence of a righthanded neutrino (ν_R) because ν_R is "inert" under $SU(2)_L \times U(1)_Y$ [i.e. the quantum numbers of ν_R would have to be $(1,0)$ under the electroweak group and these quantum numbers do not affect the chiral gauge anomaly-free constraints (see below)]; (2) the QCD scale, Λ_{QCD}, is close to the mass of the pion and explains why the crude pion theory of nuclear forces (at low energy) has not been such a poor approximation; and (3) the three chiral gauge anomaly-free constraints in four dimensions—associated with the triangular axial, global $SU(2)$ and mixed gauge-gravitational anomalies— are all consistent with the gauge symmetry structure of the standard model. The chiral gauge anomalies in a quantum field theory result from the insistence that chirality invariance and gauge invariance and/or general covariance be simultaneously maintained in the theory. The chiral gauge anomaly-free constraints impose strong conditions on the possible gauge symmetry structure of the standard model and, as such, reflect the breakdown of perturbative renormalizable quantum field theory in the presence of chiral fermions and the added constraints imposed by the topologically non-trivial gauge transformations associated with non-Abelian gauge groups. Parenthetically, the mixed gauge-gravitational anomaly is the first example of the impact of the gravitational interaction—living at the Planck scale—on the structure of the standard model.

(3) <u>Anomaly structure of the standard model:</u> under this heading, I discuss two topics: (3a) the role of the "harmless" triangular axial anomalies in contributing to the structure of the standard model; and (3b) the role of the "harmful" chiral gauge anomalies in fixing the Weyl representations of the fermions (quarks and leptons) of the standard gauge group.

(3a) Of the three chiral gauge anomalies, the triangular axial anomaly [the Adler-Bell-Jackiw (ABJ) anomaly] was first identified in connection with the $\pi^0 \rightarrow 2\gamma$ decay where the calculation of the first-order fermion loop correction to a process involving an axial vector current (whose divergence represents π^0) and two vector currents (representing the 2γ's) could not produce a unique answer in the face of the insistence on simultaneous chirality and gauge invariance of the Green's functions. The ABJ triangular axial anomaly is harmless in QED because the π^0-associated axial vector quark current is not part of the structure of QED and actually accounts completely for the $\pi^0 \rightarrow 2\gamma$ decay width. The ABJ anomaly has its useful counterpart in QCD where the "flavor-singlet" triangular axial anomaly is "harmless" because only vector quark currents interact with the gluon (gauge) fields and hence all renormalization arguments—depending on the appropriate vector Ward identities—follow through. The "harmless" triangular axial anomaly in QCD makes significant contributions to a number of QCD processes that can otherwise not be explained by perturbative QCD.

(3b) On the other hand, when an axial vector (or chiral) fermion current is coupled to a gauge field (s)—as in the case of QFD—the resulting triangular chiral gauge anomalies must be absent in order to carry through the renormalization of the gauge theory with the help of axial vector Ward identities in addition to the vector Ward identities. It turns out that the required triangular chiral gauge anomaly cancellation for QFD is made possible by the existence of both quarks and leptons in appropriate representations of the electroweak group. The quasi-non-perturbative character of the triangular axial anomaly stems from the fact that it holds to all orders of perturbation theory and hence possesses both a "short wavelength" and "long wavelength" limit; this feature introduces renormalizability problems for chiral QFD. The discovery of the triangular axial anomaly was followed by the identification of two more chiral gauge anomalies affecting the standard model, namely the non-perturbative, global $SU(2)$ anomaly and the mixed perturbative gauge-gravitational anomaly. The global $SU(2)$ anomaly is non-perturbative because it involves the topologically non-trivial "large" $SU(2)$ gauge transformation and would destroy the mathematical self-consistency of chiral QFD in the presence of an odd number of Weyl fermion doublets; fortunately, the total number of chiral quark and lepton doublets per generation in the standard model is even so that the global $SU(2)$ anomaly is absent. The mixed gauge-gravitational anomaly—is perturbative, like the triangular axial anomaly, and would create renormalizability problems for the standard model if $Tr\, Y$ summed over the quarks and leptons of one generation, did not vanish; again, $Tr\, Y$ vanishes for one generation of quarks and leptons so that the standard model is concordant with the anomaly structure constraints.

The fact that all three chiral gauge anomaly cancellations take place for the quark-lepton Weyl representations of one generation one fermions in the standard model raises the interesting obverse question as to whether obeying the three chiral gauge anomaly-free constraints carries strong structural implications for the standard model. The answer is in the affirmative since it can be shown that if one starts with the standard gauge group $SU(3) \times SU(2) \times U(1)$—without specifying the Weyl fermion representations—leads uniquely to the correct result for the minimal number of lefthanded and righthanded Weyl representations of $SU(3) \times SU(2) \times U(1)$ so that the original $SU(3) \times SU(2) \times U(1)$ gauge group (with unspecified Weyl fermion representations) becomes the phenomenologically correct standard gauge group $SU(3)_C \times SU(2)_L \times U(1)_Y$ with all quark and lepton quantum numbers properly fixed. Indeed, the structural constraints on the standard model resulting from the requirement that all three chiral gauge anomalies be absent, also yield automatic hypercharge quantization and electric charge quantization [electric charge quantization was thought to require the intervention of a non-Abelian group and not just the Abelian $U(1)_Y$]; anomaly-related structural constraints on the standard model go beyond the normal requirements of Lorentz invariance, gauge invariance and renormalizability that determine QED, and seem to encode some of the topological properties of Yang-Mills theories.

(4) <u>Topological structure of the standard model</u>: under this heading, I again discuss two topics: (4a) instantons; and (4b) sphalerons. Instantons are the "topological solitons" of relativistic (Yang-Mills) quantum field theory in Euclidean four-space. The identification of relativistic, classical and soliton-like solutions of the non-Abelian gauge equations in Euclidean four-space was a major discovery and their effect on the topological structure of the standard model in Minkowski space became a major undertaking. Instanton solutions are responsible for the non-uniqueness of the non-Abelian gauge theory vacuum because they provide a mechanism for "vacuum tunneling" between topologically inequivalent n-vacua, where n is the "winding number" of the instanton. The Minkowski interpretation of instantons as "vacuum tunneling" events leads to the complicated θ-vacuum with its associated θ parameter which, phenomenologically [because of the observed upper limit on the electric dipole moment of the neutron] must be miniscule ($\simeq 10^{-9}$). This result has created the "strong CP problem" in QCD which is still unresolved. The well-known Peccei-Quinn attempt to dispose of the "strong CP problem" by postulating additional structure—the Peccei-Quinn chiral $U(1)$ symmetry—leaves a quasi-Nambu-Goldstone boson particle (the axion) in its wake, whose presence—despite much effort over a decade—has eluded all searches. In my view, the "strong CP problem" in QCD may be indicative of our failure thus far to fully understand the topological structure of the QCD vacuum, which is a pity since the global internal symmetry and the gauge symmetry structures of the standard model are

so well in hand.

(4b) <u>Sphalerons</u>: the "color" instantons in unbroken vector-like QCD create the still-unresolved "strong CP problem". The "flavor" instantons of spontaneously-broken chiral QFD have been known to produce miniscule B- and L-violating effects (while maintaining global B-L conservation). These conclusions seemed uninteresting and were put aside, until fairly recently, when interest was revived in the possibility of suppressing the "flavor"-instanton-inhibition of B violation by means of unstable saddle-point, topological solutions (called "sphalerons") to the Yang-Mills-Higgs equations. The sphaleron barrier—related to the spontaneous symmetry breaking scale ($250\ GeV$) of the electroweak group—is about $10TeV$ and it has been found that there is a large enhancement of B violation at temperatures $T \sim 10\ TeV$. This could be of interest for "baryon asymmetry of the universe" (BAU) questions and provide a new handle on the Sakharov-type explanation for BAU. A similar enhancement of B violation (within the QFD framework) in very high energy collisions ($E \gtrsim 10\ TeV$) is also possible, provided the end products in the collision contain a high multiplicity of weak bosons ($Z, W^{\pm}$) and Higgs particles. However, these conclusions about electroweak B violation at high energies are less certain than those for high temperature because of the approximations made to the instanton solution. In sum, the topological structure of the standard model appears to possess many novel and intriguing properties but there are still too many open questions to draw final conclusions.

MECHANISMS OF COLOUR CONFINEMENT FROM LATTICE QCD

Adriano Di Giacomo

Dip. Fisica e Sezione I.N.F.N. PISA

IN HONOUR OF EDUARDO CAIANIELLO ON HIS 70th BIRTHDAY

INTRODUCTION

Quantum Chromodynamics (QCD) is widely accepted as the theory of strong interactions. It is also a popular statement that colour is confined: no free quarks or gluons have been observed in Nature.

Despite the fundamental simplicity of the QCD Lagrangean it is not clearly known if it has colour confinement built in, and, if so, by which kind of field theoretical mechanism the phenomenon takes place. The reason for that is that confinement is a large distance phenomenon, and the perturbative expansion, which is one of our main tools in field theory, is, for QCD, highly infrared singular.

Lattice formulation[1] of the theory does not rely on perturbative expansion, and can help to investigate confinement.

On a lattice QCD can be simulated in its full dynamical content. A mass scale is generated by dymensional trasmutation, which eliminates infrared problems.

In fact the evidence that colour is confined in QCD mainly comes from lattice simulations[2].

Of course it can prove to be a non trivial job to remove the cut-off, i.e. to extract continuum physics from numerical simulations on a finite, discrete array of points: theoretical ingenuity may be needed to supplement limitations in computing power. We shall illustrate how this can be done to extract information on colour confinement mechanism. Many models have been proposed for the mechanism of colour confinement. One of the most appealing from a theoretical point of view is due to t'Hooft[3] and Mandelstam[4]. The QCD vacuum behaves, according to this model, as a " Dual Superconductor of type II". Dual means that the role of electric and magnetic field is interchanged with respect to usual superconductors. The chromoelectric field penetrates the vacuum by Abrikosov[5] flux tubes, the same way as magnetic field does in normal superconductors of type II. Thus the chromoelectric field produced by a single quark is squeezed in a flux tube of constant section wich extends from the quark to infinity, whilst a quark - antiquark pair are connected by a flux tube which is a straight segment in its ground configuration. Since the energy of a flux tube is proportional to its length, it follows that an isolated quark has infinite energy, and that the $q\,\bar{q}$ potential has the form

$$V(r) = \sigma\, r \tag{1.1}$$

σ, the energy of the flux tube per unit length is known as the string tension. The flux tube joining q and $\bar{q}$ has indeed the spectrum of a string[6,7].

Usual superconductivity is produced by condensation of charged Cooper - pairs in the ground state: by analogy in QCD vacuum objects with non trivial magnetic charge should condense (monopoles or bound states of monopoles).

We have tested on the lattice the validity of such a model[8,9,10]. We have used $SU(2)$ as a gauge group, and no quarks: the mechanism of confinement should be the same for QCD, but computational difficulties lower by few orders of magnitude. We have investigated the existence of flux tubes [Sect. 2], the existence of monopoles and the colour structure of flux tubes [Sect. 3], and we are investigating monopole condensation [Sect. 4].

Beautiful physics emerges from these problems, covering aspects which are typical of particle physics and at the same time aspects which are typical of statistical mechanics and condensed matter physics.

2. FLUX TUBES

The first obvius check of the model is to look for the existence of chromo-electric flux tubes between a static $q \bar{q}$ pair. Putting such a pair on a lattice is a standard procedure[1], and the field strength in the region of space between them can be defined and measured[8,9]. Doing this by brute force would require a large montecarlo statistics, and would be prohibitively expensive, due to the short range quantum fluctuations (lattice artifacts), which have to be removed to go to continuum physics.

However, if flux tube configurations exist and are classical solutions of the equations of motion, as in usual superconductors, one can implement on the lattice the semiclassical approximation, by freezing the short range quantum fluctuations around the classical solution. This is done by a cooling procedure, which smoothly brings configurations towards a local minimum of the action[10,12]. The linear force eq.(1.1) between the pair is preserved during cooling[12], and the residual classical configuration can be exposed, with almost non fluctuations[8,9].

The result is a nice flux tube, with chromoelectric field directed along the line joining the pair, and transverse size of $\sim$ .5 fm.

Confinement is produced by flux tubes, in agreement with the model of dual superconducting vacuum.

3. MONOPOLES IN QCD. COLOUR STRUCTURE OF THE FLUX TUBES.

A more sophisticated question is to check that vacuum is indeed supercon-ductor of type II, i.e. that flux tubes originate from a (dual) Meissner effect. This means identifying the monopoles which should condense and produce a spon-taneous breaking of (dual) gauge invariance, i.e. superconductivity.

Of course the gauge invariance which must be realized à la Goldstone has nothing to do with the colour gauge group. Colour is an unbroken symmetry, and is realized à la Wigner - Weyl: condensing monopoles must be colour singlets. Another reason for that is that, contrary to ordinary superconductors which are made of e^- and nuclei, the QCD ground state must be invariant under colour

206

charge conjugation.

How can such monopoles be defined?

A hint comes from the $SU(2)$ Higgs model, with the Higgs belonging to the adjoint representation, the so called Georgi - Glashow model[13]. In this model stable monopole configurations do exist[14,15]. At large distance from the location of the monopole the $a-th$ colour component of the Higgs field behaves as (hedgehog):

$$\phi^a(\vec{x}) \simeq \frac{x^a}{|\vec{x}|} \tag{3.1}$$

and the gauge field as

$$W_0^a = 0 \qquad W_i^A \simeq -\frac{1}{e}\varepsilon_{iab}\frac{x^b}{|\vec{x}|} \tag{3.2}$$

An "electromagnetic field" can be defined[14], which is an $SU(2)$ singlet

$$F_{\mu\nu} = \hat{\phi}^a\, G^a_{\mu\nu} - \frac{1}{e}\varepsilon_{abc}\hat{\phi}^a D_\mu\hat{\phi}^b D_\nu\hat{\phi}^c$$

$$\hat{\phi}^a = \frac{\phi^a}{\sqrt{\sum_a \phi^{a\,2}}} \qquad \mathbf{G}_{\mu\nu} = \partial\mathbf{W}_\nu - \partial_\nu\mathbf{W}_\mu - e\,\mathbf{W}_\mu \wedge \mathbf{W}_\nu \tag{3.3}$$

The field $F_{\mu\nu}$ is colour singlet and even under colour charge conjugation. The field $F_{\mu\nu}$ produced by the monopole is, at large distances, the field of a point monopole.

$F_{\mu\nu}$ can also be written[16]

$$F_{\mu\nu} = M_{\mu\nu} + H_{\mu\nu} \tag{3.4}$$

with

$$M_{\mu\nu} = \partial_\mu B_\nu - \partial_\nu B_\mu; \qquad B_\mu = \hat{\phi}^a W^a_\mu \tag{3.5}$$

and

$$H_{\mu\nu} = \frac{1}{e}\varepsilon_{abc}\hat{\phi}^a\partial_\mu\hat{\phi}^b\partial_\nu\hat{\phi}^c \tag{3.6}$$

In a gauge where $\hat{\phi}^a = \delta_3^a$ $H_{\mu\nu}$ vanishes, $B_\mu = W^3_\mu$ and

$$F_{\mu\nu} = \partial_\mu W^3_\nu - \partial_\nu W^3_\mu \tag{3.7}$$

Going to that gauge is called an abelian projection. $F_{\mu\nu}$ is gauge invariant, and assumes the form (3.7) after abelian projection.

In QCD there is no Higgs field. However monopoles do exist. Indeed if (in $SU(2)$ theory)

$$\mathcal{O}(x) = \vec{\sigma} \cdot \vec{b}(x) \tag{3.8}$$

the gauge transformation which dyagonalizes $\mathcal{O}(x)$ is singular on the line

$$\vec{b}(x) = 0 \tag{3.9}$$

The singularity has the topology of a monopole[17]. For any operator $\mathcal{O}(x)$ of the form (3.8) an abelian projection can be defined as the gauge transformation which brings $\vec{b}$ along the 3 axis. An abelian projected field can be defined, and the sources of its magnetic vector will be monopoles. The number and the locations of such monopoles depend on the choice of the operator $\mathcal{O}(x)$: the guess of t'Hooft[17] is that physics is independent of it. To verify if monopoles defined by any abelian projection are the ones which produce dual superconducivity by condensing in the vacuum, one must define a field describing monopoles, or what in statistical mechanics is called a disorder operator, and check that it condenses in the ground state of the theory[10]. Let $\mu(x)$ be such an operator, having a monopole charge $q \neq 0$. Under a dual gauge transformation $\alpha(x)$

$$\mu(x) \rightarrow e^{iq\alpha(x)}\mu(x)$$

and hence a condensate

$$\langle 0 \mid \mu(x) \mid 0 \rangle \neq 0$$

signals that $|0\rangle$ is not gauge invariant. Gauge invariance is spontaneously broken, and that will produce Meissner effect[18].

A preliminary test, however, can be obtained by analyzing the colour structure of the flux tubes between $q\,\bar{q}$ pairs which were described in sect.2 The field which is subjected to Meissner effect should be the electric vector of $F_{\mu\nu}$, i.e. the electric vector of the abelian projected field eq.(3.7). After abelian projection, thereefore,

208

the field in the flux tubes should be mainly oriented along the direction 3 in colour space.

This can be tested, and has been tested for a few choices of the operator $\mathcal{O}(x)$ used to define the monopoles[19,9]. The answer is negative. This could only mean that those particular choices of $\mathcal{O}(x)$ where not giving the right monopoles. In any case this proves that not all choices of $\mathcal{O}(x)$ give the same physics.

4. MONOPOLE CONDENSATION

As anticipated in sect.3, to detect Meissner effect a disorder operator $\mu(x)$ must be defined for monopoles, and its vacuum expectation value must be measured. An alternative procedure consists in measuring the correlation at large distances

$$\langle 0|\mu(\vec{x},t)\,\mu^{\dagger}(\vec{y},t)|0\rangle \underset{|\vec{x}-\vec{y}|\to\infty}{\simeq} Ae^{-k|\vec{x}-\vec{y}|} + B \tag{4.1}$$

By cluster property and by use of tanslation invariance of $|0\rangle$

$$B = |\langle 0|\mu(\vec{x},t)|0\rangle|^2 \tag{4.2}$$

If the correlation in eq.(4.1) does not vanish at large distances there is Meissner effect.

The operator $\mu(\vec{x},t)$ can be defined by use of a method introduced by Kadanoff and Ceva[20] for the Ising model and then elaborated by other people[21,22] for different systems. μ is defined by the so called disorder algebra

$$\mu(\vec{x},t)W_i^a(\vec{y},t) = \begin{cases} \left[gW_ig^{-1} - \dfrac{1}{e}\partial_i g\, g^{-1}\right]^a \mu(\vec{x},t) & \vec{y}\notin T_{\vec{x}} \\ W_i^a(\vec{y},t)\mu(\vec{x},t) & \vec{y}\in T_{\vec{x}} \end{cases} \tag{4.3}$$

$T_{\vec{x}}$ is a volume surronding the point $\vec{x}$, e.g. a sphere centered on $\vec{x}$; g is a gauge transformation singular in $\vec{x}$, which belongs to the homotopy class of a monopole, e is the gauge coupling constant.

A form for $\mu(\vec{x},t)$ can be explicitly written[2,3]

$$\mu(\vec{x},t) = \exp\left[R(\vec{x},t) + \mathcal{L}_c(\vec{x},t)\right] \tag{4.4}$$

Here

$$R(\vec{x},t) = -\frac{i}{e} \int_{R_3 - T_{\vec{x}}} d^3\xi \left[D_i G^{i0}(\vec{\xi},t) \right]^a \alpha^a(\vec{\xi},t) \tag{4.5}$$

$\alpha^a(\vec{x},t)$ is the gauge parameter, $g(\vec{x},t) = \exp(i\alpha^a(\vec{x},t)T^a)$. $\mathbf{G}^{i0}(\vec{x},t)$ is the canonical conjugate momentum to $\mathbf{W}_i$. It is easy to check by use of the canonical commutation relations that $\mu(\vec{x},t)$ would satisfy the algebra (4.3) for $\mathcal{L}_c = 0$.

$\mathcal{L}_c$ is a kind of renormalization counterterm which must be added[20] to make the disorder operator $\mu(\vec{x},t)$ independent of the way in which the discontinuity introduced by it is propagated. Explicitely

$$\mathcal{L}_c(\vec{x},t) = -\frac{1}{4}\delta(0) \int_{R_3 - T_{\vec{x}}} d^3\xi \, \mathrm{Tr}\left\{ D_i \mathbf{A} \right\}^2 \tag{4.6}$$

$$A^a(\vec{x},t) = \frac{1}{e}\alpha^a(\vec{x},t) \int_{R_3 - T_{\vec{x}}} d^3\xi \, \delta^3(\vec{x} - \vec{\xi}) \tag{4.7}$$

$\mathcal{L}_c$ only depends on the fields $\mathbf{W}_i$ and not on their conjugate momenta, so that its introduction does not modify the algebra (4.3).

The operator $\mu(\vec{x},t)$ as given by eq.'s (4.4), (4.5), (4.6), (4.7) can be translated on the lattice.

The translation should not produce any problem, especially in view of the fact that we plan to use cooling to detect condensation, and cooling eliminates all the renormalization effects expected between lattice and continuum[23]. Work is in progress in this direction.

If we will be able to find one definition of monopoles, which gives condensation of the disorder parameter, this will be the proof that the mechanism of dual superconductivity is at work in QCD.

This would be an important step in understanding which are the relevant degrees of freedom of QCD at large distances.

Much of the work illustrated above was made possible by combined knoledge of high energy physics and condensed matter physics. It is from the example of physicists like Eduardo that we learned to enjoy different branches of physics and to appreciate how much they can have in common.

REFERENCES

1. K.G.Wilson Phys. Rev. **D10** 2445 (1974)

2. M.Creutz Phys. Rev. Lett. **43** 553 (1979)

3. G.t'Hooft in *High Energy Physics* ed. A.Zichichi, Ed.Compositori Bologna 1976

4. S.Mandelstam Phys. Rep. **23C** 245 (1976)

5. A.A.Abrikosov JETP **32** 1442 (1957) , Sov. Phys. JETP **5** 1174 (1957)

6. Y.Nambu Phys.Rev. **D10** 4262 (1974)

7. A.B. Nielsen, P.Olesen Nucl. Phys. **B 61** 45 (1973)

8. A.Di Giacomo, M.Maggiore, S.Olejnik Phys. Lett. **B 236** 199 (1990)

9. A.Di Giacomo, M.Maggiore, S.Olejnik Nucl. Phys. **B 347** 441 (1990)

10. L.Del Debbio, A.Di Giacomo, M.Maggiore, S.Olejnik Phys. Lett. **B 267** 254 (1991)

11. M.Campostrini, A.Di Giacomo, H.Panagopoulos, E.Vicari Nucl. Phys. **B329** 683 (1990)

12 M.Campostrini, A.Di Giacomo,M.Maggiore, H.Panagopoulos, E.Vicari Phys. Lett.

   B 225 403 (1989)

13. H.Georgi, S.L.Glashow Phys. Rev. Lett. **32** 438 (1974)

14. G.t'Hooft Nucl. Phys. **B79** 276 (1974)

15. A.M.Polyakov Sov. Phys. JETP Lett. **20** 194 (1974)

16. J.Arafune, P.G.O.Freund, C.J.Goebel J. Math. Phys. **16** 433 (1975)

17. G.t'Hooft Nucl. Phys. **B 190** [**FS3**] 455 (1981)

18. S.Weinberg Progr. Theor. Phys. Suppl. **86** 43 (1986)

19. J.Greensite, J.Winchester Phys. Rev. **D40** 4167 (1989)

20. L.Kadanoff, H.Ceva Phys. Rev. **B3** 3918 (1971)

21. E.C.Marino, B.Schroer, J.A.Swieca Nucl. Phys. **B200** [**FS4**] 473 (1982)

   J.Froelich, P.A.Marchetti Comm. Math. Phys. **112** 343 (1987)

22. E.C.Marino, J.Stephany Ruiz Phys. Rev. **D 39** 3690 (1989)

23. M.Campostrini, G.Curci, A.Di Giacomo, G.Paffuti Zeit. Phys. C **32** 377 (1986)

M.Campostrini, A.Di Giacomo, H.Panagopoulos Phys. Lett. **B 212** 206 (1988)

On the Structure of the Pomeron

E. Predazzi*

Abstract A longstanding problem in particle physics concerns the origin and the structure of the Pomeron. Arguments are given to relate this problem to the question of the so-called small-x physics, in particular to the burning question of the gluon distribution. A scheme based on the exchange of a dipole Pomeron with unit intercept is briefly discussed.

---

* Dipartimento di Fisica Teorica dell'Università and INFN - Sezione di Torino, ITALY

The origin of the so-called *Pomeron i.e.* of the Regge trajectory responsible for the diffractive part of an elastic reaction is one that has been very much debated but, to date, little clarified. Here, we consider the relation between high-energy, small momentum transfer hadron scattering (i.e.*soft physics*) and deep inelastic scattering (DIS), which probes the hadronic structure at small distances. This relation is important in order to understand the rise of cross-sections from the point of view of the structure of hadrons and of the interaction of their constituents. Recently, see e.g. [1], the idea of the dominance of semi-hard collisions opened new perspectives for perturbative QCD calculations: in essence, big hadronic cross-sections can result from a very large number of small partonic cross-sections. This approach, however, has problems with unitarity (violation of the Froissart bound) so that the range of applicability of perturbative QCD remains questionable.

From the DIS point of view, it is natural to attribute the growth of the hadronic cross sections to the increase of the gluon distribution function at small x. The use of current models for the structure functions and the scattering amplitude, however, leads to a contradiction with the data at a quantitative level [2].

Formally, the deep inelastic structure functions are related to the forward elastic hadron scattering amplitude by unitarity. The essential kinematical variables are the usual ones of deep inelastic scattering, the virtual photon momentum $q = k - k'$, $Q^2 = -q^2 > 0$, the virtual photon energy (in the lab system) $\nu = pq$ and the total initial energy squared $s = (p + q)^2$ (notice that our definition of ν is slightly different from the usual one).

The photon-hadron coupling is assumed to be determined via vector dominance or by $\bar{q}q$ loop calculations.

The deep inelastic lepton-hadron cross-section is expressed as usual through the structure functions W_1 and W_2 by

$$d^2\sigma/d\Omega dE' = 4e^2 E'^2/Q^4 \quad [2W_1(\nu, Q^2)sin^2(\theta/2) + W_2(\nu, Q^2)cos^2(\theta/2)]. \quad (1)$$

W_1 and W_2 are related to the forward virtual Compton scattering (i.e. to the elastic off-shell hadron scattering) in terms of the transverse and longitudinal cross sections. Since we are interested in the high-energy behaviour of the hadronic scattering amplitude, a Regge behaviour for W_1 and W_2 is appropriate. Summing over all relevant Regge contributions, we shall write

$$W_1(\nu, Q^2) \sim \nu\sigma_T \to \sum_i \beta_1^i(Q^2)(\nu/s_o)^{\alpha_i(0)} \tag{2}$$

$$\nu W_2(\nu.Q^2) \sim Q^2(\sigma_T + \sigma_L) \to \sum_i Q^2 \beta_2^i(Q^2)(\nu/s_o)^{\alpha_i(0)-1}. \tag{3}$$

We shall approximate the above equations by a leading Pomeron exchange interpreted in QCD as a reggeized gluon ladder exchange. In the Bjorken limit $(\nu \to \infty,\ Q^2 \to \infty,\ \nu/Q^2$ fixed$)$, the structure functions obey Bjorken scaling, $(x = Q^2/2\nu)$

$$W_1(\nu, Q^2) \to F_1(x),$$

$$\nu W_2(\nu, Q^2) \to F_2(x).$$

Regge and Bjorken regions do not coincide but they overlap and where they overlap we have as $x \to 0$

$$W_1(\nu, Q^2) \to x^{-\alpha_P(0)} \tag{4a}$$

$$W_2(\nu, Q^2) \to x^{1-\alpha_P(0)} \tag{4b}$$

where we have confined ourselves to the dominant Pomeron contribution (whose intercept we have denoted by $\alpha_P(0)$); even though their small-x behaviour is dominated by the gluon contribution (which are believed to form the bulk of Pomeron exchange) the structure functions may receive contributions also from valence, sea quarks and gluons. Thus, at small x, the gluon distribution will be written as

$$G(x) \sim x^{-\alpha_P(0)} \tag{5.}$$

According to perturbative QCD calculations [3]

$$\alpha_P(0) = 1 + \delta, \qquad \delta = (12/\pi)\alpha_s ln2 > 0.3. \tag{6}$$

It has become popular to use $\delta = 0.5$ to fit the data on the DIS structure functions. A singularity in $xG(x)$ as $x \approx 0$ is required also by more general arguments based on the solution of evolution equations [4]. According to these calculations,

$$xG(x) \sim \exp 2\sqrt{[A\xi(Q^2)ln(1/x)]} \tag{7}$$

which increases faster than $ln(1/x)$ at $x \to 0$ but slower than any power of $1/x$. In any case, great care should be taken when performing the limit $x \to 0$ since this region lies outside the domain of perturbative calculations and new phenomena are not unexpected.

$xG(x)$ represents the number of gluons; assuming $\hat{\sigma} \sim \alpha_s(Q^2)/Q^2$ for the cross section of each constituent, we arrive at a *Froissart bound* (here $x = Q^2/s$)

$$W \sim xG(x)\alpha_s(Q^2)/\sigma_t \tag{8}$$

which is certainly violated by eqs. (5) and (6).

We argue that violation of unitarity is prevented if a parametrization for the small-x gluon distribution

$$G(x) \approx ln(1/x)x^{-\alpha_P(0)} \tag{9}$$

with $\alpha_P(0) = 1$ is used instead of (5,6).

The form (9) is suggested by the analogy with a dipole Pomeron (DP) model (see below and ref. [5]). It is singular at $x \to 0$ as required [4], but does not necessitate shadowing (or unitarity) correction as in the case of the supercritical Pomeron, $\alpha_P(0) > 1$.

The nature of the Pomeron and the related problem of the rate of increase of total cross sections is a controversial subject.

A popular model, the so-called Lipatov Pomeron [3], is based on perturbative QCD calculations of a gluon ladder exchange which leads to a lower bound for the Pomeron intercept [see eq.(6)]. After unitarization this results in the saturation of the Froissart bound

$$\sigma_t \sim ln^2 s \tag{10}$$

and gives a value of $\sigma_t(\bar{p}p)$ at the Tevatron energy $\sqrt{s} = 1.8$ TeV which lies higher than the $E710$ and CDF preliminary values (see below).

In a series of papers [6], Donnachie and Landshoff (D-L) have treated the Pomeron intercept as a free parameter finding $\alpha_P(0) \approx 0.08$. With this small value of $\alpha_P(0)$ they get total cross sections compatible with the available data. In the D-L model, σ_t is well below the Froissart bound, which makes it possible to avoid the unitarization procedure over the range of present and next generation accelerators.

In yet another model for the Pomeron, based on a double pole exchange, [5], cross sections rise at a unit Pomeron intercept $\alpha_P(0) = 1$, which means that the Froissart bound is never violated. The Regge form of the input dipole Pomeron Ansatz [5] is

$$A(s,t) = ig^2 ln(-is/s_1)(-is/s_2)^{\alpha_P(t)-1}G(t) \qquad (11)$$

where $\alpha_P(t) = 1 + \alpha't$, $G(t) = \exp Bt$. Details of various properties of the DP model, possible modifications and generalizations can be found in the review paper [7] (see also references therein). The replacement, suggested in [7],

$$(s/s_0)^{\alpha(t)}ln(s/s_0) \longrightarrow (1 + s/s_0)^{\alpha(t)}ln(1 + s/s_0) \qquad (12)$$

and interpreted as the inclusion of a Pomeron daughter contribution avoids negative values of the amplitude for $s < s_0$. The above modification is very interesting as it is also directly related to a phenomenological parametrization of the structure functions.

We don't know of any convincing theoretical "derivation" of the DP model from first principles. It should, however, be stressed that this is true of all models, even of the simple Pomeron pole; from perturbative QCD a complicated j-plane structure emerges and a lower bound on the parameter δ of the intercept can be obtained [3].

A conventional argument against the DP used to be that the logarithmic rise predicted for the total cross section was not fast enough to fit the data. However, the recent preliminary measurements of the $\bar{p}p$ total cross section from the Tevatron for CDF and $E710$ at $\sqrt{s} = 1.8 GeV$ [8], i.e., respectively

$$\sigma_t(\bar{p}p) = 73.3 \pm 3.0$$

$$\sigma_t(\bar{p}p) = 72.0 \pm 3.6$$

favour now the logarithmic rise, rendering unlikely models based on the saturation of Froissart bound for σ_t.

Using the above arguments, we can now go back to the discussion of the gluon distribution. As the form of the latter can not be derived from first principles, there is a large literature on its phenomenological parametrization. A fairly general form for $xG(x)$ which incorporates scaling violations was recently studied in detail by Morfin and Tung [9]

$$xG(x, Q^2) = e^{A_0} x^{A_1}(1 - x)^{A_2}[ln(1 + 1/x)]^{A_3} \tag{13}$$

where

$$A_i(Q) = C_{i,0} + C_{i,1}T(Q) + C_{i,2}T^2(Q), (i = 0,...3)$$

$$T(Q) = ln[ln(Q/\Lambda)/ln(Q_0/\Lambda)], \quad Q_0 \approx 2GeV.$$

The above parametrization has been tuned to fit all the existing data for $x > 0.03$. Expression (13) as fitted in [9] (i.e. given the numerical values of the various coefficients), is singular at $x = 0$ on two counts: as a power and logarithmically.

Eq. (13) can be modified setting $\Lambda_1 = 0$ [cf. δ in cqs. (5, 6)] and using instead

$$xG(x, Q^2) = c^{B_0}(1 - x)^{B_1} ln[(1 + 1/x)]^{B_2} \tag{14}$$

the various parameters can be chosen in such a way as to fit very well the data for $x > 0.03$ and for various values of Q^2. For not too small values of x there is little difference between (13) and (14) but they deviate at very small $x < 0.1$.

As one can check, the Froissart bound $xG(x) \sim ln^2(1/x)$ is preserved by (14). Future experimental data to be collected from HERA at small x should definitely discriminate between different models for soft gluon distributions, e.g. between parametrizations (13) and (14).

In conclusion, in our approach we have related the unexplored behaviour of the structure functions at small x to the asymptotic behaviour of the total cross sections within a coherent picture of deep inelastic and soft collisions.

References

1. Proceedings of the Int. Conf. on Elastic and Diffractive Scattering; Nucl. Phys. (Proc. Suppl.) **B 12** (1990),1-142.

2. E. Leader; Phys. Lett. **B 253** (1991),457.

3. L. N. Lipatov; JETP **90**(1986) 1536.

4. J. C. Collins and J. Kwiecinski; Nucl. Phys. **B 316** (1989), 307.

5. L. L. Jenkovszky, E. S. Martynov, B. V. Struminsky; Phys. Lett. **B 249** (1990), 535.

6. A. Donnachie, P. V. Landshoff; Nucl. Phys. **B 267** (1986) 690.

7. L. L. Jenkovszky; Fortsch. d. Phys. **34** (1986), 751.

8. S. Shukla, E 710 Collab.; S. White, CDF Collab., see these Proceedings.

9. See, for instance, Wu-Ki Tunk, Proc. of the Workshop on *Parton Distribution Functions at small x*, DESY, May 1990; see also J. G. Morfin and Wu-Ki Tung, preprint Fermilab-Pub 90/74, April 1990.

10. L. L. Jenkovszky, F. Paccanoni and E. Predazzi; Proceedings on the IVth Blois Workshop, Elba (June 1992).

ABOUT SYSTEMS OF THREE PARTICLES WITH CHARGES $+1$ $+1$ -1
or
IS A "MÉNAGE À TROIS" STABLE? [1]

André Martin

Theory Division, CERN
CH-1211 Geneva 23, Switzerland

ABSTRACT

Depending on their masses, three particles of charge $+1, +1, -1$ can be bound or dissociated into two bound particles and a free particle. Jean-Marc Richard, Tai Tsun Wu and myself have shown that simple convexity considerations allow a unified treatment of this problem and make it possible to confirm, for instance, that the system proton-deuteron-muon is bound, while the system proton-electron-positron is not.

CERN-TH.6376/92
January 1992

---

[1] Talk given on the occasion of the 70th birthday of Professor Eduardo Caianiello

Primo, vorrei dire quanto felice sono di essere qui, per il compleanno di Eduardo Caianiello, et di potere dirgli "Many Happy Returns". Now, I must say that when I tried to find a subject for my talk, I was somewhat embarrassed, because my field of interest is rather far from quantum electrodynamics and from the problem of the functioning of the brain. Finally, I chose a subject which has the advantage of being simple and accessible to any person having followed an elementary course of quantum mechanics, which is that of finding out whether three particles of electric charge $+1, +1, -1$ form a bound state or not, a problem on which I have recently been working with Jean-Marc Richard and Tai Tsun Wu [1]. Everybody knows particular cases of this problem, for instance the fact that the molecular hydrogen ion is stable. Extensive calculations have been made in particular cases, but what we want to do here, which seems appropriate for this meeting, is to show that thinking with your brain may save a lot of computing time and a lot of energy.

Let me list known cases first:

- ppe^- is bound. See quantum chemistry books.

- $e^+e^-e^-$ is bound. This was first predicted by Wheeler [2] and proved experimentally by Mills [3] (nothing to do with Yang-Mills!), an experimentalist from Bell Labs, who happens to be a friend of my physicist son, Thierry.

- pe^-e^- is bound, but furthermore has only one bound state in which the two electrons form a spin zero according to Hill [4], and also only one bound state in which the two electrons form a spin one, according to Grosse and Pittner [5].

- $p\mu^-e^-$ is NOT bound. This was first proved by Wightman in his thesis [6]. Since then he has moved to axiomatic field theory.

- more generally, Grosse, Glaser, Thirring and myself [7] showed that a system made of an infinitely heavy proton, an electron, and an extra negative particle with mass $> 1.57m_e$ is unavoidably unstable.

In this list, you see that all systems in which the two particles with charges of the same sign have the same mass are stable. This has indeed been proved by Hill, using a $1S1S' + 1S'1S$ trial function [4].

What Jean-Marc Richard, Tai Wu and myself tried to do is to have a new look at the problem by considering "le cas le plus général" for masses, as Louis Michel would say, and using general properties of the Hamiltonian. The problem is that the negatively-charged particle attracts the two positively-charged particles, but they repel each other. It is a little like a "ménage à trois". Two men are in love with the same woman who attracts both of them, but they have repulsion for one another. Who will win? Will they all stay together? Those are the questions.

We will only consider the purely non-relativistic problem. The Hamiltonian of the system can be written as

$$H = \frac{p_1^2}{2m_1} + \frac{p_2^2}{2m_2} + \frac{p_3^2}{2m_3} - \frac{e^2}{r_{13}} - \frac{e^2}{r_{23}} + \frac{e^2}{r_{12}}. \tag{1}$$

The system will be stable if the energy $E_3(m_1, m_2, m_3)$ is strictly less than the energy of the lower two-body system, $(1,2)$ or $(1,3)$. If we assume

$$m_2 > m_3,$$

the two-body system with the lower energy will be $(1,2)$ with an energy

$$E_2(m_1, m_2) = -\frac{e^4}{2\left(\frac{1}{m_1} + \frac{1}{m_2}\right)} \tag{2}$$

It is more convenient to use variables entering linearly into the Hamiltonian, i.e.,

$$x_1 = \frac{1}{m_1} \quad x_2 = \frac{1}{m_2} \quad x_3 = \frac{1}{m_3}. \tag{3}$$

Then the energy of the ground state of the system possesses a concavity property: if $H(\lambda) = A + \lambda B$, the energy of the ground state $E(\lambda)$ of $H(\lambda)$ is concave in λ, i.e.,

$$\frac{d^2 E(\lambda)}{d\lambda^2} < 0. \tag{4}$$

Hence, denoting the energy by the same letter (a mathematical sin!):

$$\frac{d^2}{dx_i^2} E(x_1, x_2, x_3) < 0 \quad i = 1, 2, 3. \tag{5}$$

We also have the Feynman-Hellmann theorem, which says

$$\frac{d}{dx_i} E_3(x_1, x_2, x_3) = \langle \frac{p_i^2}{2} \rangle > 0. \tag{6}$$

Now, in these variables (2) becomes

$$E_2(x_1, x_2) = -\frac{e^4}{2(x_1 + x_2)}. \tag{7}$$

Now the problem of stability versus instability is scale invariant in the sense that we can replace m_i by λm_i or equivalently x_i by $\lambda^{-1} x_i$, and e^2 by μe^2. Therefore, without loss of generality, we can constrain the x_i's to have a sum equal to unity. Then we shall call them

$$\alpha_1, \alpha_2, \alpha_3 \quad \text{with} \quad \alpha_1 + \alpha_2 + \alpha_3 = 1. \tag{8}$$

Then, any three-body system can be represented by a point in a triangle, the distances to the sides of the triangle being the α's, as shown in Fig. 1. Since we assume $m_2 \geq m_3$, we have only to consider the left half of the triangle. From the result of Hill, previously mentioned, we have stability on the vertical line. We know that there are points with instability on the line $\alpha_1 = 0$. From Refs. [6] and [7] specifically we have instability for $\alpha_3/\alpha_2 > 1.57$.

We shall now prove that in the left half of the triangle, the domain of instability is star-shaped with respect to summit 3, which corresponds to $\alpha_2 = \alpha_1 = 0$. Indeed, consider (Fig. 2) a point P where the three-body system is unstable. We can follow the straight segment joining P to 3. Along this segment $\alpha_2/\alpha_1 = const.$, so that by rescaling we can use

$$x_1 = \frac{\alpha_1}{\alpha_1 + \alpha_2}, \quad x_2 = \frac{\alpha_2}{\alpha_1 + \alpha_2} \quad x_3 = \frac{\alpha_3}{\alpha_1 + \alpha_2}.$$

x_1 and x_2 are constants. As we move from 3 to P, x_3 increases and, by Feynman-Hellmann, the three-body binding energy should <u>increase</u> algebraically, while the binding energy of the two-body subsystem $(1,2)$ stays constant. If one already has instability in P, this remains therefore true on the whole segment.

A second property is that the <u>border</u> of the domain of instability is <u>convex</u>. Let two points, P' and P'', be on the border of the stability domain. Any point of the two-dimensional triangle defines a ray in the three-dimensional space $\vec{x}$:

$$\frac{x_1}{\alpha_1} = \frac{x_2}{\alpha_2} = \frac{x_3}{\alpha_3}$$

and we can intersect the rays associated to P' and P'' by a plane $x_1 + x_2 = const.$, defining in this way π' and π''. Naturally, $E_2(x_1', x_2') = E_2(x_1'', x_2'')$, since the two-body energy is a function of $x_1 + x_2$. On the other hand,

$$E_3(x_1', x_2', x_3') = E_2(x_1', x_2')$$

and

$$E_3(x_1'', x_2'', x_3'') = E_2(x_1'', x_2'').$$

Now, from the concavity property

$$E_3(\lambda \vec{x}' + (1 - \lambda)\vec{x}'') > E_3(\vec{x}') = E_3(\vec{x}''),$$

and from the constancy of $x_1 + x_2$,

$$E_2(\lambda x_1' + (1 - \lambda)x_1'', \lambda x_2' + (1 - \lambda)x_2'') = E_2(x_1', x_2') = E_2(x_1'', x_2'').$$

Hence any point of the segment $\pi'\pi''$ is <u>unstable</u>, and projecting back on the triangle, any point of $P'P''$ is unstable, and we get the qualitative behaviour shown in Fig. 2.

To make things quantitative, a minimum amount of calculations are needed, and what we have done is to use the calculations already done by others. First of all one can find, using concavity again, a minimum width for the stability band shown in Fig. 2, by using the fact that at many points on the diagonal $\alpha_2 = \alpha_3$, the binding energy is known or has an algebraic upper bound from variational calculations. Now concavity tells us that

$$E(\alpha_1, \frac{\alpha_2 + \alpha_3}{2}, \frac{\alpha_2 + \alpha_3}{2}) > \frac{1}{2}\left(E(\alpha_1, \alpha_2, \alpha_3) + E(\alpha_1, \alpha_3, \alpha_2)\right) = E(\alpha_1, \alpha_2, \alpha_3),$$

so that we can use the value on the diagonal to get an upper bound on the three-body energy and compare it with the two-body energy. In this way, one checks, for instance, that the system proton-deuteron-muon is bound, a fact which is important for muon catalysed fusion. One can also get a lower bound on the size of the instability region. We have already seen that we have instability if $\alpha_1 = 0, \alpha_3/\alpha_2 > 1.57$.

Another region where a calculation can be made is the vicinity of $\alpha_2 = \alpha_3 = 0$, where the two particles with positive charge are very heavy, and where the Born-Oppenheimer approximation can be used. I remind you that in the first step of the Born-Oppenheimer approximation, particles 2 and 3 are held fixed, and the energy of particle 1 is calculated or bounded (for instance by a variational method <u>and</u> by using Temple's inequality). Then, minimizing with respect to the distance between 2 and 3, one gets a lower bound for the three-body energy, since the kinetic energies of particles 1 and 2 are neglected. However, in this approach, the energy of particle 1 approaches $-\frac{m_1 e^2}{2}$, when the distance between 2 and 3 goes to infinity, instead of $-\frac{1}{\left(\frac{1}{m_1}+\frac{1}{m_2}\right)}\frac{e^2}{2}$. This makes it difficult to prove instability except in one case, which is $m_2 = \infty$, i.e., $\alpha_2 = 0$. This is what has been done (after the pioneering work of Larry Spruch) by Armour and Schrader [8], who have been able to prove instability for $\alpha_2 = 0, \alpha_1/\alpha_3 < 1.51$.

These two results allow us to find a triangle with guaranteed instability, $3XY$ in Fig. 3. This in turn allows us to settle problems which have caused much computer ink to flow, and also, for instance, to get a clear proof that the system proton-electron-positron is unstable. It is also the case for $p\mu^-\mu^+$, and in fact one sees geometrically in Fig. 3 that the system $A^+B^-B^+$ is unstable as soon as $m_{A^+} > 4.19\, m_{B^\pm}$.

In conclusion, returning to our "ménage à trois", we see that the ménage is stable if the two men have about the same weight, but unstable if one of them is much heavier. This is illustrated in Fig. 4, drawn by my climatologist son, Philippe, and sent to me by telefax from Sydney before this meeting. There is still one situation where one has stability. It is if the lady is very light with respect to the men, irrespective of their relative weights $(pd\mu^-)$. This is because she is a "femme légère".

Postscript

Upon reflection, one realizes that most of the considerations hold also for particles with different charges. The next simplest case is $Z_2 = Z_3$. Then symmetry with respect to the vertical bisector is preserved, and the star-shaped property of the instability domain, as well as its convexity, are preserved. What is remarkable is that for $Z_1 > Z_2 = Z_3$ all three-body systems, <u>for any mass</u>, are stable. This is because for α_1 and α_2 very small, particles 1 and 2 form a very compact structure with a net attractive charge at large distances. Particle 3 is attracted by a Coulomb-like potential and has one (in fact an infinity of) bound state(s). Then the star-like property shows that we have binding everywhere in the left half-triangle.

For $Z_1 < Z_2 = Z_3$, but sufficiently close to Z_2, we have the same qualitative picture as for $Z_1 = Z_2$, but when Z_1 is small enough, binding disappears both at $\alpha_1 = 0$, $\alpha_2 = \alpha_3 = \frac{1}{2}$ and at $\alpha_2 = \alpha_3 = 0$, and therefore <u>everywhere</u> by concavity. The case where all Z's are different is a little bit more complex, because what replaces the line $\alpha_2 = \alpha_3$ is the line where the binding energies of particles 1 and 2, and 1 and 3 are equal, which is given by

$$\frac{(Z_1 Z_2)^2}{1 - \alpha_3} = \frac{(Z_1 Z_3)^2}{1 - \alpha_2}$$

This line goes through the symmetric point of summit 1 with respect to the line $\alpha_1 = 0$. Many qualitative features remain, and for instance one has stability for any mass if Z_1 is larger than Z_2 and Z_3, but some numerical calculations are unfortunately needed to know the evolution of the stability domain when Z_1 decreases.

References

[1] A. Martin, J.M. Richard and T.T. Wu, preprint CERN-TH.6227/91 (revised version).

[2] J.A. Wheeler, *Ann. N.Y. Acad. Sci.* **48** (1946) 219.

[3] A.P. Mills, *Phys. Rev. Lett.* **46** (1981) 717.

[4] R.N. Hill, *J. Math. Phys.* **18** (1977) 2316.

[5] H. Grosse and L. Pittner, *J. Math. Phys.* **24** (1982) 1142.

[6] A.S. Wightman, Thesis, Princeton University (1949).

[7] V. Glaser, H. Grosse, A. Martin and W. Thirring, in "Studies in Mathematical Physics", E.H. Lieb, B. Simon and A.S. Wightman eds., (Princeton University Press, 1976) p. 169.

[8] E.A.G. Armour and D.M. Schrader, *Can. J. Phys.* **60** (1982) 581.

ppe⁻
pdμ⁻
STABLE
pe⁺e⁻
$\alpha_3=0$
α_3
α_2
e⁻e⁺e
pp̄e⁻
α_1
pμ⁻e⁻   pe⁻e⁻
$\alpha_1=0$

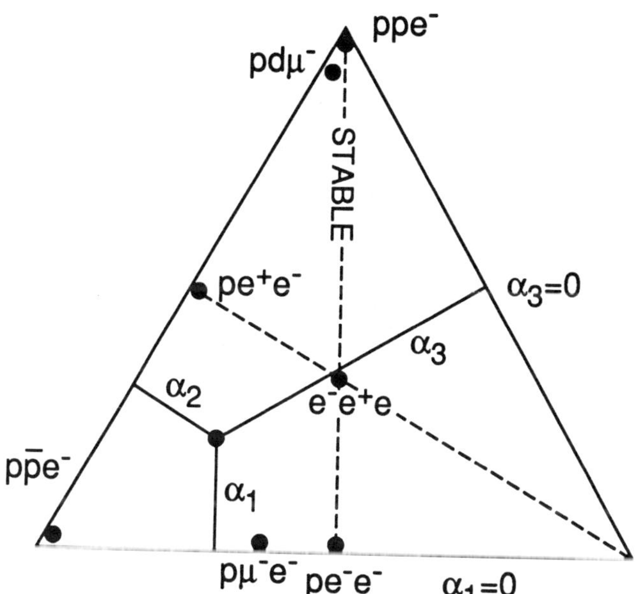

Fig. 1

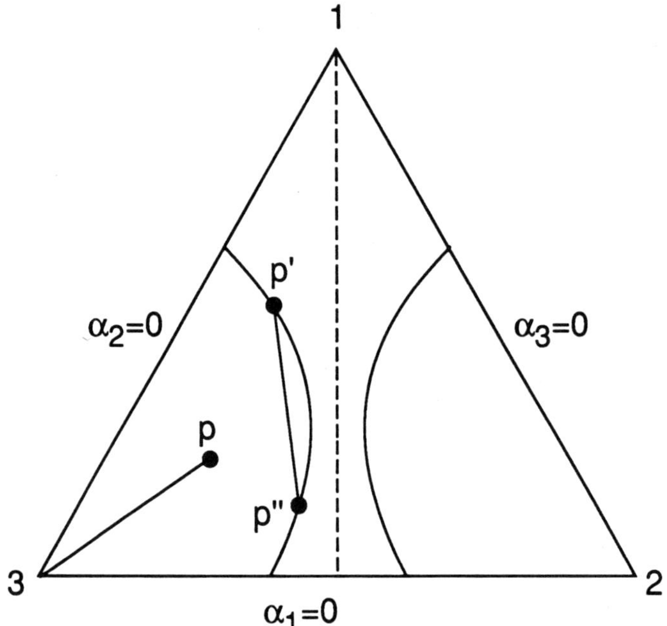

Fig. 2

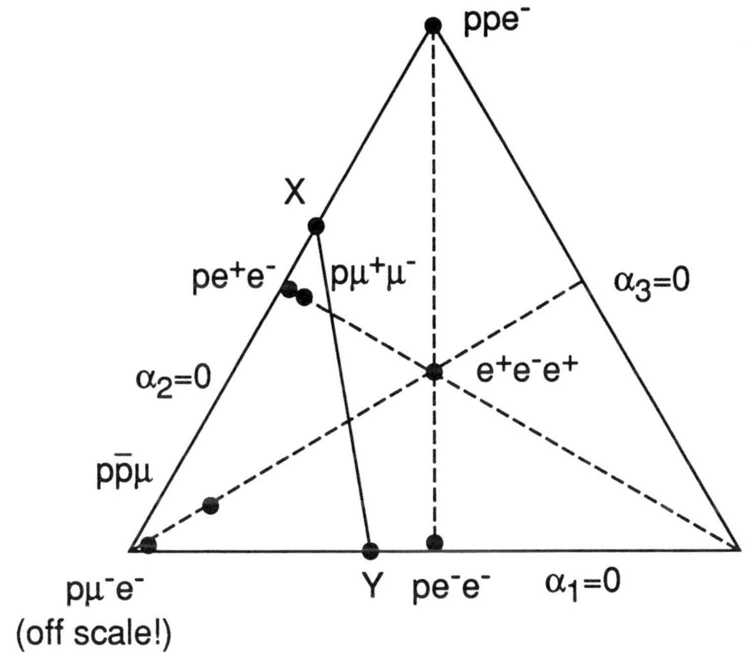

Fig. 3

Fig. 4

SIMPLE BEHAVIOUR OF MICROSCOPIC SYSTEMS

S. Fubini[1,2]   and   A. Molinari [2,3]

[1] *Theory Division CERN, Geneva, Switzerland*

[2] *Dipartimento di Fisica Teorica dell'Università di Torino, Torino, Italy*

[3] *Ministero Affari Esteri - Consolato Generale d'Italia - Boston, USA*

We dedicate this short essay to our old friend Eduardo Caianiello for his 70th birthday. We express our admiration for his wide and important contribution to science and for the rôle he has played in promoting international collaboration.

1.

The deepest level in the understanding of nature is achieved when it becomes possible to derive from general theoretical principles simple physical laws which are experimentally verified to a high degree of precision.

A very successful trend in physics is indeed to focus on *simple* microscopic systems, ranging from atoms to elementary particles, and to explore whether their behaviour can be described in the above mentioned conceptual framework. The wonderful experimental agreement of the theoretical predictions with the hydrogen atom spectrum, with the magnetic moments of the electron and of the muon and, more recently, the outstanding acheivement of electroweak unification are examples forcefully indicating that we are on the right track in carrying out this programme. This situation lends strong support to the reductionist philosophy, which can be summarized in the sentence *small is beautyful.*

However it came samewhat as a surprise at least to some of us, that simple physical laws in wonderful agreement with the experimental data are not an exclusive feature of the microscopic systems. Although this recognition was already achieved, at least partially, many years ago in the field of superfluidity and superconductivity, yet the number of examples of macroscopic systems exhibiting surprisingly simple behaviour appears to be rapidly increasing.

The instance of amazing agreement between simple theory and experiment were first met, as mentioned above, in the domain of superconductivity, e.g. in the Meissner effect, in the quantization of the magnetic flux and in the Josephson effect. In superfluidity as well, the quantization of the circulation of the superfluid current around a vortex line provided a similar example.

More recently the magic (10^{-8}) experimental agreement of the simple theoretical prediction [1].

$$\sigma = n\frac{e^2}{h}\,(n \text{ integer})$$

for the electrical conductivity in the quantum (integer) Hall effect, together with the discovery of the fractional Hall effect represent admirable and still somewhat surprising (at least in the eyes of the non-experts) analogous cases. Finally the new challenge for both theorists and exsperimentalists appears to be the unravelling of possible simple physical laws in the field of high T_c superconductivity.

In line with these considerations in this short essay we shall revisit in the framework of a simple bosonic model, the basic concepts (such as the spontaneously broken symmetry and the macroscopic occupation of a quantum level) which are believed to be the source of the exceptional success achieved by theoretical models in predicting some of the simple features macroscopically displayed by many-body systems.

Furthermore since at a microscopic level a many-body system is dealt with in a framework quite similar to the one adopted here, both approaches, e.g., heavily relaying on a still non completely understood ground state (*the vacuum*), it might be possible that macroscopic and microscopic simplicity share a common basis after all.

We hope that our elementary field theoretical treatment will shed light on this issue, providing at the same time a firmer ground for the understanding of some of the many problems we are still facing in the field of condensed matter physics.

2.

Let us consider a scalar boson field ψ, associated with particles of mass m, charge q and enclosed in a volume Ω. This fields of course interacts with the electromagnetic field $(A_i, V)(i = 1, 2, 3,)$ according to the model Lagrangian, valid in the non relativistic approximation†,

$$\mathcal{L} = \frac{1}{2m}(D_i\psi)(D_i\psi) + \\ -\frac{i}{2}\left[\psi^+(D_t\psi) - (D_t\psi)^+\psi\right] + U(\psi^+\psi) \tag{1}$$

D_i and D_t being the standard covariant derivatives

$$D_i = \frac{\partial}{\partial x_i} - iqA_i \\ D_t = \frac{\partial}{\partial t} + iqV \tag{2}$$

and $U(\psi^+\psi)$ the potential term (we set $\hbar = c = 1$).

In this Section, for the sake of simplicity, we shall consider, semi-classically, (A_i, V) as an *external field*, while in the next Section the electromagnetic field will be restored to its natural rôle of a fully dynamical variable, on the same footing as the field ψ.

---

† in the following we shall take $\hbar = c = 1$

When the occupation number of one (or of a few) quantum levels is macroscopically large our model is more conveniently treated by expressing the complex field ψ in terms of the *polar* coordinates χ and θ, both being of course function of space and time, according to

$$\psi = \chi e^{i\theta}. \tag{3}$$

The Lagrangian then reads

$$\mathcal{L} = \frac{1}{2m}\left(\partial_i \chi\right)\left(\partial_i \chi\right) + \chi^2 \left[\frac{1}{2m}q^2 \hat{A}_i^2 + q\hat{V}\right] + U\left(\chi^2\right) \tag{4}$$

where the fundamental *gauge invariant potentials* $\hat{A}_i$ and $\hat{V}$ are given by*

$$\hat{A}_i = A_i - \frac{1}{q}\frac{\partial\theta}{\partial x_i} \tag{5a}$$

$$\hat{V} = V + \frac{1}{q}\frac{\partial\theta}{\partial t} \tag{5b}$$

We can now introduce the momentum canonically conjugated to θ, namely

$$p_\theta = \frac{\partial\mathcal{L}}{\partial\dot\theta} = \chi^2 = \rho \tag{6}$$

and the current

$$j_i = \frac{\partial\mathcal{L}}{\partial A_i} = \frac{iq}{2m}\left\{\psi^+ D_i\psi - \psi(D_i\psi)^+\right\} = q^2\frac{\rho}{m}\hat{A}_i \ . \tag{7}$$

The above relation reduces to the familiar London equation of superconductivity when ρ can be viewed as a constant.

The canonical Poisson bracket of the variables ρ and θ is given by

$$[\rho(\vec{x}), \theta(\vec{y})] = i\delta^3(\vec{x} - \vec{y}) \tag{8}$$

---

* Note the characteristic field combination which appears in our Abelian Higgs mechanism. This result is sometimes referred to by saying that the photon field $\left(\vec{A}, V\right)$ *eats up* the Goldstone boson $\theta(x)$. In connection with this point see also ref. [1].

By introducing next the constant of the motion

$$N = \int_\Omega \rho(\vec{x})d\vec{x} \tag{9}$$

we can write

$$[N, \theta(\vec{y})] = i \quad , \tag{10}$$

a relation whose importance in the context of many-body physics has been much emphasized by P. Anderson [2] and which also plays a crucial rôle in quantum optics [3].

If we now want to introduce quantization and accordingly interpret the particle number N and the phase θ as operators, so that equations (8) and (10) become commutation relations, we have to be careful especially when the eigenvalues of N are small.

This point has been strongly emphasized in the simple example of an harmonic oscillator [4] [5] *.

In the case of macroscopic values of N the situation, however, is quite clear. Indeed equation (10) leads then to the inequality

$$\frac{\Delta N}{N}\Delta\theta \geq \frac{1}{2N} \quad . \tag{11}$$

In particular when the root mean square deviation ΔN and $\Delta\theta$ in (11) are computed in the coherent states framework their product is minimized. Correspondingly, for large N, the use of semiclassical considerations is warranted and very accurate values for both N and θ can be obtained.

Precisely this fact is the basis of the astonishing success achieved by simple theoretical predictions in superconductivity and superfluidity.

Indeed for a macroscopically large value of N/Ω one is naturally led to write

$$\rho(\vec{x}) = f + \sigma(\vec{x}) \tag{12}$$

---

* We wish to thank Marco Roncadelli for an illuminating discussion on this issue.

234

with

$$f = \frac{N}{\Omega} \tag{13}$$

and to condsider a *large N expansion* in powers of $\sigma(\vec{x})/f$.

In order to appreciate better how the latter is realized and to get a deeper insight into the meaning of $\sigma(\vec{x})$, a suitable way is to cast our formalism into the language of the Feynman path integrals. This approach will be outlined in the next Section.

3.

The programme referred to at the end of the previous Section is conveniently implemented in the grand canonical ensemble where the grand Hamiltonian

$$\hat{H} = H - \mu N \tag{14}$$

is introduced, μ being the chemical potential.

The dynamical variables relevant to the problem are of course the complex field ψ and the electromagnetic field (A_i, V). We furthermore assume, for sake of illustration, the potential term in the Lagrangian (1) to be of the form

$$U\left(\psi^+\psi\right) = \frac{b}{2}\left(\psi^+\psi\right)^2 \tag{15}$$

with b positive.

We can then write the path integral generating functional

$$\begin{aligned}
Z &= \int d\psi d\psi^+ \exp\left[-i\int \hat{\mathcal{L}} dt\right] \\
&= \int d\rho d\theta \exp\left[-i\int \hat{\mathcal{L}} dt\right]
\end{aligned} \tag{16}$$

where, in corrispondence to (4), the grand canonical Lagrangian density reads

$$\hat{\mathcal{L}} = \frac{1}{8m}\frac{\partial_i\rho\partial_i\rho}{\rho} + \rho\left[\frac{1}{2m}q^2\hat{A}_i^2 + q\hat{V}\right] - \mu\rho + \frac{b}{2}\rho^2 - \frac{1}{8\pi}(\vec{E}^2 - \vec{B}^2) \quad , \tag{17}$$

$\hat{A}_i$ and $\hat{V}$ being given by (5).

We next evaluate the generating functional using the stationary phase method. In searching for the saddle point we only keep in the Lagrangian (17) the binomial $-\mu\rho + (b/2)\rho^2$.

Accordingly we find that the main contribution to the functional integral (16) arises from

$$\rho \cong f = \frac{\mu}{b} \quad . \tag{18}$$

One thus sees that large values of the chemical potential μ, in connection with a moderate strenght b of the potential U, on the one hand ensure a high accuracy of the stationary phase evaluation of the generating functional and on the other entail a macroscopic occupation of some quantum levels (since, when μ is very large, so is f).

Furthermore by inserting (18) into (17), recalling (12) and retaining only the leading terms in the expansion in $\sigma(\vec{x})/f$, we obtain

$$\hat{\mathcal{L}} \Longrightarrow \hat{\mathcal{L}}_0 = \mathcal{L}_{e.m.} + \mathcal{L}_\sigma \tag{19}$$

where

$$\mathcal{L}_{e.m.} = f\left[\frac{1}{2m}q^2\hat{A}_i^2 + q\hat{V}\right] - \frac{\mu^2}{2b} - \frac{1}{8\pi}(\hat{E}^2 - \hat{B}^2) \tag{20}$$

and

$$\mathcal{L}_\sigma = \frac{1}{8m}\frac{\partial_i\sigma\partial_i\sigma}{f} + \frac{b}{2}\sigma^2 \quad . \tag{21}$$

Now, looking at the term (20), the key point to be realized is that, due to the extremely large value of f, all possible corrections arising from additional terms in the potential U, from quantum fluctuations around the saddle point or from the existence of impurities in the system are completely negligible. Indeed all these terms are not expected to exhibit the factor ρ (i.e. f), which indeed *protects* only the terms explicitly appearing in eq. (20).

This outcome turns out to be quite pleasant because the Meissner law and the flux quantization in superconductivity just stem from the term $\hat{A}_i^2$ in (20), whereas the elemetary laws of the Josephson effect are implied by the $\hat{V}$ term*

Concerning (21), it highlights remarkably the connection of our approach with the concept of ODLRO [2] (off-diagonal long range order). Indeed consider the boson propagator

$$G(\vec{x} - \vec{y}) = \langle \rho(\vec{x}) \rho(\vec{y}) \rangle \tag{22}$$

together with (12).

The since clearly

$$\langle \sigma(\vec{x}) \rangle = 0 \tag{23}$$

it follows that

$$\langle \rho(\vec{x}) \rangle = f \quad ; \tag{24}$$

therefore we obtain

$$G(\vec{x} - \vec{y}) = f^2 + \langle \sigma(\vec{x}) \sigma(\vec{y}) \rangle \tag{25}$$

Now (21) allows us to compute the last term on the RHS of the above relation yielding

$$\langle \sigma(\vec{x}) \sigma(\vec{y}) \rangle \propto e^{-r/\lambda} (r = |\vec{x} - \vec{y}|) \tag{26}$$

where the quantity $\lambda = \left(2\sqrt{mfb}\right)^{-1}$ is obviously related to the coherence lenght.

Equation (25) and (26) express the ODLRO as discussed in ref. [2].

In conclusion the aim of this contribution has **been to promote further thinking** on these fascinating aspects of nature: we hope that the simple approach we have developed, largely inspired by the work of Weinberg, Feynman and Anderson, might be of some help in the attempt to achieve a deeper understanding of some of the macroscopic quantum phenomena.

---

* On this point we received much ispiration by reading the article of S. Weinberg *Superconductivity for particle theorists* [6]. One of us (S. F.) is grateful to Val Telegdi for having drawn his attention to his beautyful paper.

References

[1] R.P. Feynman *Statistical Mechanics* , W.A. Benjamin, Inc., (1972) p. 303

[2] P.W. Anderson *Basic notions of condensed matter physics*, Addison-Wesley Publishing Company, (1984) p. 229

[3] R. London *The quantum theory of light*, Clarendon Press, (1983)

[4] L. Susskind and J. Glogower, Physics **1**, 49 (1964)

[5] P. Carruthers and M.M. Nieto, Rev. Mod. Phys. **40**, 411 (1968)

[6] S. Weinberg, Progr. Theor. Phys. Supplement **86**, 43 (1986)

ORDER CRITERIA IN THE FRACTIONAL QUANTUM HALL EFFECT[†]

G. Cristofano - G. Maiella - R. Musto - F. Nicodemi

Dipartimento di Scienze Fisiche - Università di Napoli

and INFN Sezione di Napoli

Mostra d'Oltremare Pad.19 - 80125 Napoli - Italy

Abstract

A natural description of a 2D system of electrons in a transverse magnetic field is achieved by using Coulomb gas conformal vertex operators. They provide an understanding of the Fractional Quantum Hall Effect in terms of quantized vortex excitations with finite energy and fractional charge, which represent the ordered structure at the basis of the phenomenon.

[†] It is a pleasure to dedicate this paper to E.R. Caianiello in honour of his 70th birthday.

§ 1. Introduction

As it is well known, the quantized Hall effect[1], namely the appearance of plateaux in the Hall conductance, σ_H, and of corresponding zeros in the longitudinal resistance ρ_{xx}, for a system of electrons confined to a 2D region and in the presence of an intense perpendicular magnetic field, occurs both for integral (IQHE) and for rational (FQHE) values of the electron's filling factor, ν. However, there appears to be a substantial difference beetwen the two phenomena, both on the experimental and on the theoretical side. Indeed, samples of lower mobilities show no indications of quantization at *any* noninteger value of ν. Instead, such samples develop extremely wide plateaux in σ_H and wide zero-resistance minima in ρ_{xx} at integral ν, which persist for a much wider interval in temperature than those present in the FQHE. The IQHE has a rather simple explanation in terms of non interacting electrons and of the filling of the single particle Landau levels, which, although broadened into a band, appear as the main energy structure of the system in the presence of an intense magnetic field. The occurence of the plateaux is then attributed to the location of the Fermi energy in a zone of localized states, while a change occurs when the Fermi energy is in a zone of conduction states. Indeed it has been shown that when the Fermi level is in a gap ρ_{xx} vanishes, while σ_H is simply proportional to the derivative with respect to the magnetic field, of the density of states below the Fermi level[2]. The Landau levels provide then the "order principle" responsible for the behaviour of the sample.

On the other hand, the FQHE is present only in very pure samples and at very low temperatures (of the order of the mK) implying a structure much more sensitive to the disorder and a much smaller energy gap for the excitations. It is also clear that the FQHE cannot be explained in terms of the single particle behaviour of the electrons and that it requires the formation of some kind of "collective excitations" (quasiparticles).

In this paper we discuss a proposal[3] on the role of 2D-Conformal Field Theory (2D-CFT) in describing the characteristic features of the FQHE, based on the use of Coulomb Gas Vertex[4] Operators. The main idea we will exploit is that the description of the 2D electron system based on 2D-CFT has to be valid not only on the plane but also on the torus and in general on an arbitrary Riemann surface. In doing so we can exhibit the topological properties that are implicit in the conformal approach and that are relevant for the FQHE. The main results that arise from this point of view[3], and that will be discussed in the following, are:

- the existence of a finite set of anyonic particles that, for fractional filling $\nu = 1/m$ leads to the m-fold degeneracy of the ground state on the torus; this degeneracy is required in order to implement modular covariance of the underlying 2D-CFT;

- the Laughlin-Jastrow (LJ) wave function[5] for the electrons and its natural generalization to the case of doubly periodic boundary conditions[6], are obtained as appropriate expectation value of Vertex operators; in principle this procedure can be extended to an arbitrary Riemann surface;

- the m-fold degenerate set of ground state wave functions on the torus carry an irreducible representation of a discrete subgroup of the magnetic translation group; in this context the vacuum degeneracy is a signal of the topological order of the quantum Hall fluid;

- the Hall conductance has the expected value $\sigma_H = 1/m$, in natural units e^2/h, corresponding to the ratio of the "electric" to the "magnetic" charge associated to the Vertex operator.

§ 2. The Coulomb gas Vertex Operators

The fractional QHE has provided the new exciting possibility of having direct experimental evidence of one of the more specific features of two dimensional physics, namely the existence of "anyons", i.e. particles obeying fractional statistics[7].

The point we wish to stress here is that in two spatial dimensions not only there exists the possibility of new values of the spin but also that fields of *any* statistics can be obtained out of a *scalar* field thanks to a *bosonization* procedure.

The main tool to be used in this procedure is the Vertex Operator, first introduced in the framework of the dual resonance model[8], and that has wide application in string theory and, more in general, in the study of 2D-CFT[9]. It is defined as:

$$V_\alpha(z) =: e^{i\alpha\phi(z)} :$$

(1)

where the field ϕ has the standard mode expansion:

$$\phi(z) = q - ip\ln z + \sum_{n\neq 0} \frac{a_n}{n} z^{-n}$$

(2)

and the coefficients satisfy the commutation relations:

$$[a_n, a_{n'}] = n\delta_{n,-n'} \qquad [q,p] = i$$

(3)

It is easy to see that V_α transforms in general as a non integer spin operator under rotation. Indeed the CR with the generator of rotations

$$[L_0, V_\alpha] = \left(z\frac{d}{dz} + \frac{\alpha^2}{2} \right) V_\alpha(z)$$

(4)

implies for a 2π rotation

$$e^{i2\pi L_0} V_\alpha(z) e^{-i2\pi L_0} = V_\alpha(z) e^{i\alpha^2\pi}$$

(5)

so that $V_\alpha(z)$ describes a particle of spin $\alpha^2/2$, corresponding to a boson (fermion) only for α^2 even (odd).

In order to show the relation between spin and statistics we start with the basic reduction formula

$$V_{\alpha_1}(z_1)V_{\alpha_2}(z_2) = (z_1 - z_2)^{\alpha_1\alpha_2} : V_{\alpha_1}(z_1)V_{\alpha_2}(z_2) :$$

(6)

obtaining the "braiding relation"

$$V_{\alpha_1}(z_1)V_{\alpha_2}(z_2) = e^{i\pi\alpha_1\alpha_2}V_{\alpha_2}(z_2)V_{\alpha_1}(z_1)$$

(7)

By taking $\alpha_1 = \alpha_2 = \alpha$ one finds

$$V_\alpha(z_1)V_\alpha(z_2) = e^{i\pi\alpha^2}V_\alpha(z_2)V_\alpha(z_1) \tag{8}$$

that, compared with eq. (5), shows the correct spin-statistics relation. The generalization of eq. (4) to arbitrary conformal transformations:

$$[L_n, V_\alpha(z)] = z^n \left(z\frac{d}{dz} + \frac{\alpha^2}{2}(n+1) \right) V_\alpha(z) \tag{9}$$

shows that $V_\alpha(z)$ is a primary field of conformal weight $\alpha^2/2$. As we shall see in details in the next paragraph a consistent description of the system at filling $\nu = 1/m$ is achieved by assuming that the chiral scalar field $\phi(z)$, given by eq. (2), is defined on S_1, i.e. it is compactified on a circle of radius R, with:

$$R^2 = m \tag{10}$$

As a consequence we are lead to introduce a set of m Vertex operators:

$$V_{\alpha_l}(z) =: e^{i\alpha_l\phi(z)} : \tag{11}$$

where $\alpha_l = l/\sqrt{m}, \ l = 1, 2, \ldots, m$, which realize the primary fields of a 2D-CFT, with action $\mathcal{A} = \dfrac{1}{2\pi} \int (\partial_z\phi)^2$ and central charge $c = 1$.

The conformal weights of the basic fields can be expressed in terms of an "electric", q_e, and a "magnetic", q_m, charge as[10] $\Delta + \bar{\Delta} = q_e^2 R^2/2 + q_m^2/2R^2$ and $\Delta - \bar{\Delta} = q_e q_m$.

Our vertices have conformal weights $\Delta = l^2/2m, \bar{\Delta} = 0$ and spin $s = \Delta$. Then, the consistency of our chiral conformal theory requires that $q_e/q_m = 1/R^2$. The choice of a rational value for R^2 ensures that the chiral theory can be realized with only a finite number of conformal blocks.

Writing again the general braiding relation (7) for the set of Vertex operators given by eq.(11):

$$V_{\frac{l}{\sqrt{m}}}(z_1)V_{\frac{p}{\sqrt{m}}}(z_2) = e^{i\pi lp/m} V_{\frac{p}{\sqrt{m}}}(z_2)V_{\frac{l}{\sqrt{m}}}(z_1) \tag{12}$$

we see that the braiding factor $exp[ilp\pi/m]$, can be thought as due either to a charge l/m going around a magnetic flux p or viceversa. To a generic anyonic vertex it is then associated an electric charge l/m and a magnetic charge l. In particular to the electron, corresponding to $l = m$, i.e. $\alpha = \sqrt{m}$, will be associated an electric charge 1 and a magnetic flux m.

§ 2. Quantum Hall Effect on a torus

The relevance of the Vertex operators for the FQHE[11], and more generally for anyon physics[12], is easily realized when one builds "many -body amplitudes", by taking appropriate expectation values of Vertex operators. Indeed, defining the state $|k>$ and its dual $< k|$ as:

$$< k|p = k < k| \qquad < k|a_n^+ = 0 \ , \quad p|k >= k|k > \qquad a_n|k >= 0 \tag{13}$$

$(n > 0)$, and using the reduction formula eq.(6), it is easy to see that, if $k = \sum_i \alpha_i$,

$$< k|V_{\alpha_1}(z_1)....V_{\alpha_M}(z_M)|0 > = \prod_{i<j=1}^{M} (z_i - z_j)^{\alpha_i \alpha_j} \tag{14}$$

When $\alpha_i = m$, $\forall i$, with m odd in order to have a fermionic operator, one obtains

$$< k|V_{\sqrt{m}}(z_1)....V_{\sqrt{m}}(z_{N_e})|0 > = \prod_{i<j=1}^{M} (z_i - z_j)^{m} \tag{15}$$

which is the *analytic part of the Laughlin wave function at filling* $\nu = 1/m$.

As we have mentioned in the previous section the requirement of a consistent description of the FQHE on an arbitrary Riemann surface fixes the possible values of the compactification radius for the ϕ field, and, as a consequence, the correct set of Vertex operators and corresponding anyonic particles. Here we will limit ourselves to the case of the torus.

In the framework of 2D-CFT the many-body correlation functions on a genus g surface, Σ, may be cast in a very simple form. In fact they can be expressed in terms of conformal blocks as:

$$< V_{\alpha_1}(w_1)....V_{\alpha_M}(w_M) >_l^g =< k|V_{\alpha_1}(w_1)....V_{\alpha_M}(w_M)|g;l > \tag{16}$$

where w_i are a local holomorphic parametrization of Σ and the genus g "vacuum" states[13], $|g;l >$, are annihilated by all destruction operators on Σ, which are defined locally by

$$a[f] = \oint_Q f(z)\partial\phi(z)$$

where $f(z)$ are holomorphic on $\Sigma - Q$.

We will not derive here this result, but we recall that it can be obtained using a technique, well known both in string and 2D-CFT, based on a sewing procedure, which allows to build any N-point correlation function on a Riemann surface of genus g (i.e. a compact surface with g handles) starting from a $(N+2)$ correlation function on a $g-1$ genus Riemann surface.

The explicit form of an arbitrary anyonic ground state wave function on a square torus of side L is then found to be[3,13]:

$$< V_{\frac{p_1}{\sqrt{m}}}(w_1)\ldots V_{\frac{p_M}{\sqrt{m}}}(w_M) >_l^{g=1} = \prod_{i<j=1}^{M} \left[\frac{\Theta_1(w_{ij}|i)}{\Theta'_1(0|i)}\right]^{\frac{p_i p_j}{m}} \Theta\left[\begin{array}{c} l/m \\ 0 \end{array}\right]\left(\frac{Wm}{L}\Big|im\right) , \tag{17}$$

where w_i are torus variables related to the plane variables z_i by means of $w_i = (L/2\pi i)\, ln\, z_i$, $w_{ij} = (w_i - w_j)/L$ and $W = \sum_{i=1}^{M} \frac{p_i w_i}{m}$ is the center of "charge" coordinate. For the description to be consistent the total flux pearcing the surface of the torus must be an integral multiple, N_s, of the elementary flux $\phi_0 = hc/e$ or, in magnetic units $\lambda^2 = \hbar c/eB$, $L^2 = 2\pi N_s$.

The N_e-electron wave function is then obtained by taking all p_i equal to m in eq.(17). This leads to the amplitude:

$$f_l(w_1, w_2 \ldots w_{N_e}) = \prod_{i<j=1}^{N_e} \left[\frac{\Theta_1(w_{ij}|i)}{\Theta'_1(0|i)}\right]^m \Theta\left[\begin{array}{c} l/m \\ 0 \end{array}\right]\left(W\frac{m}{L}\Big|im\right) . \tag{18}$$

recovering the proposal made in ref.[5]. As it could have been expected, the quantity entering the N_e-electron wave function, eq.(18), is the prime form[15]:

$$E(w_i, w_j) \equiv \frac{\Theta_1(w_{ij}|i)}{\Theta'_1(0|i)} \tag{19}$$

whose logarithm gives the singular part of the Coulomb Green function $G(w_{ij}) = <\phi(w_i)\phi(w_j)>$ on the torus. This is the analog of the factor $(z_i - z_j)$, appearing in eq.(14), whose logarithm gives the Coulomb Green function on the plane.

Finally we observe that the compactification radius $R^2 = m$ is the correct one only at filling $\nu = N_e/N_s = 1/m$. In fact this value of the filling is required in order that the set of functions given by eq.(18) satisfy the correct periodicity conditions in each electron variable. Furthermore, only at this filling[3] the m-fold degenerate ground state wave functions transform covariantly under the modular transformations $\tau \to \tau + 2$ and $\tau \to -1/\tau$, reflecting the gauge invariance of the problem.

§ 3. The magnetic translation group

A better understanding of the m-fold degeneracy of the ground state is obtained observing that, when the correct filling condition $N_s = mN_e$ is fulfilled, the wave functions, given by eq. (18), provide an m-dimensional irreducible representation of a discrete subgroup of the *total* magnetic translation group[15]. This group concerns charged particles moving in a constant uniform magnetic field. In that case the physical situation is clearly translation invariant, but, since in the Hamiltonian enters the vector potential, a translation of the wave function must be supplemented by a gauge transformation. The symmetry operations corresponding to both transformations are known as "magnetic translations". Their crucial feature is that magnetic translations along different directions do not commute. In the gauge $\vec{A} = By\hat{x}$ the action of the total magnetic group on the holomorphic part of the wave function is generated by:

$$S_a = \prod_{i=1}^{N_e} S_a^i \qquad T_b = \prod_{i=1}^{N_e} T_b^i \tag{20}$$

where S_a^i and T_b^i are defined as[15,16]:

$$S_a^i f(w_1, \ldots w_i, \ldots w_{N_e}) = f(w_1, \ldots w_i + a, \ldots w_{N_e})$$

$$T_b^i f(w_1, \ldots w_i, \ldots w_{N_e}) = e^{-b^2/2 + ibw_i} f(w_1, \ldots w_i + ib, \ldots w_{N_e}) . \tag{21}$$

It is then easy to show that for transformations with finite steps $a = b = \dfrac{L}{N_s}$ one has:

246

$$S_{\frac{L}{N_s}} \, f_l(w_1, \ldots w_{N_e}) = e^{i2\pi \frac{l}{m}} \, f_l(w_1, \ldots w_{N_e})$$

$$T_{\frac{L}{N_s}} \, f_l(w_1, \ldots w_{N_e}) = f_{l+1}(w_1, \ldots w_{N_e}) . \tag{22}$$

Therefore the m-fold degeneracy of the N_e-electron wave function on the torus is due to the non-commutative structure of the magnetic translation group:

$$S_{\frac{L}{N_s}} \, T_{\frac{L}{N_s}} = e^{i\frac{2\pi}{m}} \, T_{\frac{L}{N_s}} \, S_{\frac{L}{N_s}} . \tag{23}$$

The phase factor present in eq.(23) can be thought as the change of phase obtained when the N_e electrons are taken around the elementary plaquette of side L/N_s.

In the language of the underlying conformal theory the operators S_{L/N_s} and T_{L/N_s} can be realized in terms of the chiral field $\phi(z)$ as[3,13]

$$S_{L/N_s} = \exp\left(\frac{1}{R} \oint_A \partial\phi(z)\right) \qquad T_{L/N_s} = \exp\left(\frac{1}{R} \oint_B \partial\phi(z)\right) , \tag{24}$$

where A and B are the two homology cycles of the torus and R is the compactification radius.

Finally, we can see[3] that the Hall conductance is given by $\sigma_H = 1/m$, a result implicit in our formalism as the ratio of electric to magnetic charge is exactly $1/m$. Indeed, in order to evaluate the response of the system to an external electric field let us imagine to insert parallel to the B cycle of the torus a solenoid and to change adiabatically its magnetic flux by $\Phi = \alpha \, \phi_0$. In the new situation, the periodicity condition of the wave function along the A cycle is changed, as it must acquire a phase factor $e^{i2\pi\Phi/\phi_0}$ in each variable, and the ground state of the system will then be given by

$$f_{l,\alpha}(w_1, w_2 \ldots w_{N_e}) = \prod_{i<j=1}^{N_e} \left[\frac{\Theta_1(w_{ij}|i)}{\Theta'_1(0|i)}\right]^m \Theta\left[\begin{array}{c} (l+\alpha)/N_s \\ 0 \end{array}\right]\left(W\frac{m}{L}\Big|im\right) =$$

$$= T_{\alpha\frac{L}{N_s}} \, f_l(w_1, w_2 \ldots w_{N_e}) \tag{25}$$

showing that the insertion of a flux $\alpha\phi_0$ along the B-cycle is equivalent to a T-translation of a step $\alpha\frac{L}{N_s}$. This statement allows us to evaluate the conductance.

As the translation is along the B cycle, while the induced electric field is along the A cycle, we have zero longitudinal conductance. The Hall conductance can then be obtained by the Faraday law $\Delta\Phi = \sigma_H^{-1}Q$. Indeed for $\Delta\Phi = \alpha\phi_0$ a charge $Q = \alpha N_e/N_s$ crosses the A cycle of the torus, so that $\sigma_H = 1/m$ in natural units e^2/h.

§ 4. Final Remarks

In this paper we have been trying to show that an explicit realization of 2D-CFT in terms of Coulomb gas-like Vertex operators provide a natural language for describing the FQHE and that it incorporates in a simple and consistent formalism many theoretical ideas that have been recently advocated. We wish to stress that the success of an approach based on 2D-CFT can be traced back to the presence of a novel kind of long range order hidden in the $\nu = 1/m$ LJ wave function. In fact, it is possible to exhibit the order structure of the system, by allowing a singular gauge transformation that maps the fermionic wave function into a bosonic wave function[17]. This order is related to the peculiar nature of two dimensional objects entering the description of the FQHE as "particles" carrying both electric and magnetic charge, a property that is automatically encoded, as we have seen, in the Vertex operator formalism. Our approach leads naturally to a description of 2D electrons in a transverse magnetic field as an ordered system. It is then no surprise that we find as a consistency condition the relation $N_s = mN_e$. As m is the "magnetic charge" of the statistical field attached to the the electron vertex such relation realizes algebraically the average balance between the external magnetic field and the statistical one that can be taken as the origin of the gap in the excitation spectrum at filling $\nu = 1/m$. In our language the origin of the gap can be traced back to the fact that, in an average field sense, the "elementary" anyons of charge $1/m$ and statistical factor $\theta = \pi/m$ fill up exactly m Landau levels. There is then a definite energy gap (of the order of $\hbar\omega$) before the next lot of available states which mast be in the next level.

We hope that our discussion of the FQHE based on 2D-CFT will help to elucidate the order principle underlying this exciting new physics.

References

1. For the general aspects of Q.H.E., see e.g.:

   The Quantum Hall Effect, R.E. Prange and S.M. Girvin, eds., Springer, New York 1987;

   A. H. MacDonald, *Quantum Hall Effect: a Perspective*, Jaca Book, Milano 1989.

   G. Morandi, *The Quantum Hall Effect*, Bibliopolis, Napoli, 1988

2. P. Streda, *J. Phys. C: Solid State Phys.* **15** (1982), 207

3. G. Cristofano, G. Maiella, R. Musto and F. Nicodemi, *Phys. Lett.* **B262** (1991) 88; *Mod. Phys. Lett.* **A6** (1991) 1779; *Mod. Phys. Lett.* **A6** (1991) 2985.

4. V.S. Dotsenko, V.A. Fateev, *Phys. Lett.* **B134** (1985) 291.

5. R.B. Laughlin, *Phys. Rev. Letters* **50** (1983) 1395.

6. F.D.M. Haldane and E.H. Rezayi, *Phys. Rev.* **B31** (1985) 2529.
   R.B. Laughlin, *Ann. of Phys.* **191** (1989) 163.

7. F. Wilczek, *Phys. Rev. Lett.* **49** (1982) 957; F. Wilczek and A. Zee, *Phys. Rev. Lett.* **51** (1983) 2250.
   For a review see: F. Wilczek, *States of Anyon Matter*, IASSNS-HEP-90/29 preprint.

8. S. Fubini and G. Veneziano, *Nuovo Cimento* **67A** (1990) 29, *Ann. of Phys.* **63** (1970) 12.

9. For a review on the subject see: P. Ginsparg, *Les Houches*, (1988), Vol. **XLIX**, D. Brezin and J. Zinn-Justine eds.

10. P. Di Francesco, H. Saleur and J.B. Zuber, *J. Stat. Phys.* **49** (1987) 57.

11. See e.g. S. Fubini and C.A. Lütken, *Mod. Phys. Lett.* **A6** (1991) 487.

12. G.V. Dunne, A. Lerda and C.A. Trugenberger, MIT preprint CPT # 1938 (1991).

13. R. Dijkgraaf, E. Verlinde and H. Verlinde, *Comm. Math. Phys.* **115** (1988) 649.
    See also G. Cristofano, R. Musto, F. Nicodemi and R. Pettorino, *Phys. Lett.* **B211** (1988) 417.

14. G. Cristofano, G. Maiella, R. Musto and F. Nicodemi, *Phys. Lett.* **237B** (1990) 379;

15. Yong-Shi Wu, *Topological Aspects of the Quantum Hall Effect* in Proceedings Banff 1989-Edited by H.C.Lee

16. D. Mumford, *Tata Lectures on Theta* I, Birkhäuser, 1983.

17. S.M. Girvin and A.H. Mac Donald, Phys. Rev. Lett. **58** (1987) 1252; S.C. Zhang, T.H. Hansen and S. Kivelson, Phys. Rev. Lett. **62** (1989) 82; N. Read, Phys. Rev. Lett. **62** (1989) 86.

FERMION SYSTEMS WITHOUT FERMION VARIABLES ?

Sergio De Filippo

Dipartimento di Fisica Teorica
Università di Salerno
Baronissi, 84081, Salerno, Italy
and
I.N.F.N. Napoli

Abstract

The pseudo-classical limit of fermion systems is considered. The corresponding super-Hamiltonian systems are shown to be equivalent to hierarchies of ordinary Hamiltonian systems. Space supersymmetries are correspondingly recast in terms of hierarchies of ordinary Hamitonian symmetries.

1. Introduction and motivations

In spite of the astonishing success of the standard model in organizing and partially explaining an enormous amount of experimental data, non-perturbative aspects of QCD are mainly the object of conjectures and speculations. This is not only true for non-perturbative quantities like the masses and spettroscopies even of the 'hydrogen atoms' of QCD, namely pions, kaons, charmonium, and so on, but also for the crucial qualitative hypothesis of colour confinement. Numerical methods, namely Monte Carlo simulations, should then play a new fundamental role, since they are expected to substantiate the claims of QCD and *prove* its very consistency [1].

As well known, Monte Carlo methods in quantum field theory are used to approximate (Euclidean) path integrals for lattice versions of the considered model. In fact, for a pure gauge theory, one is confronted with the evaluation of the mean value of functionals $F[A]$, (where A denotes the gauge field, and omitted colour, vector and space-time indices are implied), given by multiple integrals in a very large number of variables, which, according to standard notation, read

$$\langle F[A] \rangle \equiv \frac{\int \mathcal{D}[A] F[A] \exp\{-S_{p.g.}[A]\}}{\int \mathcal{D}[A] \exp\{-S_{p.g.}[A]}\tag{1}$$

Here $S_{p.g.}[A]$ may be either the Wilson gauge invariant action, or a naively discretized variant. These expressions can be generalized to fermion fields ψ [2], according to

$$\langle F[\overline{\psi}, \psi, A] \rangle \equiv \frac{\int \mathcal{D}[A] \mathcal{D}[\overline{\psi}] \mathcal{D}[\psi] F[\overline{\psi}, \psi, A] \exp\{-S[\overline{\psi}, \psi, A]\}}{\int \mathcal{D}[A] \mathcal{D}[\overline{\psi}] \mathcal{D}[\psi] \exp\{-S[\overline{\psi}, \psi, A]\}}\tag{2}$$

although fermion integration is algebraic in character, and then eq. (2), as it stands, is not suitable for numerical simulations. One then exploits the feature of the QCD action, (and, due to the requirement of perturbative renormalizability, of most of the proposed models of fundamental interactions,) of being quadratic in fermion fields, i.e.

$$S[\overline{\psi}, \psi, A] = \overline{\psi} \Delta[A] \psi + S_{p.g.}[A],\tag{3}$$

where the fermion kernel $\Delta[A]$ is matrix valued and summation is implied on colour, flavour, spinor and space-time indices. This structure allows for a preliminary *gaussian integration* on fermion variables, in terms of well known formulas, like

$$\int \mathcal{D}[\overline{\psi}] \mathcal{D}[\psi] \exp\{-\overline{\psi} \Delta[A] \psi\} = \det \Delta[A]\tag{4a}$$

$$\int \mathcal{D}[\overline{\psi}] \mathcal{D}[\psi] \overline{\psi}_j \psi_k \exp\{-\overline{\psi} \Delta[A] \psi\} = \{\Delta[A]\}_{jk}^{-1} \det \Delta[A]\tag{4b}$$

and so on, which are the algebraic generalization of the gaussian integral formulas for ordinary (i.e. bosonic) variables, the presence of the determinant instead of its inverse being obvious just on purely dimensional grounds, (fermionic integration acts as derivation). Once fermion integration is performed, one is reduced to an ordinary bosonic functional integral, which, looking for simplicity at the denominator in eq. (2), acquires a seemingly very similar form to its analogue in eq. (1), the only difference being the replacement of the pure gauge action $S_{p.g.}[A]$ by the effective action

$$S_{eff}[A] \equiv S_{p.g.}[A] - \text{Tr} \ \ln \ \Delta[A], \tag{5}$$

at least if $\Delta[A]$ is positive definite or its eigenvalues all have even multiplicity. This similarity however is misleading as to the actual evaluation of the boson integral, since this effective action is highly non-local, which makes Monte Carlo simulations exceedingly more time-consuming than for the pure gauge case. In spite of the ingenious tricks devised to replace the numerical simulations directly corresponding to such non-local Euclidean actions, with more tractable ones [3], reliable Monte Carlo simulations for QCD are still out of reach for the available processing speed.

The difficulties mentioned above can be traced back to the lack of a measure-theoretic notion of Feynman integral for fermion degrees of freedom, which in turn is connected with the long-standing problem of defining a more or less formal classical limit for fermion systems [4]. In particular the emergence of super-string theory [5] made very popular the concept of a (pseudo-)classical super-dynamical system, as a system with both traditional bosonic and anticommuting fermionic variables [6].

In the present paper a possible scheme replacing pseudo-classical super-Hamiltonian fermion theories with hierarchies of ordinary Hamiltonian theories, is proposed. The possibility to apply it in order to devise an alternative path integral quantization, hopefully leading to effective Monte Carlo simulation methods, is under investigation. Technicalities and detailed proofs [7] will be skipped, while physical motivations and possible applications will be stressed.

2. Pseudo-classical mechanics

Consider a quantum system with both boson and fermion degrees of freedom, whose operator algebra is generated by two families, respectively of boson

$$(b_j)_{j\epsilon B}, \quad (b_j^\dagger)_{j\epsilon B}, \quad [b_j, b_k^\dagger]_- = \hbar\delta_{jk}, \quad [b, b]_- = [b^\dagger, b^\dagger]_- = 0, \tag{6}$$

(where here and henceforth omitted indices are arbitrary) and fermion operators

$$(f_j)_{j\epsilon F}, \quad (f_j^\dagger)_{j\epsilon F}, \quad [f_j, f_k^\dagger]_+ = \hbar\delta_{jk}, \quad [f, f]_+ = [f^\dagger, f^\dagger]_+ = 0, \tag{7}$$

which, according to the usual convention, are assumed to commute between themselves,

$$[b, f]_- = [b^\dagger, f]_- = 0 \tag{8}$$

The associative algebra generated by these operators is endowed with a Z_2 grading g, recursively defined by

$$\mathbf{g}(b) = \mathbf{g}(b^\dagger) = 0, \quad \mathbf{g}(f) = \mathbf{g}(f^\dagger) = 1, \quad \mathbf{g}(AB) = \mathbf{g}(A) + \mathbf{g}(B)(mod2), \tag{9}$$

and with a super-commutator, defined according to

$$[A, B] = AB - (-1)^{\mathbf{g}(A)\mathbf{g}(B)}BA \tag{10a}$$

for pure elements [1] , i.e. of definite grading, and linear w.r. to both arguments,

$$[\alpha A + \beta B, C] = \alpha[A, C] + \beta[B, C],$$
$$[C, \alpha A + \beta B] = \alpha[C, A] + \beta[C, B], \quad \alpha, \beta\epsilon C \tag{10b}$$

The supercommutator then is graded antisymmetric, i.e.

$$[A, B] = -(-1)^{\mathbf{g}(A)\mathbf{g}(B)}[B, A], \tag{11}$$

satisfies the graded Jacobi identity

$$(-1)^{\mathbf{g}(A)\mathbf{g}(C)}[A, [B, C]] + (-1)^{\mathbf{g}(C)\mathbf{g}(B)}[C, [A, B]] + (-1)^{\mathbf{g}(B)\mathbf{g}(A)}[B, [C, A]] = 0 \tag{12}$$

---

[1] Henceforth pure elements are always considered implicitly in otherwise meaningless equations.

and is a c-map, i.e.

$$g([A, B]) = g(A) + g(B)(mod2); \qquad (13)$$

otherwise stated the operator algebra is a graded Lie algebra. Furthermore the super-commutator is a graded derivation, namely

$$[AB, C] = A[B, C] + (-1)^{g(B)g(C)}[A, C]B, \qquad (14)$$

this property establishing a link between the associative and the graded-Lie products.

The (reverse) Dirac prescription, inducing a Poisson structure on the classical commutative associative algebra in the purely bosonic case, can be generalized to the present case, leading to a super-Poisson structure in the corresponding pseudo-classical graded-commutative associative algebra, i.e such that

$$AB = (-1)^{g(A)g(B)}BA \qquad (15)$$

To be specific, a super-Poisson bracket is defined according to

$$\{a_j, a_k\} \equiv (i\hbar)^{-1}[a_j, a_k], \qquad (16)$$

where a_j, a_k are either boson or fermion generators, sharing with the super-commutator the same formal properties of being a graded antisymmetric bilinear c-map fulfilling the graded derivation property and graded Jacobi identity

As a class of specific examples, fermion extensions of the KdV equation will be used here in order to illustrate this algebraic pseudo-classical setting. The KdV equation [8] is quite popular due to its relevance in the quantization of the Liouville equation, which in turn is relevant to the quantization of the Polyakov string below the traditional critical dimension [9]. The generalization of the known relation between the KdV equation and the Virasoro algebra [10] to the integrable space-supersymmetric KdV equation and the super-Virasoro algebra aroused considerable interest in super-KdV equations [11].

The proposed fermion extensions of the KdV eq. all have the general form

$$u_t = 6uu_x - u_{xxx} + e\varphi\varphi_{xx} \ , \ \varphi_t = f\varphi_{xxx} + gu_x\varphi + hu\varphi_x, \qquad (17)$$

where t and x subscripts, (which are replaced in the following, where notationally convenient, by dot ($\dot{}$) and prime superscripts ($'$)), respectively denote time and space derivatives, while u and φ are a bosonic and fermionic field respectively, and e, f, g, h are constant parameters (ordinary numbers). In particular the choice

$$e = 3, \quad f = -4, \quad g = 3, \quad h = 6 \tag{18}$$

gives the earliest version, which admits two different super-Hamiltonian realizations [12]. One of them corresponds to the super-Poisson brackets

$$\{u(x), u(y)\}_1 = \delta'(x - y), \{u(x), \varphi(y)\}_1 = 0, \{\varphi(x), \varphi(y)\}_1 = a\delta(x - y) \tag{19}$$

with $a = 1$, and the super-Hamiltonian functional

$$H_1[u, \varphi] = \int dx(u^3 + u_x^2/2 + su\varphi\varphi_x + t\varphi\varphi_{xxx}), \tag{20}$$

with $s = 3$, $t = -2$. The alternative Hamiltonian realization is given by the super-Poisson brackets

$$\{u(x), u(y)\}_2 = -\delta'''(x - y) + 2u'(x)\delta(x - y) + 4u(x)\delta'(x - y)$$

$$\{u(x), \varphi(y)\}_2 = \varphi'(x)\delta(x - y) + 3\varphi(x)\delta'(x - y);$$

$$\{\varphi(x), \varphi(y)\}_2 = c(-\delta''(x - y) + u(x)\delta(x - y)) \tag{21}$$

with $c = 4$ and, for $b = 1$, by the super-Hamiltonian

$$H_2[u, \varphi] = \int dx \frac{\left(u^2 + b\varphi\varphi_x\right)}{2} \tag{22}$$

The two space-supersymmetric versions of the KdV eq. are given by eq. (17) with

$$e = -3, \quad f = -1, \quad g = 3, \quad h = 3 \tag{23a}$$

$$e = -2, \quad f = -1, \quad g = 2, \quad h = 4 \tag{23b}$$

respectively [11,13]. The former is super-Hamiltonian with super-Poisson brackets given by eq.s (21) with $c = -1$ and super-Hamiltonian functional corresponding to H_2 in eq. (22) with $b = -1$, while the latter admits a super-Hamiltonian realization with super-Poisson brackets and super-Hamiltonian functional given respectively by eq.s (19) with $a = -1$ and H_1 in eq. (20) with $s = -2$, $t = 1/2$.

3. A constructive approach to pseudo-classical mechanics

As to the generalization of classical dynamics mentioned above, several attitudes are possible. The simplest choice is to work in a purely algebraic setting in analogy to the Berezin approach to the Feynman path integral on fermion variables. In this context the formulation of classical dynamics in terms of derivations on the ring of smooth functions on a given phase manifold, or related settings, can be taken as the starting point $[14]^2$. It is then quite natural to generalize this notion of the classical dynamical system, just taking more general non-Abelian rings as dynamical variable sets. In particular a Z_2 graded ring $\mathcal{F}$, is presumably the most general arena for super-dynamics. However this extremely general axiomatic context has several limitations, the most severe being the lack, without further specifications, of the notion of flow corresponding to a given super-vector field (derivation). Furthermore, as mentioned above, it deprives path integral quantization of its strongly advocated probabilistic interpretation, which is already present in original Feynman conception [16] and, strictly speaking, is rigorously established for its Euclidean version [17].

A less abstract setting, which can be considered in a sense the opposite constructive viewpoint, is presented below; it will also be shown, by a specific example, that super-Hamiltonian supersymmetries can be recast as ordinary Hamiltonian symmetries.

To fix language and notation, consider the traditional context in which super-dynamics is usually formulated. The implied graded ring $\mathcal{F}$ is generated by two families $(u_\alpha)_{\alpha \in \beta}$, $(\varphi_\beta)_{\beta \in F}$ of pure elements

$$\mathbf{g}(u_\alpha) = 0, \ \alpha \in B; \quad \mathbf{g}(\varphi_\beta) = 1, \ \beta \in F, \tag{24}$$

---

^{2} A related aspect consists of the algebraic characterization of tangent and cotangent bundles and correspondingly of their Lagrangian and Hamiltonian vector fields [15].

respectively corresponding to bosonic and fermionic degrees of freedom. The families B and F of bosonic and fermionic indices, (which are in principle independent, except for super-symmetric theories), can be either finite or infinite; in particular they are formally considered non-denumerably infinite if field theories are involved. Time evolution is then defined by putting

$$\dot{u}_\alpha = X_\alpha(\mathbf{u},\varphi), \ \alpha \in B, \quad \dot{\varphi}_\beta = Y_\beta(\mathbf{u},\varphi), \ \beta \in F \tag{25}$$

where X_α, Y_β are in general formal series in their arguments and mostly just polynomials, while here and henceforth bold characters $\mathbf{u}$ and φ denote the whole families of Bose and Fermi variables. The above equations are usually considered as transcriptions in local coordinates of a dynamical equation for a vector field globally defined (or in principle globally definable) in some suitable supermanifold [18]. Here local coordinates are used, but a coordinate independent approach is possible [19].

The r.h.s.s of eq. (25) are assumed to be local components of a super-Hamiltonian vector field corresponding to a super-Poisson structure and a super-Hamiltonian function H, i.e. an even (i.e. $\mathbf{g}(H) = 0$) element of $\mathcal{F}$. To be specific, $\mathcal{F}$ is endowed with a super-Poisson bracket, i.e a graded antisymmetric c-map

$$(f,g) \in \mathcal{F} \times \mathcal{F} \mapsto \{f,g\} \in \mathcal{F} \tag{26}$$

satisfying, just like the super-commutator, the graded Jacobi identity and being a derivation with respect to both arguments, i.e.

$$\{g,f\} = -(-1)^{\mathbf{g}(f)\mathbf{g}(g)}\{f,g\}, \tag{27a}$$

$$\mathbf{g}(\{f,g\}) = \mathbf{g}(fg) = \mathbf{g}(f) + \mathbf{g}(g)(mod\ 2), \tag{27b}$$

$$(-1)^{\mathbf{g}(f)\mathbf{g}(h)}\{f,\{g,h\}\} + (-1)^{\mathbf{g}(h)\mathbf{g}(g)}\{h,\{f,g\}\} +$$
$$(-1)^{\mathbf{g}(g)\mathbf{g}(f)}\{g,\{h,f\}\} = 0, \tag{27c}$$

$$\{f,gh\} = (-1)^{\mathbf{g}(f)\mathbf{g}(g)}g\{f,h\} + \{f,g\}h \tag{27d}$$

The super-hamiltonian character of eq.s (25) then means in particular that

$$X_\alpha(\mathbf{u},\varphi) = \{u_\alpha, H\}, \qquad Y_\beta(\mathbf{u},\varphi) = \{\varphi_\beta, H\} \tag{28}$$

and in general that, for a generic element f of $\mathcal{F}$,

$$\dot{f} = \{f, H\}. \tag{29}$$

In order to recast super-Hamiltonian dynamics in an ordinary Hamiltonian setting, consider explicitly the u's and φ's as local coordinates on a supermanifold locally modelled on $\mathbb{C}_c^{\#(B)} \times \mathbb{C}_a^{\#(F)}$, $\#(B)$ and $\#(F)$ respectively denoting the (possibly infinite) number of boson and fermion degrees of freedom [18]; here $\mathbb{C}_c$ denotes the subalgebra of commuting supernumbers, $\mathbf{g}(\mathbb{C}_c) = 0$, and $\mathbb{C}_a$ the subspace of anticommuting ones, $\mathbf{g}(\mathbb{C}_a) = 1$. This means that the u's and φ's can be represented as a follows

$$u_\alpha = u_{\alpha,B} + \sum_k \frac{1}{(2k)!} u_{\alpha,(j_1,j_2,\ldots,j_{2k})} \varsigma_{j_1} \varsigma_{j_2} \cdots \varsigma_{j_{2k}} \tag{30a}$$

$$\varphi_\beta = \sum_k \frac{1}{(2k-1)!} \varphi_{\beta,(j_1,j_2,\ldots,j_{2k-1})} \varsigma_{j_1} \varsigma_{j_2} \cdots \varsigma_{j_{2k-1}} \tag{30b}$$

where $u_{\alpha,B}$ is the body of u_α, while the sum in (30a) is its soul $u_{\alpha,S}$ and summation is implied on repeated Grassmann indices. Here $\varsigma_1, \varsigma_2, \ldots, \varsigma_k, \ldots$ with $\varsigma_i \varsigma_j + \varsigma_j \varsigma_i = 0$, are a family of generators of the Grassmann algebra Λ, whose dimensionality is irrelevant in what follows. If in particular $\Lambda = \Lambda_\infty$, which is always implicitly assumed in field theory in order for the corresponding quantum theory to be able to accommodate the whole Fock space, expressions given in eq.s (30) are to be meant as formal series, thus avoiding any notion of convergence in Λ [18]. Finally power series coefficients in eq.s (30) are ordinary complex variables, completely antisymmetric in their Grassmann indices. Here for simplicity they are assumed to be real, which does not imply that u and φ are real supernumbers, since, although Grassmann generators are taken as usual to be real, i.e. $\varsigma_k^* = \varsigma_k$, their products $\varsigma_1 \varsigma_2 \ldots \varsigma_k$ are real or imaginary according to k.

As to functions X_α and Y_β appearing in eq.s (25), it is worth remembering that they are defined as power series in φ and in the souls of u [18]; once expressions given in eq.s (30) are substituted for u and φ in these series, X_α, Y_β are given as (formal) power series in Grassmann generators as follows

$$X_\alpha(\mathbf{u}, \varphi) = X_{\alpha,B}(\mathbf{u}_B) + \sum_k \frac{1}{(2k)!} X_{\alpha(j_1,j_2,\ldots,j_{2k})}(\mathbf{u}_{[c]}, \varphi_{[d]}) \varsigma_{j_1} \varsigma_{j_2} \cdots \varsigma_{j_{2k}} \tag{31a}$$

$$Y_\beta(\mathbf{u},\varphi) = \sum_k \frac{1}{(2k-1)!} Y_{\beta,(j_1,j_2,\ldots,j_{2k-1})}(\mathbf{u}_{[c]},\varphi_{[d]})\varsigma_{j_1}\varsigma_{j_2}\cdots\varsigma_{j_{2k-1}} \tag{31b}$$

where power series coefficients are ordinary complex variables, fulfilling the same antisymmetry and reality (for consistency) requirements as for coefficients in eq.s (30), and $[c],[d]$ denote generic (respectively even and odd) Grassmann multi-indices, i.e. ordered subsets of $\{j_1,j_2,\ldots,j_{2k}\}$ in eq. (10a) and $\{j_1,j_2,\ldots,j_{2k-1}\}$ in eq. (10b). Once expressions (30),(31) are substituted for left- and right-hand sides in equations (25), an equivalent ordinary differential system for real coefficients of the expansions (30), from now on called component variables, is obtained

$$\dot{u}_{\alpha,B} = X_{\alpha,B}(\mathbf{u}_B) \tag{32a}$$

$$\dot{\varphi}_{\alpha,(i)} = Y_{\alpha,(i)}(\mathbf{u}_B,\varphi_{(i)}) \tag{32b}$$

$$\dot{u}_{\alpha,(i,j)} = X_{\alpha(i,j)}(\mathbf{u}_B,\mathbf{u}_{(i,j)},\varphi_{(i)},\varphi_{(j)}) \tag{32c}$$

$$\dot{\varphi}_{\alpha(i,j,h)} = Y_{\alpha(i,j,h)}\left(\mathbf{u}_B,\mathbf{u}_{(i,j)},\mathbf{u}_{(i,h)},\mathbf{u}_{(j,h)},\varphi_{(i)},\varphi_{(j)},\varphi_{(h)},\varphi_{(i,j,h)}\right) \tag{32d}$$

$$\vdots$$

which for the super-dynamical system defined in eq. (17) reads

$$\dot{u}_B = 6u_B u_B' - u_B''' \tag{33a}$$

$$\dot{\varphi}_{(k)} = f\varphi_{(k)}''' + gu_B'\varphi_{(k)} + hu_B\varphi_{(k)}' \tag{33b}$$

$$\dot{u}_{(h,k)} = 6u_B u_{(h,k)}' + 6u_{(h,k)}u_B' - u_{(h,k)}''' +$$
$$e\left(\varphi_{(h)}\varphi_{(k)}'' - \varphi_{(k)}\varphi_{(h)}''\right) \tag{33c}$$

$$\dot{\varphi}_{(j,h,k)} = f\varphi_{(j,h,k)}''' + gu_B'\varphi_{(j,h,k)} + g\left(u_{(j,h)}'\varphi_{(k)} + c.p.\right)$$
$$+ hu_B\varphi_{(j,h,k)}' + h\left(u_{(j,h)}\varphi_{(k)}' + c.p.\right) \tag{35d}$$

If eq.s (25) are assumed to define a super-Hamiltonian system according to eq.s (28), then coefficients in the expansion of the super-hamiltonian function

$$H(\mathbf{u},\varphi) = H_B(\mathbf{u}_B) + \sum_k \frac{1}{(2k)!} H_{(j_1,j_2,\ldots,j_{2k})}\varsigma_{j_1}\varsigma_{j_2}\cdots\varsigma_{j_{2k}} \tag{34}$$

are constants of motion for system (32). For the specific examples H_1 and H_2 in eq.s (20),(22) these coefficient are given by

$$H_{1,B} = H_{2,B} = \int dx \left(u_B^3 + u'^2_B/2 \right) \qquad (35a)$$

$$H_{1,(1,2)}\left[u_B, u_{(1,2)}, \varphi_{(1)}, \varphi_{(2)} \right] =$$
$$\int dx \left[3u_B^2 u_{(1,2)} + u'_B u'_{(1,2)} + s u_B \left(\varphi_{(1)}\varphi'_{(2)} - \varphi_{(2)}\varphi'_{(1)} \right) + 2t\varphi_{(1)}\varphi'''_{(2)} \right] \qquad (35b)$$

$$H_{2,(1,2)}\left[u_B, u_{(1,2)}, \varphi_{(1)}, \varphi_{(2)} \right] = \int dx \left[2u_B u_{(1,2)} + b \left(\varphi_{(1)}\varphi'_{(2)} - \varphi_{(2)}\varphi'_{(1)} \right) \right] /2 \qquad (35c)$$

and so on.

It can be proved [7] that suitable closed subsystems of eq.s (32) are ordinary Hamiltonian systems with respect to ordinary Poisson structures defined in terms of the original super Poisson brackets. The ordinary Hamiltonian functions are given correspondingly by complex coefficients in expansion (34), which, consistently with the assumed reality of component variables, are taken to be real. Consider in fact a finite subset of $2n$ Grassmann indices, which, owing to the completely equivalent role played by all of them, can be identified without loss of generality with $S = \{1, 2, ..., 2n\}$. The closed subsystem of eq.s (32) including the evolution equations for component variables whose Grassmann multi-indices only contain indices up to $2n$, is Hamiltonian, with $H_{(1,2,...,2n)}$ as the Hamiltonian function, with respect to a suitable Poisson structure on the Abelian ring generated by the fixed set of component variables.

To be specific, the ordinary Poisson brackets mentioned above are given by

$$\left\{ v_{[g]}, w_{[g']} \right\} = 0, [g], [g'] \subset \{1, 2, \ldots, 2n\}, \quad \text{if}[g] \cup [g'] \neq \{1, 2, \ldots, 2n\} \quad (36a)$$

$$\left\{ v_{[g]}, w_{[g']} \right\} = (-1)^{P_1}(-1)^{P_2} \{v, w\}_{[g] \cap [g']} \quad \text{if}[g] \cup [g'] = \{1, 2, \ldots, 2n\} \quad (36b)$$

where v, w are arbitrary super-variables, $[g], [g'] \subset S$ are Grassmann multi-indices, $[g] \cap [g']$ denotes the corresponding naturally ordered subset and $p_1, p_2 = 0, 1$ respectively the parity of the permutations

$$(1, 2, \ldots, 2n) \rightarrow ((1, 2, \ldots, 2n)\backslash[g], [g]) \qquad (37a)$$

$$[g'] \to ([g']\backslash[g], [g'] \cap [g]) \tag{37b}$$

Here and henceforth the same symbol is used to denote both the original super-Poisson and their corresponding ordinary Poisson brackets. What is meant is clear from their arguments. For the simplest non-trivial case, $n = 1$, the ordinary Poisson brackets corresponding, according to eq.s (36),(37), to the super-Poisson brackets in eq.s (19),(21) are given respectively by

$$\left\{u_B(x), u_{(1,2)}(y)\right\}_1 = \delta'(x-y), \left\{\varphi_{(1)}(x), \varphi_{(2)}(y)\right\}_1 = -a\delta(x-y), \quad h < k,$$

$$\left\{u_B, u_B\right\}_1 = \left\{u_{(1,2)}, u_{(1,2)}\right\}_1 = \left\{\varphi_{(1)}, \varphi_{(1)}\right\}_1 = \left\{\varphi_{(2)}, \varphi_{(2)}\right\}_1 = \left\{u, \varphi\right\}_1 = 0 \tag{38}$$

and

$$\left\{u_B, u_B\right\}_2 = \left\{u_B, \varphi\right\}_2 = \left\{\varphi_{(j)}, \varphi_{(j)}\right\}_2 = 0, \quad j = 1, 2$$

$$\left\{u_B(x), u_{(1,2)}(y)\right\}_2 = -\delta'''(x-y) + 4u_B(x)\delta'(x-y) + 2u'_B(x)\delta(x-y)$$

$$\left\{u_{(1,2)}(x), u_{(1,2)}(y)\right\}_2 = 4u_{(1,2)}(x)\delta'(x-y) + 2u'_{(1,2)}(x)\delta(x-y)$$

$$\left\{u_{(1,2)}(x), \varphi_j(y)\right\}_2 = 3\varphi_j(x)\delta'(x-y) + \varphi'_j(x)\delta(x-y), \quad j = 1, 2$$

$$\left\{\varphi_{(1)}(x), \varphi_{(2)}(y)\right\}_2 = c[\delta''(x-y) - u_B(x)\delta(x-y)], \tag{39}$$

while for the case $n = 2$ only the Poisson brackets corresponding to the super-Poisson brackets in eq. (19) are given for simplicity below:

$$\left\{\varphi_{(r)}, \varphi_{(s)}\right\}_1 = \left\{\varphi_{(r,s,v)}, \varphi_{(w,y,z)}\right\}_1 = \left\{\varphi, u\right\}_1 = 0; \quad r, s, v, w, y, z = 1, 2, 3, 4;$$

$$\left\{u_B(x), u_B(y)\right\}_1 = \left\{u_{(r,s)}(x), u_{(r,s)}(y)\right\}_1 =$$

$$= \left\{u_{(1,2,3,4)}(x), u_{(1,2,3,4)}(y)\right\}_1 = 0; r, s = 1, 2, 3, 4;$$

$$\left\{\varphi_{(1)}(x), \varphi_{(2,3,4)}(y)\right\}_1 = -\left\{\varphi_{(2)}(x), \varphi_{(1,3,4)}(y)\right\}_1 = \left\{\varphi_{(3)}(x), \varphi_{(1,2,4)}(y)\right\}_1 =$$

$$= -\left\{\varphi_{(4)}(x), \varphi_{(1,2,3)}(y)\right\}_{(1)} = -\delta(x-y),$$

$$\left\{u_B(x), u_{(1,2,3,4)}(y)\right\}_1 = \left\{u_{(1,2)}(x), u_{(3,4)}(y)\right\}_1 = -\left\{u_{(1,3)}(x), u_{(2,4)}(y)\right\}_1 =$$

$$= \left\{u_{(1,4)}(x), u_{(2,3)}(y)\right\}_1 = \delta'(x-y) \tag{40}$$

A relevant consequence of the present setting is the possibility of shedding new light on supersymmetry transformations [20]. Consider in fact the two supersymmetric versions of the KdV eq. given by eq.s (17) with parameters chosen according to eq.s (23a) or (23b). In order to rewrite the space-supersymmetry (infinitesimal) transformation

$$\delta u(x) = \eta \varphi'(x), \qquad \delta \varphi(x) = \eta u(x) \tag{41}$$

in terms of the component fields, develop the anticommuting parameter η, as an odd supernumber, according to the chosen basis of the Grassmann algebra under consideration, namely

$$\eta \equiv \sum_k \frac{1}{(2k-1)!} \eta_{(j_1 j_2 \dots j_{2k-1})} \zeta_{j_1} \zeta_{j_2} \cdots \zeta_{j_{2k-1}} \tag{42}$$

meaning that it corresponds to a whole (possibly infinite) set of real parameters. The above supersymmetry transformation can then be represented in terms of a multi-parameter family of transformations for component fields, as follows

$$\delta u_B = 0, \delta \varphi_{(i)} = \eta_{(i)} u_B, \delta u_{(i,j)} = \eta_{(i)} \varphi'_{(j)} - \eta_{(j)} \varphi'_{(i)} \tag{43a}$$

$$\delta \varphi_{(i,j,h)} = \eta_{(i)} u_{(j,h)} - \eta_{(j)} u_{(i,h)} + \eta_{(h)} u_{(i,j)} + \eta_{(i,j,h)} u_B;$$
$$\delta u(i,j,h,k) = \eta_{(i)} \varphi'_{(j,h,k)} - \eta_{(j)} \varphi'_{(i,h,k)} + \eta_{(h)} \varphi'_{(i,j,k)} - \eta_{(k)} \varphi'_{(i,j,h)}$$
$$+ \eta_{(i,j,h)} \varphi'_{(k)} - \eta_{(i,j,k)} \varphi'_{(h)} + \eta_{(i,h,k)} \varphi'_{(j)} - \eta_{(j,h,k)} \varphi'_{(i)}, \dots \tag{43b}$$

It is easy to check directly that the transformations above are infinitesimal symmetries for eq.s (17) with parameters chosen either according to eq. (23a), or (23b). To show how supersymmetry transformations can be realized as Hamiltonian symmetries in terms of component fields, consider the closed subsystem for two selected Grassmann indices corresponding to the supersymmetric version defined by eq. (23b). It is then straightforward to check that the corresponding two-parameter family of transformations obtained from eq.s (43a) for $i, j = 1, 2$ is Hamiltonian with Hamiltonian generators given by

$$Q_1\left[u_B, \varphi_{(1)}, \varphi_{(2)}, u_{(1,2)}\right] = \int dx \left(\varphi_{(2)} U_B\right),$$
$$Q_2\left[u_B, \varphi_{(1)}, \varphi_{(2)}, u_{(1,2)}\right] = \int dx \left(-\varphi_{(1)} u_B\right) \tag{44}$$

and that invariance now simply reduces to the fact that Q_1 and Q_2 above Poisson commute with $H_{1(1,2)}$:

$$\{Q_1, H_{1,(1,2)}\}_1 = \{Q_2, H_{1,(1,2)}\}_1 = 0 \tag{45}$$

4. Concluding remarks

The most relevant applications of the presented result doubtless reside in its quantum counterparts, namely in the corresponding quantization methods for systems with fermions, in terms of component variables. The present setting gives in particular the possibility of devising a new kind of approximation where a Grassmann algebra with a finite set of generators replaces the original infinite dimensional algebra implied in fermion field theories, this corresponding to a reduced Fock space with a maximum total occupation number. It should be remarked however that it is relevant at purely *classical* level too, with reference to both the general initial value problem and the more specific question of integrability [7]. A further bonus of the proposed setting consists in recasting supersymmetry superalgebras in terms of ordinary (Hamiltonian) symmetry algebras [20] as suggested by the example of supersymmetric KdV equation. Finally a byproduct of the proposed approach is given by the emergence of whole new classes of ordinary Poisson structures.

References

[1] M. Creutz, *Quarks, gluons and lattices*, C. U. Press, Cambridge, 1983.

[2] F. A. Berezin, *The Method of Second Quantization*, Academic Press, New York, 1966.

[3] F. Fucito, E. Marinari, G. Parisi, C. Rebbi, Nucl. Phys. B180[FS2] (1981) 369

[4] J. Schwinger, Phil. Mag. 44 (1953) 1171. I. L. Martin, Proc. Roy. Soc. A251 (1959) 536.

[5] P. Ramond, Phys. Rev. D13 (1971) 2415. A. Neveu, J. H. Schwarz, Nucl. Phys. B31 (1971) 86. M. B. Green, J. H. Schwarz, Phys. Lett. 109B (1982) 444.

[6] R. Casalbuoni, Nuovo Cim. 33A (1976) 389. F. A. Berezin, M. S. Marinov, Ann. Phys. 104 (1977) 336.

[7] S. De Filippo, Phys. Lett. B265 (1991) 127; J. Phys. A25 (1992) 949.

[8] S. De Filippo, Phys. Lett. B131 (1983) 65, references therein.

[9] J. L. Gervais, Phys. Lett. B 160 (1985) 277, and ref.s quoted therein.

[10] J. L. Gervais and A. Neveu, Nucl. Phys. B 209 (1982) 125.

[11] Yu. I. Manin, A. O. Radul, Commun. Math. Phys. 98 (1985) 65. M. Chaichian, P. P. Kulish, Phys. Lett. B 183 (1987) 169. P. Mathieu, Phys. Lett. B 203 (1988) 287.

[12] B. A. Kupershmidt, Phis. Lett. 102A (1984) 213.

[13] P. Mathieu, J. Math. Phys. 29 (1988) 2499.

[14] S. De Filippo, G. Landi, G. Marmo, G. Vilasi, Phys. Lett. B220 (1989) 576, and ref.s therein.

[15] S. De Filippo, G. Landi, G. Marmo, G. Vilasi, Ann. Inst. H. Poincare' 50 (1989) 205

[16] R. P. Feynman, Rev. Mod. Phys. 20 (1948) 367.

[17] K. Symanzik, J. Math. Phys. 7 (1966) 510.

[18] B. DeWitt, Supermanifolds, Cambridge University Press, New York (1984). F. A. Berezin, Introduction to Superanalysis, Reidel, Dordrecht (1987).

[19] S. De Filippo, *A Bosonization Procedure for super-Hamiltonian Theories. Global Approach.* Preprint, Salerno.

[20] S. De Filippo, Phys. Lett. B279 (1992) 336

THE CLASSICAL TRAJECTORIES ARE NOT EQUIPROBABLE

M. Fusco Girard

*Dipartimento di Fisica Teorica e Istituto Nazionale di
Fisica della Materia
Università di Salerno - Italy*

Abstract

A probability for each classical path connecting given points in the same
time can be obtained by considering the path's contribution to the semiclassical
propagator, in the configuration space. If the initial and the final points are
conjugate with respect to a particular path, that path's probability dominates
the others'. The case of an one-dimensional quartic oscillator is in particular
investigated, and it is shown that in the generic case in which the initial and
the points are non conjugate each other, the classical paths' probabilities become
asymptotically the same, and the series for the semiclassical propagator can be
resummed.

*My first scientific paper was written together with Prof. Eduardo Caianiello [1]. It
is therefore an honour and a pleasure for me to have the opportunity of contributing
this paper to the volume celebrating the 70th birthday of Prof. Caianiello, a man
who gave so relevant contributions to the development of Physics and of the Theory
of Neural Networks in this region of Italy.*

As well known, according to the Feynman's path integral propagator approach [2] the total amplitude $K(\underline{x}_B, t_B; \underline{x}_A, t_A)$ for a quantum particle to go from an initial point $\underline{x}_A$ at time t_A to the point $\underline{x}_B$ at time t_B is given by a sum to which all the paths $\underline{x}(t)$ connecting the given points in the same time $T = t_B - t_A$ contribute by equal amounts, but with a phase which is proportional to the action $S(\underline{x}(t))$ for the path. This provides a very natural bridge between quantum and classical mechanics: in the semiclassical (also called WKB) limit, the action's extremals, i.e. the classical paths, emerge via stationary phase principle from the totality of curves connecting the given points, while the contributions of non-classical paths cancel by interference; in this limit therefore the path integral for the propagator reduces to a sum of terms, one for each of the classical paths [3]:

$$K_{WKB} = \frac{1}{(2\pi\hbar)^{\frac{m}{2}}} \sum_{\alpha} \left| \frac{\partial^2 S_\alpha}{\partial \underline{x}_A \partial \underline{x}_B} \right|^{\frac{1}{2}} e^{\left[\frac{1}{\hbar} \left(S_\alpha - n_\alpha \frac{\pi}{2} \right) \right]} \tag{1}$$

where the index α labels the classical paths, m is the degree of freedoms and the determinant is known as the Van Vleck's determinant [4]; S_α is the action for the path, i. e. the integral of the Lagrangian:

$$S_\alpha = \int_{t_A}^{t_B} \mathcal{L}\left[\underline{x}_\alpha(t), \underline{\dot{x}}_\alpha(t), t\right] dt \tag{2}$$

and n_α denotes the number of the focal points along the $\alpha-$th path. Let us recall that a point x_c, T_c, is called a focal point or a conjugate point to the initial one x_A, t_A, with respect to the $\alpha-$th path, if there it results

$$\underline{\delta}(t_c) = 0 \tag{3}$$

where $\underline{\delta}(t)$ denotes the solution of the Jacobi equation, which gives the linearised time evolution of the path $\underline{x}(t) = \underline{x}_\alpha + \underline{\delta}(t)$, infinitely near to $x_\alpha(t)$, and leaving the same initial point.

The presence of the index α in Eq. (1) is due to the circumstance that in the generic case of a non-linear system, there are many classical paths connecting the given points in time T; all these paths give in principle different contributions to the semiclassical propagator, and it is therefore interesting to investigate how they cooperate to the formation of the total quantum amplitude. An immediate consequence of Eq. (1) is that if the final point is conjugate to the initial one with respect to a particular path, that path's contribution to the total semiclassical propagator will be dominant with respect to the others'. Indeed, as well known, [4] the Jacobi field $\underline{\delta}(t)$ is proportional to the inverse of the matrix formed with the second derivatives of the action

$$\underline{\delta}(t) \sim \left| \frac{\partial^2 S_\alpha}{\partial \underline{x}_A \partial \underline{x}_B} \right|^{-1} \tag{4}$$

so that in a focal point the van Vleck's determinant is infinite. In the Feynman's language the various classical paths present interfering alternatives to a semiclassical particle, and the modulus squared of the amplitude gives the probability of the alternative: if the final point is conjugate to the initial one for a path $x_i(t)$, the probability of that path therefore is of overmingly importance with respect to the others'.

This point has been investigated in a recent paper [5], where the case of a quartic one-dimensional oscillator has been in particular analysed. The Lagrangian is

$$\mathcal{L} = \frac{1}{2}\dot{x}^2 - \frac{1}{2}\omega^2 x^2 - \frac{1}{4}\lambda x^4 \tag{5}$$

where, in order to have only bound states, the parameter λ is assumed positive. The inclusion of the quartic term in the potential strongly modifies the structure of the extremals of the action with respect to the case of a purely harmonic potential: indeed, it is easy to see that for the anharmonic oscillator there are denumerably infinite classical paths connecting given points in time T, while for the corresponding "unperturbed" system, the harmonic oscillator, there is in general only one of such paths. These paths can be classified into four classes, according to the four possibilities for the signs of the initial and the final momenta: for the first class the signs are $(+,+)$, for the second class $(+,-)$, for the third $(-,+)$, and finally, for the fourth class, the signs are $(-,-)$. Each class including a shortest

268

member, plus all the paths which are obtained by adding to it an integer number n of complete oscillations. The energies of the paths are given by all the solutions of the equation which expresses the energy's conservation, i.e.

$$T = \sigma_i t^K(E) - t^A(E) + (n + \delta_i)P(E) \tag{6}$$

where $t^K(E)$ is the time for the particle to go from the origin to x, along the direct path of energy E, and $P(E)$ is the period at that energy; $\sigma_i = +1$ for the first and the third classes, and -1 in the other cases, and $\delta_i = 0$ for the first class, $\frac{1}{2}$ for the second and the third classes, and $\delta_i = 1$ for the fourth class. Quantities occurring in the last equation are expressed in terms of elliptic integrals as follows:

$$t^K = t^K(E) \equiv \int_0^{x_K} \frac{dx}{D(x)} = \frac{1}{\sqrt{\frac{1}{2}\lambda R}} F(\gamma_K, r) \tag{7}$$

$$P(E) = 4 \int_0^b \frac{dx}{D(x)} = \frac{4}{\sqrt{\frac{1}{2}\lambda R}} K(r) \tag{8}$$

where

$$D(x) = \sqrt{2E - \omega^2 x^2 - \frac{1}{2}\lambda x^4} \tag{9}$$

$F(\varphi, k)$ and $K(k)$ are the elliptic integral and the complete elliptic integral of first kind, respectively, and the quantities occurring in eq. (7) are given by:

$$R = a^2 + b^2 \;, \tag{10}$$

$$a^2 = \frac{\omega^2}{\lambda} + \left[\frac{\omega^4}{\lambda^2} + \frac{4E}{\lambda}\right]^{\frac{1}{2}} \tag{11}$$

$$b^2 = \frac{\omega^2}{\lambda} + \left[\frac{\omega^4}{\lambda^2} + \frac{4E}{\lambda}\right]^{\frac{1}{2}} \;, \tag{12}$$

$$\gamma_K = Arcsin\frac{x_K}{b}\left[\frac{R}{a^2 + x_K^2}\right]^{\frac{1}{2}} \;, \tag{13}$$

$$r = \frac{b^{\frac{1}{2}}}{R} \,. \tag{14}$$

The period $P(E)$ and the times $t_B(E)$, $t_A(E)$, are monotonically decreasing functions of the energy E, so that the right hand side in eq. (6) is a monotonically decreasing function of E, starting from a finite value (which increases with n) at $E = E\min$, which is the minimum value of the energy such that the particle can reach the point x_B; therefore, the equation (6) has a unique solution $E_n(x_A, x_B, T, \sigma_i, \delta_i)$ for each integer n greater than a minimum value $n_{\min}$. This solution of the equation can be only numerically found. From the knowledge of the energy of the path, it is possible to compute the action $S = W - Et$, where W is given by

$$W = \sigma_i \, w^B - w^A + (n + \delta_i) J(E) \tag{15}$$

where $J(E)$ is the action variable,

$$J(E) = \oint p dq \tag{16}$$

and

$$W^K = \int_0^{x_K} p dq \tag{17}$$

the expression of $J(E)$ and w can be found in ref. [5].

It is then possible to compute the second derivative of the action S, which gives the weight of the path's contribution to the total semiclassical propagator. After a straightforward calculation one obtains that

$$S_{x_A x_B} = -W_{x_A, E} \cdot E_{x_B} \tag{18}$$

where the suffixes denote derivatives (see Ref. [5] for details). The analysis of the divergences of $S_{x_A x_B}$ permits the individuation of the focal points, and if one follows the oscillator along a path, by choosing the initial point as the origin, it is easy to see that given the two points x_A and x_B, there exist two countable sequences of times T_n, such that the two points are conjugate with respect to a path $x_n(t)$. The two sequences correspond to paths belonging to the second and to the fourth class, respectively.

Interesting conclusions can be obtained when one analyses the asymptotic behaviour of the various quantities involved in Eq. (6), for large energies, which corresponds to the case $n >> 1$. It is then easy to see that the energies of the paths increase like n^4, and by using this result in eq. (18) one finds the surprising result that in the general case the amplitude of the paths contribution to the total semiclassical propagator become asymptotically the same; the various path's therefore are asymptotically equiprobable. This result also implies that the van Vleck's series for the semiclassical propagator does non converge, but it is possible to see that in the general case it can be Cesaro resummed.

These conclusions can extended to a large class of potentials [6]. In conclusion, there are two possible outcomes for an experience in which the times of flight and the energies of semiclassical particles subjected to such potentials, are recorded: if the particles are ejected with equally distributed energies from a source in x_A, in the generic case, particles arriving after time T in x_B will have with the same frequency (for paths with a large number n of complete oscillations), all the energies solutions of eq. (1); otherwise, if the time T is equal to one of the times T_n for which the final point is conjugate to the initial one with respect to a path $x_n(t)$, particles will arrive preferably with the energy E_n of that path.

REFERENCES

[1] E. R. Caianiello, M. Fusco Girard, M. Marinaro, Lett. Nuovo Cim. $\underline{20}$, 431 (1977).

[2] R. P. Feynman, Rev. Mod. Phys. $\underline{20}$, 367 (1948); R. P. Feynman and A.R. Hibbs, *Quantum Mechanics and Path Integrals*, Mc Graw-Hill, New York, 1965.

[3] L. S. Shulman, *Techniques and Applications of Path Integration*, John Wiley and Sons, New York, 1981, in particular chapt. 13 and references therein.

[4] M. C. Gutzwiller, J. Math. Phys. $\underline{8}$, 1979 (1967); L.S. Shulman, Ref. [3].

[5] M. Fusco Girard, "*On The Semiclassical Propagator for the Anharmonic Oscillator*", Preprint Univ. Salerno 1991.

[6] M. Fusco Girard, in preparation.

ELECTROMAGNETISM, RELATIVITY AND QUANTUM MECHANICS

A. Campolattaro

Department of Physics,
The University of Maryland, Baltimore Country Campus,
Baltimore, Maryland 21228, U.S.A.

1. Introduction

In 1873 appeared in London the treatise by James Clerk Maxwell on the Electricity and Magnetism which established a mile-stone in Physics, not only because it systematized the whole of electro-magnetic phenomena but also because, as it appeared at the time, together with Newtonian mechanics, completed the knowledge of all physical phenomena.

The situation was in effect, much less open to optimism than as expected. In fact immediately shewed an enormous conflict between Newtonian dynamics and electro-magnetism quite unexpected which, as we all well know, was resolved only in 1905 with Einstein's work on the theory of Relativity. This theory represented a drastic departure from the newtonian vision of the natural laws while at the same time, left unchanged the Maxwell theory.

The problems however were not at all over. In 1900 Max Planck introduced the notion, quite revolutionary, of the quantization of energy, opening in this way a completely new research on the basis of our knowledge of the physical world which was going to bring Quantum Mechanics. The official birth date for this new Mechanics is: December 14 1900.[1]

One has thence:

Special theory of relativity with the relativistic invariant:

$$ds^2 = \eta_{\mu\nu} dx^\mu ds^\nu \qquad \mu, \nu = 0\ldots 3 \qquad x^0 = ct \tag{1}$$

and $\eta^{\mu\nu}$ the Minkowski tensor defined by

$$\eta^{\mu\nu} = \begin{pmatrix} 1 & 0 & 0 & 0 \\ 0 & -1 & 0 & 0 \\ 0 & 0 & -1 & 0 \\ 0 & 0 & 0 & -1 \end{pmatrix} \tag{2}$$

Electro-magnetism with Maxwell's equations in terms of the electro-magnetic field tensor $F^{\mu\nu}$ given by:

$$F^{\mu\nu},_\nu = j^\mu$$
$$^*F^{\mu\nu},_\nu = 0 \tag{3}$$

where with a comma followed by an index has been indicated the partial derivative with respect to the variable with that index,

$$^*F^{\mu\nu} = \frac{1}{2}\varepsilon^{\mu\nu\sigma\tau} F_{\sigma\tau} \tag{4}$$

with $\varepsilon^{\mu\nu\sigma\tau}$ the Ricci pseudosensor with entry $+1$ if the parity of the permutation $\mu\nu\sigma\tau$ of the indices 0,1,2,3 is even, and -1 if odd, and entry zero if two or more indices are equal, and

$$F_{\mu\nu} = \eta_{\mu\sigma}\eta_{\nu\tau} F^{\sigma\tau} = \Phi_{\mu'\nu} - \Phi_{\nu'\mu} \tag{5}$$

and where j^μ is the tetra current, and Φ_μ is the tetra-potential, all defined in any book concerned with the covariant formulation of electro-magnetism.

Relativistic Quantum Mechanics with the Dirac equation for the electron in the electro-magnetic field described by the tetra-potential Φ_μ given by:

$$[\gamma^\mu(\partial_\mu + i\Phi_\mu) + im_0]\,\Psi = 0 \tag{6}$$

where Ψ is the Dirac spinors and the γ's the Dirac gamma matrices by:

$$\gamma^0 = \begin{pmatrix} 1 & 0 & 0 & 0 \\ 0 & 1 & 0 & 0 \\ 0 & 0 & -1 & 0 \\ 0 & 0 & 0 & -1 \end{pmatrix} \qquad \gamma^1 = \begin{pmatrix} 0 & 0 & 0 & 1 \\ 0 & 0 & 1 & 0 \\ 0 & -1 & 0 & 0 \\ -1 & 0 & 0 & 0 \end{pmatrix} \tag{7}$$

$$\gamma^2 = \begin{pmatrix} 0 & 0 & 0 & -i \\ 0 & 0 & i & 0 \\ 0 & i & 0 & 0 \\ -i & 0 & 0 & 0 \end{pmatrix} \qquad \gamma^3 = \begin{pmatrix} 0 & 0 & 1 & 0 \\ 0 & 0 & 0 & -1 \\ -1 & 0 & 0 & 0 \\ 0 & 1 & 0 & 0 \end{pmatrix}$$

in the units $\hbar = c = 1$.

However while the theory of Relativity and Maxwell equations represent a close body of knowledge, the third one epitomizes the dícotomy between deterministic and undeterministic Physics.

Louis de Broglie asks himself: La Physique Quantique restera-t-elle indeterministe? [2]

I am not able to give to this question an answer. But in the following pages I will show that quite possibly the answer is no.

2. Spinor representation of Maxwell's equations.

In two papers [3,4] I succeeded in demonstrating that given an electro-magnetic field tensor $F^{\mu\nu}$ it is always possible to find a spinor Ψ with components Ψ_μ such that

$$F^{\mu\nu} = \overline{\Psi} S^{\mu\nu} \Psi \tag{9}$$

with

$$S^{\mu\nu} = \frac{i}{2}\gamma^{[\mu}\gamma^{\nu]} = \frac{i}{4}(\gamma^\mu\gamma^\nu - \gamma^\nu\gamma^\mu) \tag{10}$$

On the other hand by introducing the matrix γ^5 defined by

$$\gamma^5 = \gamma^0\gamma^1\gamma^2\gamma^3 \tag{11}$$

and the property

$$\varepsilon^{\mu\nu\sigma\tau}\eta_{\sigma\alpha}\eta_{\tau\beta}\gamma^{[\alpha}\gamma^{\beta]} = 2\gamma^5\gamma^{[\mu}\gamma^{\nu]} \tag{12}$$

one has

$$^*F^{\mu\nu} = \overline{\Psi}\gamma^5 S^{\mu\nu}\Psi \tag{13}$$

Maxwell's equation read

$$(\overline{\Psi} S^{\mu\nu}\Psi)_{,\nu} = j^\mu \tag{14}$$

$$(\overline{\Psi}\gamma^5 S^{\mu\nu}\Psi)_{,\nu} = 0$$

Although the magnetic monopole has until now eluded all experiments intended for its determination, for reason of symmetry and for possible physical implications which will be discussed later, Maxwell's equations in the presence of a monopolar current g^μ and in the spinorial formalism, read

$$(\overline{\Psi}S^{\mu\nu}\Psi)_{,\nu} = j^\mu \tag{14}$$

$$(\overline{\Psi}\gamma^5 S^{\mu\nu}\Psi)_{,\nu} = g^\mu$$

3. Two gauge conditions.

The spinor Ψ defined by equation (9) is not unevocally determined because the electro-magnetic tensor $F^{\mu\nu}$ admits six independent components there were the spinor Ψ admits four components, then eight independent parameters leaving in this way two free conditions one can impose on it. I chose for reasons which will appear clear in a while, the following:

$$Im[\overline{\Psi}\square\Psi] = 0 \tag{17}$$

$$Im[\overline{\Psi}\gamma^5\square\Psi] = 0$$

that I call "gauge conditions" and which are equivalent in stating that the vector $\eta^{\mu\nu}Im(\overline{\Psi}_{,\nu}\,\psi)$ and the pseudo-vector $\eta^{\mu\nu}Im(\overline{\Psi}_{,\nu}\,\gamma^5\psi)$ are solenoidal i.e.

$$[\eta^{\mu\nu}Im(\overline{\Psi}_{,\nu}\,\Psi)]_{,\mu} = 0 \tag{19}$$

$$[\eta^{\mu\nu}Im(\overline{\Psi}_{,\nu}\,\gamma^5\Psi)]_{,\nu} = 0$$

conditions which will be used in a while.

4. The Dirac-like equation equivalent to Maxwell's equations.

In [3,4] it has also been possible to show that the generalized Maxwell equations (15) are equivalent to the single equation

$$\gamma^\mu\Psi_{,\mu} = -i\gamma^\mu\frac{e^{\gamma^5\alpha}}{\rho}\{Im(\overline{\Psi}_{,\mu}\,\Psi) - j_\mu - \gamma^5[Im(\overline{\Psi}_{,\mu}\,\gamma^5\Psi) - g_\mu]\}\Psi \tag{21}$$

276

where ρ is given by

$$\rho = \sqrt{(\overline{\Psi}\Psi)^2 + (\overline{\Psi}\gamma^5\Psi)^2} \tag{22}$$

and α is the angle of duality rotation defined by

$$\tan\alpha = \frac{\overline{\Psi}\Psi}{\overline{\Psi}\gamma^5\Psi} \tag{23}$$

The gauge conditions 19 read then that the two quantities $Im(\overline{\Psi}_{,\mu}\Psi)$ and $Im(\overline{\Psi}_{,\mu}\gamma^5\Psi)$ are two currents solenoidal like the two currents j_μ and g_μ.

In this formalism the duality trasformation [5,6] with complexion α defined by

$$\bar{F}^{\mu\nu} = F^{\mu\nu}\cos\alpha +{}^* F^{\mu\nu}\sin\alpha \tag{24}$$

is equivalent to a Touscheck-Nishijima transformation [7,8] for the spinor Ψ into the spinor Ψ' given by

$$\Psi' = e^{\frac{1}{2}\gamma^5\alpha}\Psi \tag{25}$$

In this representation several algebraic properties of the electro-magnetic field are deduced [5,6,9,10,11].

5. The neutrino equation.[12,13]

Of particular interest in the case of an electro-magnetic field generated by two currents given by

$$j_\mu = Im(\overline{\Psi}_{,\mu}\Psi) \tag{26}$$

and

$$g_\mu = Im(\overline{\Psi}_{,\mu}\gamma^5\Psi) \tag{27}$$

in fact in this case equation 21 reads:

$$\gamma^\mu\partial_\mu\Psi = 0$$

which is the celebrated equation for the neutrino that in this formalism reads as the electro-magnetic field generated by the two currents 24 and 25.i.e.

$$(\overline{\Psi}S^{\mu\nu}\Psi)_{,\nu} = \eta^{\mu\sigma}Im(\overline{\Psi}_{,\sigma}\Psi) \tag{29}$$

and

$$(\overline{\Psi}\gamma^5 S^{\mu\nu}\Psi),_\nu = \eta^{\mu\sigma} Im(\overline{\Psi},_\sigma \gamma^5 \Psi) \quad . \tag{30}$$

6. Dirac's equation for the free electron.

If one performs the substitution

$$j_\mu \rightarrow j_\mu + Im(\Psi,_\mu \Psi)$$
$$g_\mu \rightarrow g_\mu + Im(\Psi,_\mu \gamma^5 \Psi) \tag{31}$$

equation (21) reads

$$\gamma^\mu \Psi,_\mu - i\gamma^\mu \frac{e^{\gamma^5 \alpha}}{\rho}(j^\mu - \gamma^5 g_\mu)\Psi = 0 \tag{32}$$

If one introduces two currents

$$j^\mu = m\overline{\Psi}\gamma^\mu\Psi$$
$$g^\mu = n\overline{\Psi}\gamma^5\gamma^\mu\Psi \tag{33}$$

with m and n two scalars such that one has

$$j^\mu,_\mu = g^\mu,_\mu = 0 \tag{34}$$

then in is shown [12] that one has that equation (32) reduces to

$$\gamma^\mu \Psi,_\mu - i(m - in)\Psi = 0 \tag{35}$$

and therewith it is also shown that equations (34) are satisfied by

$$m = \frac{i}{2}\frac{2iA_\mu(\overline{\Psi}\gamma^\mu\Psi) - (\overline{\Psi}\gamma^5\Psi),_\mu}{\overline{\Psi}\gamma^5\Psi} \tag{36}$$

$$n = \frac{1}{2}\frac{2iA_\mu(\overline{\Psi}\gamma^5\gamma^\mu\Psi) + (\overline{\Psi}\gamma^\mu\Psi),_\mu}{\overline{\Psi}\Psi} \tag{37}$$

with A_μ given by

$$A_\mu = \frac{1}{\rho}(j_\mu \sin\alpha - g_\mu \cos\alpha) \tag{38}$$

In particular for $n = 0$ the first of the (34) is satisfied with m constant and equation (35) becomes the Dirac equation for the free electron which is therefore equivalent to the Maxwell equations for an electro-magnetic field given by

$$(\overline{\Psi} S^{\mu\nu} \Psi)_{,\nu} = \eta^{\mu\nu} Im(\overline{\Psi},_\nu \Psi) + m\overline{\Psi}\gamma^\mu\Psi \tag{39}$$

$$(\overline{\Psi}\gamma^5 S^{\mu\nu} \Psi)_{,\nu} = \eta^{\mu\nu} Im(\overline{\Psi},_\nu \gamma^5\Psi) \tag{40}$$

7. Conclusions.

The correspondence which has been shown between the Dirac equation and the Maxwell equations has been eleborated upon and various other results will be presented in forthcoming papers. In the present one it has only been indicated that possibly a deterministic interpretation of Relativistic quantum Mechanics is indeed obtainable once one has seen that the Dirac equation for the free electron is equivalent to a set of non-linear Maxwell equations involving some suitable currents. This, however, is true only if there are no magnetic manoples. If these exist the free particle obeys still an equation of Dirac but the mass term is a complex quantity and it is not constant. A puzzling interpretation of these results has been offered by the author [12,13] who is at present at work elaborating upon it. The consequences will be presented in a forthcoming paper.

References.

[1] Max von Laue. Eulogy for Max Planck. Read in Albani Church in Göttingen on October 7, 1947.

[2] L. de Broglie, *La Physique Quantique restera-t-elle indeterminste?* Gauthier-Villars. Paris 1953.

[3] A.A. Campolattaro,*New spinor representation of Maxwell's equations I, Generalities.* Intern. J. Theor. Physics. 19,2,99,126 (1980).

[4] A.A. Campolattaro, *New spinor representation of Maxwell's equations II, Algebraic properties.* Intern. J. Theor. Physics. 19,2,127-138 (1980).

[5] G.Y. Rainich. Trans. Amer. Math. Soc. 27,106 (1925).

[6] C.W. Misner and J.A. Wheeler. Ann. of Phys. 2, 525 (1957).

[7] B.F. Touscheck, Nuovo Cimento 5, 1281 (1957).

[8] K. Nishijima, Nuovo Cimento 5, 1349 (1957).

[9] H.S. Ruse, Proc. London Math. Soc. 41, 302 (1936).

[10] L. Witten, Phys, Rev. 115, 206 (1959).

[11] L. Witten, *Gravitation. An Introduction to current research.* 382 L. Witten, ed. John Wiley aqnd Sons, New York (1962).

[12] A.A. Campolattaro, Intern. J. Theor. Physics. 29, 2 ,141-155 (1990).

[13] A.A. Campolattaro, Intern. J. Theor. Physics. 29, 5, 477-482 (1990).

Quantum Groups, Coherent States and Squeezing

E. Celeghini [b] *, M. Rasetti* [♮] *, and G. Vitiello* [♯]

ABSTRACT

We show how the two concepts of quantum groups and squeezing can be bridged in the simple setting provided by a set of states which are a modified version of the Biedenharn and McFarlane states. We conjecture that the notion of quantum group coherent state is in general the natural candidate to describe squeezed quantum states of matter.

[b] Dipartimento di Fisica and Sezione I.N.F.N., Universitá di Firenze, Italy

[♮] Dipartimento di Fisica and Unitá I.N.F.M., Politecnico di Torino, Italy

[♯] Dipartimento di Fisica, Universitá di Salerno, and I.N.F.N., Sezione di Napoli, Italy

It has been recognized that squeezed coherent states [1],[2] constructed as generalized coherent states for suitable algebras may describe a wide class of systems characterized by reduced quantum fluctuations. Instances are given by $su(2)$ and $su(1,1)$ (both in the single- and the multi-boson realizations [3]) as well as by $g\ell(\infty)$ (soliton [4]) squeezing. Applications have been suggested ranging from gravitational wave detection and polariton theory [5] to low-noise optical communications, to the inhibition of atomic phase decay [6].

On a different side, quantum groups have shown to be an exceptionally promising and rich structure whereby one can expect a growing wealth of new results in statistical mechanics and quantum field theory [7]. Stemmed out of the algebraic structure dictated by integrability conditions (quantum Yang-Baxter (q-Y.-B.) equations) for a class of integrable systems, quantum groups can be intuitively thought of as the deformation of the universal enveloping algebra of some given Lie algebra $\mathcal{L}$ of dynamical variables induced by replacing the Jacobi identities with the q-Y.-B. equations, yet preserving the associativity properties of $\mathcal{L}$. As usual, quantum groups is synonimous of Hopf algebra.

In a recent paper [8], we have shown how the two concepts of quantum groups and squeezing can be bridged in the simple setting provided by a set of states which are a modified version of the states introduced by Biedenharn and McFarlane [9] (B.-McF.). We conjecture that the notion of quantum group coherent state is in general the natural candidate to describe squeezed quantum states of matter. In this report we closely follow ref. 8 to which the reader is addressed for more details.

The *quantum* version of the Jordan-Wigner map is generated by the set of operators $\{a_q, \bar{a}_q, N_q; q \in \mathbb{C}\}$, with relations

$$[N_q, a_q] = -a_q \quad ; \quad [N_q, \bar{a}_q] = \bar{a}_q \ , \tag{1}$$

$$a_q \bar{a}_q - q^{\frac{1}{2}} \bar{a}_q a_q = q^{-\frac{1}{2} N_q} \ . \tag{2}$$

The structure lying behind (1) and (2) has been shown to be a *quantum* superalgebra [10] $\mathcal{S} \sim \mathrm{osp}_q(1|2)$.

282

As we intend to discuss the property of squeezing of the generalized coherent states $(G.C.S)_q$ associated with S, we have to equip first the realization (1), (2) of S with the notion of hermiticity. We work in the customary Fock space $\mathcal{F}$, which will make the physics of squeezing much more transparent. This also implies that $N_q^\dagger = N_q \equiv N$, where $N = N_{q=1}$ is the usual boson number operator.

Conjugation of (1) and (2) immediately leads then to

$$a_q^\dagger = \bar{a}_{q^*} [\chi_q(N)]^{-1} \quad , \quad \bar{a}_q^\dagger = \chi_q(N) a_{q^*} \quad , \tag{3}$$

where $\chi_q(N)$ is a function to be determined by suitable self-consistency conditions. A convenient, natural assumption to fix it is that $a_q, a_q^\dagger$ are canonically conjugate with each other, $i.e.$ that in the corresponding sector S be identical with the usual Weyl-Heisenberg algebra, namely $[\,a_q, a_q^\dagger\,] = 1$. This condition is implemented by the choice

$$\chi_q(N) = \frac{([\,N+1\,]_{q^*}[\,N+1\,]_q)^{\frac{1}{2}}}{N+1} \quad , \tag{4}$$

where $[\,n\,]_q \equiv \dfrac{q^{\frac{1}{2}n} - q^{-\frac{1}{2}n}}{q^{\frac{1}{2}} - q^{-\frac{1}{2}}} = \dfrac{\sinh\left(\frac{1}{2}nz\right)}{\sinh\left(\frac{1}{2}z\right)}$, with $z \equiv \ln q = x + iy$. One has as well

$$a_{q^*} = a_q \left(\frac{[\,N\,]_{q^*}}{[\,N\,]_q}\right)^{\frac{1}{2}} \text{ and } \bar{a}_{q^*} = \left(\frac{[\,N\,]_{q^*}}{[\,N\,]_q}\right)^{\frac{1}{2}} \bar{a}_q.$$

Notice that as $q \to 1$, $a_q^\dagger \to \bar{a}_q \to a^\dagger$, $\bar{a}_q^\dagger \to a_q \to a$, $a_q^\dagger a_q \to a^\dagger a \equiv N$ (recall that $a_q^\dagger a_q = \dfrac{N}{[\,N\,]_q}\bar{a}_q a_q$), $a, a^\dagger$ denoting the generators of the customary bosonic realization of the Weyl-Heisenberg algebra over $\mathcal{F}$.

We denote by $\{|n > |n \text{ integer} \geq 0\}$ the usual basis of $\mathcal{F}$. In view of (3), (4),

$$a_q|n> = \left(\frac{[\,n\,]_q}{[\,n\,]_{q^*}}\right)^{\frac{1}{4}} n^{\frac{1}{2}}|n-1> \quad ,$$

$$\bar{a}_q|n> = \left(\frac{[\,n+1\,]_q}{[\,n+1\,]_{q^*}}\right)^{\frac{1}{4}} \left(\frac{[\,n+1\,]_q[\,n+1\,]_{q^*}}{n+1}\right)^{\frac{1}{2}}|n+1> \quad , \tag{5}$$

whence $a^\dagger a|n> = n|n>$, $\bar{a}_q a_q|n> = [\,n\,]_q|n>$. $|0>$ is the highest weight vector of S: $a|0> = 0$.

If one defines now the coherent states $\{|\alpha; q> |\alpha, q \in \mathbb{C}\}$ by

$$a_q|\alpha; q> = \alpha|\alpha; q> \quad , \tag{6}$$

it is straightforward to check that

$$|\alpha; q> = \mathcal{N}(|\alpha|) \exp_q (\alpha \bar{a}_q)|0> \quad , \tag{7}$$

where $\exp_q (x) \equiv \sum_{n=0}^{\infty} \dfrac{x^n}{[n]_q!}$ is the *quantum* version of the exponential function $([n+1]_q! \equiv [n+1]_q [n]_q!)$, and $\mathcal{N}(|\alpha|)$ is a normalization factor which with the above choice turns out to be independent of q : $\mathcal{N}(|\alpha|) = \exp\left(-\frac{1}{2}|\alpha|^2\right)$.

Then also,

$$|\alpha; q> = \exp\left(-\frac{1}{2}|\alpha|^2\right) \sum_{n=0}^{\infty} \frac{\alpha^n}{\sqrt{n!}} \left(\frac{[n]_{q^*}!}{[n]_q!}\right)^{\frac{1}{4}} |n> \quad . \tag{8}$$

We notice that in the limit $q \to 1$, $|\alpha; q>$ turns into a customary Glauber coherent state [11].

We can thus define the *quantum* analog of position (Q_q) and momentum (P_q) operators :

$$P_q \equiv i \left(\frac{m\hbar\omega}{2}\right)^{\frac{1}{2}} \left(\bar{a}_q - \bar{a}_q^\dagger\right) \quad ,$$
$$Q_q \equiv \left(\frac{\hbar}{2m\omega}\right)^{\frac{1}{2}} \left(\bar{a}_q + \bar{a}_q^\dagger\right) \quad . \tag{9}$$

Q_q and P_q are hermitian and have commutation relation

$$[Q_q, P_q] = i\hbar \left(\frac{[N+1]_q [N+1]_{q^*}}{N+1} - \frac{[N]_q [N]_{q^*}}{N}\right) = i\hbar_q \quad . \tag{10}$$

Operators (10) give rise to a *quantum* version of the quantized harmonic oscillator (q.h.o.), with hamiltonian

$$H_q \equiv \frac{1}{2m} P_q{}^2 + \frac{1}{2} m\omega^2 Q_q{}^2 = \frac{1}{2}\hbar\omega \left(\bar{a}_q^\dagger \bar{a}_q + \bar{a}_q \bar{a}_q^\dagger\right)$$
$$= \frac{1}{2}\hbar\omega \left(\frac{[N+1]_q [N+1]_{q^*}}{N+1} + \frac{[N]_q [N]_{q^*}}{N}\right) \quad . \tag{11}$$

Eq. (11) suggests a possible experimental test of the q-effects through the spectrum of the q.h.o.for large N. If one considered indeed eq. (10) − in the low z limit − as a *canonical* form with $\hbar_q = \hbar \left(1 + \frac{1}{4} \mathrm{Re}\,(z^2) N(N+1) + \mathcal{O}(z^4)\right)$, the q.h.o.hamiltonian would then be written, up to the same order, as $H_q = \hbar_q \omega_q$

$\left(N + \dfrac{1}{2}\right)$, with $\omega_q = \omega \left(1 - \dfrac{1}{6}\mathrm{Re}\,(z^2)N(N+1) + \mathcal{O}(z^4)\right)$. In this perspective the spectrum of H_q can be thought of as that of the usual harmonic oscillator in which the level spacing is deformed in such a way that it increases with n (up to $\mathcal{O}(z^4)$ as $\hbar\omega \left(1 - \dfrac{1}{2}\mathrm{Re}\,(z^2)(n+1)^2\right)$).

In order to show that the $(G.C.S)_q$ $\{|\alpha; q >\}$ are squeezed, we proceed now to evaluate the variances of Q_q and P_q. Denoting, as customary,

$$(\Delta Q_q)^2 = < Q_q^2 > - < Q_q >^2 \quad , \quad (\Delta P_q)^2 = < P_q^2 > - < P_q >^2 \quad , \tag{12}$$

where $< \bullet > \equiv < \alpha; q| \bullet |\alpha; q >$, one can readily check that one can write $(\Delta Q_q)^2 = \dfrac{\hbar}{2m\omega}\,(\mathcal{A}_q + \mathcal{B}_q)$, $(\Delta P_q)^2 = \dfrac{1}{2}m\hbar\omega\,(\mathcal{A}_q - \mathcal{B}_q)$, with

$$\mathcal{A}_q \equiv < \bar{a}_q^\dagger \bar{a}_q > + < \bar{a}_q \bar{a}_q^\dagger > - 2 < \bar{a}_q >< \bar{a}_q^\dagger > , \; \mathcal{B}_q \equiv \left(\Delta \bar{a}_q^\dagger\right)^2 + \left(\Delta \bar{a}_q\right)^2 \; . \tag{13}$$

For simplicity of notation, we shall henceforth use units such that $m\omega = 1$.

Due to (10), we say that one has *strong squeezing* in P_q (respectively Q_q) if $\Delta P_q < \left(\dfrac{1}{2}|< \hbar_q >|\right)^{\frac{1}{2}}$ (respectively $\Delta Q_q < \left(\dfrac{1}{2}|< \hbar_q >|\right)^{\frac{1}{2}}$). Analogously we say that one has *weak squeezing* if $\hbar_q$ is replaced by $\hbar$ in the above inequalities. It should be pointed out that the $\{|\alpha; q >\}$'s are <u>not</u> minimum uncertainty states[‡]. Thus, not surprisingly, the locus of points $\mathcal{C}$ for which $\mathcal{B}_q = 0$ does not necessarily separate, in the (α, q) space (isomorphic to $\mathbb{C} \times \mathbb{C}$) regions characterized by weak squeezing in P_q from regions with weak squeezing in Q_q, as one has with conventional squeezed states[12].

The calculations are lengthy but straightforward. Notice that $|\alpha|^2 = < N >$, therefore in our numerical analysis we confine our attention to relatively low $|\alpha|$ ($< \sim 10$). One finds

$$< \bar{a}_q >< \bar{a}_q^\dagger > = \frac{e^{-2|\alpha|^2}}{4|\alpha|^2\,|\sinh\left(\frac{1}{2}z\right)|^2}\left|e^{|\alpha|^2\exp\left(\frac{1}{2}z\right)} - e^{|\alpha|^2\exp\left(-\frac{1}{2}z\right)}\right|^2 \quad , \tag{14.1}$$

[‡] This is related to the fact that also the states (7), as the B.-McF. states are not coherent states for the q-$W.H.$ group, in that they do not have the right transformation properties under the quantum group action.

$$< \bar{a}_q^\dagger \bar{a}_q + \bar{a}_q \bar{a}_q^\dagger > = \frac{e^{-|\alpha|^2}}{4|\alpha|^2 \, |\sinh\left(\frac{1}{2}z\right)|^2}$$

$$\left\{ 2 \left[e^{|\alpha|^2 \cosh x} \cosh\left(|\alpha|^2 \sinh x\right) - e^{|\alpha|^2 \cos y} \cos\left(|\alpha|^2 \sin y\right) \right] \right. \tag{14.2}$$

$$\left. + |\alpha|^2 \left[\mathrm{Ei}\left(|\alpha|^2 e^x\right) + \mathrm{Ei}\left(|\alpha|^2 e^{-x}\right) + 2\,\mathrm{Re}\,\mathrm{E1}\left(-|\alpha|^2 e^{iy}\right) \right] \right\},$$

$$\mathcal{B}_q = \frac{e^{-|\alpha|^2}}{2|\alpha|^2}\,\mathrm{Re}\left\{ \frac{e^{-2i\varphi_\alpha}}{\{\sinh\left(\frac{1}{2}z\right)\}^2}\left[\exp\left(-\frac{1}{2}z + |\alpha|^2 e^z\right) + \exp\left(\frac{1}{2}z + |\alpha|^2 e^{-z}\right) \right.\right.$$

$$\left.\left. - 2\cosh\left(\frac{1}{2}z\right) e^{|\alpha|^2} - e^{-|\alpha|^2} \left(\exp\left(|\alpha|^2 e^{\frac{1}{2}z}\right) - \exp\left(|\alpha|^2 e^{-\frac{1}{2}z}\right) \right)^2 \right] \right\};$$

$$\tag{14.3}$$

where φ_α denotes the phase of α, whereas $\mathrm{Ei}(\varsigma)$ and $\mathrm{E1}(\varsigma)$ denote the exponential integral functions. Numerical results show that there is a rich structure with interesting squeezing phenomena. In ref.8 plots in the α-plane are reported which show squeezing regions.

For different q's near 1 we find regions in the α-plane of strong squeezing and curves C along which $\Delta P_q = \Delta Q_q = \left(\frac{1}{2}\hbar \mathcal{A}_q\right)^{\frac{1}{2}}$, crossing which one switches from a regime with $\Delta P_q > \Delta Q_q$ to the one where $\Delta P_q < \Delta Q_q$.

For q not too close to the unit circle, it is always $\hbar_q > \hbar$, which corresponds to the property that along the curves C $|\alpha|$ grows monotonically and smoothly with φ_α.

The regions of both weak and strong squeezing manifest the feature of symmetry under exchange of Q_q and P_q, which geometrically corresponds to a rotation in the α-plane by $\frac{1}{2}\pi$.

The relevance of the states for which $|q| \approx 1$ but the phase of q is nonvanishing, which exhibit such a rich structure of squeezing features, is worth stressing. Moreover, in view of the physical applications of q-coherent states , the combined notions of *deformed* Planck constant $\hbar_q$ and of weak and strong squeezing may make the $\{|\alpha; q >\}$'s the appropriate tools for a phenomenological understanding of the deformation parameter q.

References

[1] D. Stoler, Phys. Rev. **D 1**, 3217 (1970)

[2] H.P.Yuen, Phys. Rev. **a 13**, 2226 (1976)

[3] J. Katriel, A.I. Solomon, G. D'Ariano, and M. Rasetti, Phys. Rev. **D 34**, 2332 (1986); J. Opt. Soc. Am. **B 4**, 1728 (1987)

[4] G. D'Ariano, and M. Rasetti, Phys. Lett. **107 A**, 291 (1985)
P.D. Drummond, and S.J. Carter, Phys. Rev. Lett. **58**, 1841 (1987); J. Opt. Soc. Am. **4**, 1565 (1987)

[5] M. Artoni, and J.L.Birman, Quantum Opt. **1**, 91 (1989)

[6] C.W. Gardiner, A.S.Parkins, and M.J. Collett, J. Opt. Soc. Am. **4**, 1683 (1987)

[7] M. Jimbo, Lett. Math. Phys. **10**, 63 (1985); **11**, 247 (1986)

[8] E. Celeghini, M. Rasetti, and G. Vitiello, Phys. Rev. Lett. **66**, 2056.(1991)

[9] L.C. Biedenharn, J. Phys. A: Math. Gen. **22**, L873 (1989)
A.J. MacFarlane, J. Phys. A: Math. Gen. **22**, 4581 (1989)

[10] E. Celeghini, T.D. Palev, and M. Tarlini, Mod. Phys. Lett. **B5**, 2056 (1991)

[11] R.J. Glauber, Phys. Rev. **130**, 2529 (1963); **131**, 2766 (1963)

[12] M.N. Nieto, in *Non-Equilibrium Quantum Statistical Physics*, G. Moore, and M.O. Scully, eds.; Plenum Press, New York, 1985

SPECTRAL MASS SUM RULES IN QUANTUM FIELD THEORY*

Salvatore De Martino   Silvio De Siena

and

Olga Mannella

Dipartimento di Fisica dell' Università di Salerno

I 84081 Baronissi (Salerno), Italy

and

Istituto Nazionale Fisica Nucleare

Sezione di Napoli

Abstract

We describe spectral mass sum rules which allows to obtain informations and bounds on physical masses and order parameters in quantum field theory models.

Introduction

We are glad that we can contribute to the proceedings of this celebrative meeting with a paper which contains as a fundamental tool the branching equations for correlation functions,namely one of the most relevant contribution of Professor E. R. Caianiello to quantumfield theory [1].

This paper deals with spectral mass sum rules (SMSR) in quantum field theoriy models [2,3,4,5,6]. The fundamental ingredients which we use are the Umezawa-Kamefuchi-Källen-Lehmann (UKKL) spectral representations of the two-point

---

*   Work supported in part by M.U.R.S.T

288

truncated correlation functions [7], and recursive equations among propagators [1,8].

SMSR are a set of equations linking spectral densities with parameters of the theory (e.g. coupling constants, bare masses, order parameters etc.). If the theory contains more fields (boson non-gauge or gauge fields,fermion fields), a set of SMSR can be written for each field.

For boson fields SMSR take the form

$$\langle m^2 \rangle = \langle V^{(2)} \rangle, \tag{1}$$

where on the l.h.s. appears the mean value,with respect to the boson spectral measure,of the squared spectral mass,and on the r.h.s. the expectation value of the polynomial of fields obtained by deriving twice the interaction potential with respect to the boson field.

For fermion fields the form is

$$\langle m \rangle = \langle V^{(2)} \rangle, \tag{2}$$

where now the l.h.s. must be interpreted as the mean value of the spectral mass with respect to a suitable combination of the fermion spectral measures, while the meaning of r.h.s. is equivalent to that of (0.1). The difference between the SMSR in the boson and in the fermion case is due to the difference between Klein-Gordon and Dirac equation.

The form of (1) and (2) suggests that these expressions represent the exact implementations of mean-field relations ; the mean field approximations can be obtained from (1) and (2) by approximating the spectral measures with delta-functions peaked on dominant masses. In fact,for the bosonic scalar quartic self-interaction this relation can be explicitly proved [9].

As a consequence of the finite character of the branching equations, the r.h.s. of (1) and (2) contains the renormalization terms which are needed for compensating the ultraviolet divergences of the spectral measures; moreover,if the interaction potential is at most quartic in each field ($g(\varphi)^4$,Yukawa,scalar and spinor quantum electrodynamics etc), it can be expressed in terms of the parameters of the theory and of mean values of spectral functions of the spectral masses,by giving then a completely spectral form to mass sum rules.

As we have remarked,SMSR can be written for a large class of field theories; however,as we are interested to obtain rigorous results (in particular,rigorous bounds on physical parameters),we limit ourself to consider super-renormalizable boson and fermion theories without gauge fields. In particular,$g(\varphi)^4$ theory with discrete or continous symmetry and Yukawa scalar and pseudoscalar interaction in two and three space-time dimensions.

1.Branching equations and spectral representations

In this section we introduce the basic tools from which we extract SMSR .We write the branching equations only for $(\varphi)^4$ model;the Yukawa model is treated in the quoted references [5,6].However,here we give for any model the spectral representations. We assume the Euclidean version of quantum field theory; the reader is referred to [10,11,12] for the more detailed notations and definitions.

1) $(\varphi)^4$ **theory.**

We consider the O(n) invariant polinomial interaction $g(\varphi_i\varphi_i)^2$ in two and three space-time dimensions defined through the Euclidean action

$$\mathcal{U} = g \int : (\varphi_i\varphi_i)^2 : dz + \frac{\sigma}{2} \int : (\varphi_i\varphi_i) : dz + \mathcal{R}, \tag{1.1}$$

where the Wick subtractions terms are denoted by :....:, and R is the suitable mass renormalization counterterm (which is zero in two dimensions).

Eq. (1.1) represents a perturbation on the free field of bare mass m_0^2.A change in the bare mass is connected to the point loop, which enters in a natural way in the SMSR equations. It is defined as

$$\Delta(m'^2_0, m_0^2) = \lim_{|x-y|\to 0} \left(S_{m'^2_0}(x-y) - S_{m_0^2}(x-y)\right). \tag{1.2}$$

and in two and three dimensions it results as

$$\Delta(m'^2_0, m_0^2) = (4\pi)^{-1} \lg \frac{m_0^2}{m'^2_0} \qquad d = 2 \tag{1.3a}$$

and

$$\Delta(m'^2_0, m^2_0) = (4\pi)^{-1}(m_0 - m'_0) \qquad d = 3. \tag{1.3b}$$

respectively.

As we said in the introduction, one of the main tool used by deriving SMSR are recursive equations among the correlation functions [1,8] ; they are formal solution of the equations of motion, and arise from an integration by part on the functional integral.

For the $(\varphi_i \varphi_i)^2$ theory defined by (1.1) we have

$$\langle \varphi_{i_1}(x_1) \cdots \varphi_{i_n}(x_n) \rangle = \sum_{k=2}^{n} \delta_{i_1 i_k} S_{m^2_0}(x_1 - x_k) \langle \not{\varphi}_{i_1}(x_1) \cdots \not{\varphi}_{i_k}(x_k) \cdots \varphi_{i_n}(x_n) \rangle +$$

$$-(\sigma + \delta m^2_0) \int \delta_{i_1 l} S_{m^2_0}(x_1 - y) \langle \varphi_l(y) \varphi_{i_2}(x_2) \cdots \varphi_{i_n}(x_n) \rangle dy +$$

$$-4g \int \delta_{i_1 l} S_{m^2_0}(x_1 - y) \langle : \varphi_l(y) \varphi_j(y) \varphi_j(y) : \varphi_{i_2}(x_2) \cdots \varphi_{i_n}(x_n) \rangle, \tag{1.4}$$

where $\delta_{ij} S_{m^2_0}(x - y)$ represents the free two-point correlation among the i and j component of the field, δm^2_0 is the infinite mass counterterm (only in 3 dimensions), while $\not{\varphi}_{i_k}(x_k)$ means that the field $\varphi_{i_k}(x_k)$ doesn't appear in the correlation function. The equations (1.4) describe entirely the theory.

The other important device which we need is the spectral representation. For d=3 and $n \geq 2$, it is well-known that spontaneous magnetization can appear, and Goldstone excitations (transverse-field) can be generated in a suitable zone of the phase space (multiple-phase region). We must then distinguish between the single-phase region and the multiple phase region. In the single phase region the UKKL representation is

$$\langle \varphi_i(x) \varphi_j(y) \rangle = \delta_{ij} \int_0^\infty dm^2 \rho(m^2) S_{m^2}(x - y), \tag{1.5}$$

where $\rho(m^2)$ is a positive, normalized spectral measure

$$\int_0^\infty dm^2 \rho(m^2) = 1, \tag{1.6}$$

which contains the effect of interaction on the two-point function; in the multiple-phase region we have

$$\langle \varphi_i(x)\varphi_j(y)\rangle = n_i n_j \mathcal{M}^2 +$$

$$+(\delta_{ij} - n_i n_j)\int_0^\infty dm^2 \rho_\perp(m^2)S_{m^2}(x-y)+$$

$$+n_i n_j \int_0^\infty dm^2 \rho_\|(m^2)S_{m^2}(x-y), \tag{1.7}$$

where n_i is a unit vector in internal space, M is the spontaneous magnetization defined through

$$n_i M = \langle \varphi_i \rangle, \tag{1.8}$$

and $\rho_\|$ and $\rho_\perp$ are,respectively,the normalized spectral densities related to the excitations along the magnetization and to the transverse- field excitations in internal space. It is well-known that $\rho_\perp(m^2)$ has a support extending to zero and that the tansverse susceptibility

$$\chi_\perp = \int_0^\infty dm^2 \int dy \rho_\perp(m^2)S_{m^2}(x-y) = \langle \frac{1}{m^2}\rangle_\perp \tag{1.9}$$

must be infinite because of the occurrence of the Goldstone-like excitations.

From now on, $\langle ...\rangle$, $\langle ...\rangle_\|,\langle ...\rangle_\perp$ indicate the mean value of that function with respect to ρ, $\rho_\|$, $\rho_\perp$.

If, however, we consider the case with d=2 and $n \geq 2$, Mermin and Wagner' s theorem prevents to have spontaneous magnetization and transverse excitations; then, a spectral representation of form (1.5) holds in all the phase space.

Finally, in the n=1 case, for d=2,3 spontaneous magnetization is generated without transverse excitations, so that we have

$$\langle \varphi(x)\varphi(y)\rangle = \mathcal{M}^2 + \int_0^\infty dm^2 \rho(m^2)S_{m^2}(x-y). \tag{1.10}$$

Also in these last two cases (1.5) holds.

2) Scalar and pseudoscalar Yukawa theory.

In two and three dimensions, Yukawa interaction in the Schwinger real representation is

$$U = -\frac{\lambda}{2} \int \; : \psi(z)\alpha\psi(z) : \varphi(z)dz + \mathcal{R}$$
$$\equiv +\frac{\lambda}{2}Tr\alpha \int \; : \psi(z)\psi(z) : \varphi(z) + \mathcal{R},$$

(1.11)

where $\alpha = \beta$ for the scalar case, $\alpha = \beta\alpha_1(1 - \alpha_6\alpha_7)$ for the pseudoscalar one, and R represents the suitable renormalization mass counterterms (in two dimensions only bosons, in three dimensions bosons and fermions).

We have two kinds of recursive equations for this model, involving the more general boson-fermion correlation functions.

Eq. (1.5) holds for the spectral representations for the bosonic part in Yukawa theory. For the two-point fermion correlation function, we have [7]

$$S^F(x - y) = \int_0^\infty dm^2 \left(\rho_+(m^2)S_{+m}(x - y) - \rho_-(m^2)S_{-m}(x - y)\right), \qquad (1.12)$$

where $S_{\pm m}$ are Fermi free Schwinger functions with masses $\pm m$ respectively.

In theories where the norm of any state is positive, the two spectral densities $\rho_\pm$ are positive and [7]

$$\rho_+(m^2) - \rho_-(m^2) \geq \rho_-(m^2). \qquad (1.13)$$

Furthermore, as we shall derive in the following section in the framework of the spectral mass sum rules, the spectral densities $\rho_\pm$ satisfy the "normalization" condition

$$\int_0^\infty dm^2\left(\rho_+(m^2) - \rho_-(m^2)\right) = 1, \qquad (1.14)$$

which is the fermionic analogue of (1.5), and expresses the anticommutativity of Fermi fields.

We add now other definitions and notations that will be used in the following. Equations (1.13), (1.14) give us

$$Z_- \equiv \int_0^\infty dm^2\rho_-(m^2) \leq 1, \qquad (1.15)$$

and then

$$1 \leq Z_+ \equiv \int_0^\infty dm^2 \rho_+(m^2) \leq 2. \tag{1.16}$$

We can also define the normalized total spectral measure

$$d\rho_F(m^2) \equiv Z_F^{-1}\left(\rho_+(m^2) + \rho_-(m^2)\right)dm^2, \tag{1.17}$$

where

$$Z_F = \int dm^2\left(\rho_+(m^2) + \rho_-(m^2)\right) \tag{1.18}$$

clearly satisfies the condition

$$1 \leq Z_F \leq 3. \tag{1.19}$$

In the following, we will denote with $\langle \cdots \rangle_F$ the average with respect to the normalized spectral measure (1.17).

2.Mass sum rules and bounds

In this section we write our SMSR.

The method is the following . One writes the branching equations for two point correlations,insert the spectral representations,apply the free equations of motion,separate the spectral from the singular part and obtains normalization conditions on the spectral densities and a spectral representation for other correlations ; on this last objects one repeat the procedure and obtain other spectral representations for correlations and SMSR [2-6]. We now specialize the results to the various models.

1) $(\varphi)^4$ **theory.**

The mass sum rules are

$$m_0^2 + \sigma + 4g(n+2)\int dm^2\rho(m^2)\Delta(m^2,m_0^2) = \langle m^2 \rangle - \delta m_0^2, \tag{2.1}$$

in the single-phase region, and

$$m_0^2 + \sigma + 4g\mathcal{M}^2 + 4g(n+1)\int dm^2\rho_\perp(m^2)\Delta(m^2,m_0^2)+$$

$$+4g \int dm^2 \rho_{||}(m^2)\Delta(m^2, m_0^2) = \langle m^2 \rangle_\perp - \delta m_0^2, \tag{2.2}$$

$$8g\,M^2 = \langle m^2 + \frac{2}{\pi}gm \rangle_{||} - \langle m^2 + \frac{2}{\pi}gm \rangle_\perp, \tag{2.3}$$

in the multiple-phase region.

The method gives also new spectral representations. If we define the mean zero field $\chi_i(x)$ (the "invariant field") as

$$\chi_i(x) = 4g : \varphi_i(x)\varphi_j(x)\varphi_j(x) : +(m_0^2 + \delta m_0^2 + \sigma)\varphi_i(x), \tag{2.4}$$

we have

$$\langle \chi_i(x)\varphi_j(y) \rangle = (\delta_{ij} - n_i n_j) \int_0^\infty dm^2 m^2 \rho_\perp(m^2) S_{m^2}(x-y)+$$

$$+n_i n_j \int_0^\infty dm^2 m^2 \rho_{||}(m^2) S_{m^2}(x-y), \tag{2.5}$$

$$\langle \chi_i(x)\chi_j(y) \rangle = (\delta_{ij} - n_i n_j) \int_0^\infty dm^2 m^4 \rho_\perp(m^2) S_{m^2}(x-y)+$$

$$+n_i n_j \int_0^\infty dm^2 m^4 \rho_{||}(m^2) S_{m^2}(x-y). \tag{2.6}$$

Notice that,in d=3,the spectral divergence in (2.1),(2.2) is exactly balanced by the infinite counterterm δm_0^2, so that we have a finite result.

Is also interesting to observe how the the longitudinal and the transverse modes work in (2.2) and (2.3). In particular, we can indirectly, but clearly, read the occurrence of Goldstone excitations in (2.3), which shows that the average distribution corresponding to $\rho_\perp(m^2)$ is lower. Actually, as it is well known, Goldstone bosons are related to the transverse part, and then must occupy the region near the origin $m^2 = 0$ in the spectral function $\rho_\perp(m^2)$.

One now search for quantitative informations,as bounds on physical masses or on order parameters;unfortunatly,in d=3 the infinite mass renormalization term makes that hard.

In d=2,however,this complication does not arise;spectral distribution reduces to the longitudinal one only,and SMSR take the form

$$m_0^2 + \sigma + 12g\mathcal{M}^2 + 4g(n+2)\int dm^2 \rho(m^2)\Delta(m^2, m_0^2) = \langle m^2 \rangle, \qquad (2.7)$$

where,obviously,the magnetization term vanishes for $n \geq 2$.

Relation (2.7),together with convexity and decreasing properties of the point loop, gives in all the phase space for $n \geq 2$,and in the single phase (zero magnetization) region for n=1,the mass bounds

$$\langle m^2 \rangle \geq \tilde{m}^2, \qquad (2.8)$$

$$\langle m^2 \rangle \geq m_{phys}^2, \qquad (2.9)$$

$$\tilde{m}^2 \geq m_{phys}^2, \qquad (2.10)$$

where m_{phys} is the physical mass, and $\tilde{m}$ is the "mean field" mass, defined through the equation

$$m_0^2 + \sigma + 4g(n+2)\Delta(\tilde{m}^2, m_0^2) = \tilde{m}^2. \qquad (2.11)$$

For n=1,finally,GHS inequalities give us

$$8g\mathcal{M}^2 \leq \langle m^2 \rangle, \qquad (2.12)$$

which relates the spontaneous magnetization with the squared mass spectral average.

2) Yukawa theory

2.1) Scalar model.

We obtain here two SMSR. If we define, as for the φ^4 theory, an "invariant field" for the bosonic part

$$\chi(x) \equiv \frac{\lambda}{2}Tr\beta : \psi(x)\psi(x) : +(m_0^2 + \delta m_0^2)\varphi(x) \qquad (2.13)$$

we have the spectral representations

$$\langle \chi(x)\varphi(y) \rangle = \int_0^\infty dm^2 m^2 \rho(m^2) S_{m^2}(x-y), \qquad (2.14)$$

$$\langle\chi(x)\chi(y)\rangle = \int_0^\infty dm^2 m^4 \rho(m^2) S_{m^2}(x-y), \tag{2.15}$$

,

the normalization (1.5) and the bosonic SMSR

$$\langle m^2 \rangle - \delta m_0^2 = m_0^2. \tag{2.16}$$

We can see that the bosonic spectral average of squared mass is nothing but the renormalized boson mass; this is not surprising,and can be put in relation with the linear dependence of the interaction from the boson field.

For the fermionic part we prove the spectral representations

$$\lambda\langle\varphi(x)\psi(x)\psi(y)\rangle = \int dm^2 \big(\rho_+(m^2)(M_0 + \delta M_0 - m)S_{+m}(x-y)+$$

$$-\rho_-(m^2)(M_0 + \delta M_0 + m)S_{-m}(x-y)\big), \tag{2.17}$$

$$\lambda^2\langle\varphi(x)\psi(x)\varphi(y)\psi(y)\rangle = \int dm^2 \big(\rho_+(m^2)(M_0 + \delta M_0 - m)^2 S_{+m}(x-y)+$$

$$-\rho_-(m^2)(M_0 + \delta M_0 + m)^2 S_{-m}(x-y)\big), \tag{2.18}$$

, the "normalization" condition (1.14) and the SMSR

$$Z_F \int d\rho_F(m^2) m = M_0 + \delta M_0 +$$

$$+\frac{\lambda^2}{2\langle m^2\rangle} Tr\beta\left(\int dm^2 \big(\rho_+(m^2)S_{+m}(0) - \rho_-(m^2)S_{-m}(0)\big) - S_{M_0}(0)\right) \tag{2.19}$$

where $S_{M_0}(0)$ is due to the Wick regularization,and M_0 is the fermionic bare mass. In obtaining equation (2.19) also the branching equation for the one-point boson correlation is used.

Moreover,in two dimensions one can extract from (2.19) a relation which does not contain the infinite mass counterterm (*finite SMSR*); introducing the adimensional spectral mass

$$\mu \equiv \frac{m}{M_0}, \tag{2.20}$$

this last can be written

$$\frac{1}{Z_F} - \frac{\lambda^2}{\pi m_0^2}\langle\mu \lg \mu\rangle_F = \langle\mu\rangle_F, \tag{2.21}$$

with the obvious meaning of the symbols.

By defining now a "fermion mean field mass" $\tilde{\mu}$ as solution of the equation

$$\tilde{\mu} = \frac{1}{Z_F} - \frac{\lambda^2}{\pi m_0^2}\tilde{\mu}\lg\tilde{\mu}, \tag{2.22}$$

we are able to prove bounds on the mean spectral mass and on the physical adimensional mass

$$\langle\mu\rangle_F \leq \tilde{\mu}, \tag{2.23}$$

$$\mu_{phys} \leq \frac{Z_F}{Z_+}\tilde{\mu} \tag{2.24}$$

This last relation provides us a (*mean field like*) bound for the physical mass.

2.2)Pseudoscalar model.

In this case,the most relevant remark [6] is that,as a simple consequence of the branching equations,of the spectral fermion representation and of the traceless property of the Dirac matrices,the expectation of the scalar field (the "magnetization") is identically zero for any value of the coupling,in disagreement with some previous calculations [13] ; then,by the same procedure as in the above case 2.1,we find that the mean spectral masses coincide with the renormalized ones, both for the scalar and for the fermion field.The model,howewer, becomes more interesting by adding a $(\varphi)^4$ term,because a magnetization *can* then be developed,and the fermion mass structure is consequently modified.

Conclusions

The problem is left open,of the extension of mass bounds to more than two-dimensional models;in such a case, the main difficulty relies in getting a *finite* SMSR without infinite mass counterterms. More general is the question of the possibile extension of these tecniques to more realistic models, such as gauge theories and expecially QCD ; in this cases complications arise in connection with the gauge invariance properties and the related renormalization difficulties.

REFERENCES

[1] E.R. Caianiello: *Combinatorics and Renormalization* (Benjamin, London 1973).

[2] J. Glimm, A. Jaffe: *Phys. Rew. D* **11** (1975) 2816.

[3] A. Kishimoto: *Comm. Math. Phys.* **47** (1976) 117.

[4] S. De Martino,S. De Siena,F. Guerra: *Lett. Nuovo Cimento* **31** (1981) 607.

[5] S. De Martino,S. De Siena,O. Mannella: *IL Nuovo Cimento* **104A** (1991) 1219.

[6] S. De Martino,S. De Siena,O. Mannella: *On the critical behaviour of the pseudoscalar Yukawa model*, to appear in *Phys. Lett. B*.

[7] S. Schweber: *An Introduction to Relativistic Quantum Field Theory* (Harper-Row Publishers,Bussum, 1964).

[8] For a review on branching equations see the contribution of F. Guerra in these Proceedings.

[9] S. De Martino,S. De Siena,F. Guerra,P. Sodano: *Lett. Nuovo Cimento* **16** (1976) 569.

[10] F. Guerra,L. Rosen,B. Simon: *Ann. Math.* **101** (1975) 111.

[11] B. Simon: *The $P(\varphi)_2$ Euclidean Quantum Field Theory* (Princeton University Press,Princeton, N.J., 1974).

[12] D. de Falco,F. Guerra: *J. Math. Phys.(N.Y.)*, **21** (1980) 1111.

[13] T. Balaban,K. Gawedzki: *Ann. Inst. H. Poincare* **36** (1982) 271.

ON THE POSSIBILITY TO "SEE" NOVEL HIGH ENERGY PHENOMENA AT TeV SCALE WITH NEUTRINO TELESCOPES

Michele Guida

Dipartimento di Fisica Teorica e S.M.S.A.

Universitá di Salerno

84081 Baronissi (Salerno), Italy

and

I.N.F.N., Sezione di Napoli, Gruppo Collegato di Salerno, Italy

Abstract

The possibility to "observe" High Energy processes, like multi–gauge bosons production at TeV energies and existence of Supersymmetric Dark Matter, with underground and aboveground $\nu - telescopes$ is discussed.

Also, it is pointed out that a *muon tracking RPC* detector can be a concrete example for realizing apparati for neutrino astronomy out of an underground laboratory.

1. Introduction

In the last few years there has been a growing interest in the High Energy Physics community about the possibility of *"observing"* novel and rare high energy phenomena, at the moment, not yet reproducible with the present generation of particle accelerators, but that, certainly, could be investigated when the next huge colliders, like LHC, SSC, etc. would move their first steps.

In fact, according to many authors [1–3] there exists the chance, worth to be taken into consideration, to exploit the conditions of extremely high energies realized in cosmic–ray interactions provided *"free"* to us by Nature and *"see"*, in the signals recorded by Neutrino Telescopes, evidence of events, otherwise not observable with other means.

Therefore, in the framework of a scientific project aimed at the construction and the operation of an International Laboratory dedicated to the study of the High Energy Cosmic Ray radiation, christened *"Southern Italy Neutrino and Gamma Astrophysical Observatory" (S.I.N.G.A.O.)*[4], in this note we want to concentrate our attention just upon two of such interesting phenomena, whose investigation in this laboratory would, certainly, shed new light to Particle Physics, Astrophysics and Cosmology still open problems.

They are:

(i) the occurring, above TeV energies, of electroweak phenomena characterized by the production of many–gauge bosons;

(ii) the possibility that Supersymmetry would provide Cosmology a natural Dark Matter candidate.

2. Multi–gauge bosons production at TeV energies.

Recent and detailed theoretical investigations have been dealing with novel Standard Model Electroweak phenomena consisting in the multi–particle production of W, Z and *Higgs* Bosons in the energy range of $\simeq (4 - 20)TeV$ [2,5–7]. If we restrict ourselves to the inclusive production of W's we can schematically represent it by the following hard subprocess:

$$A + B \rightarrow n_W W \tag{1}$$

where A and B are weakly interacting quarks or leptons and n_W is the W's molteplicity predicted to be around 30–100.

The production of many–electroweak gauge bosons would be originated by either Instanton–induced Barion–number plus Lepton–number $(B + L)$ violation processes [2,5-7] or Breakdown of the perturbation scheme of the Electroweak theory because of its infrared structure [8].

In both the cases the most striking features can be the following ones [9]:

(a) **large cross sections**: $(\approx 1pb$ for the $(B + L)$ violation phenomena and up to $1\mu b$ for the broken perturbative scheme case);

(b) **high multiplicity final states**: (i.e. Multi–W events with $< n_W >\simeq 30 - -100$);

(c) **energy thresholds in the multi–TeV range.**

Processes with these characteristics are called Geometric Flavour Interactions (GFI)[7].

Also, there have been indications according to which the mass scale of the (B+L) breaking could even be as low as 1 TeV if we consider a minimal extension of the Standard Model based on the gauge group $SU(3)_C \times SU(2)_L \times U(1)_{\gamma'}$ contained in the unified gauge group SO(10), with one extra Z' boson (with mass of the order of 1 TeV) where additional fermions, heavy Majorana neutrinos, having mass ranging from few tens to few hundreds GeV) are predicted [10].

From the investigations of some authors, among them Farrar and Meng [11], it can be inferred that with LHC, SSC, etc., multi–W (B+L)–violating processes characterized by a cross section of 1.5 pb are, in principle, observable with threshold energies below 11TeV for LHC and 28 TeV for SSC.

But, due to the huge financial and human resources necessary to realize them, these machines will not, presumably, begin to operate before the end of this century. Therefore, it is certainly worth drawing our attention to the cosmic radiation realm where almost certainly multi–W phenomena can be initiated in very high energy cosmic ray interactions by either protons interacting in the Earth's atmosphere or by neutrino's interactions with the atmosphere or inside the Earth. In both cases, the signal of interest is due to multiple penetrating muons, or muon bundles, that would be produced by the decay of the many–gauge bosons.

But, on one hand, for what protons are concerned, unfortunately, the signal due to proton–initiated multi–W events can be, surely, considered negligible because these events are buried under a very large background produced by standard hadronic cosmic ray showers in the atmosphere where the competing QCD cross sections are $\sim 10^3$ times larger [2].

On the other hand, for the neutrino–induced multi–W processes it should be pointed out that the majority of them occurs in the Earth's rock. That makes it possible that a neutrino telescope, dedicated to the observation of High Energy Neutrinos coming from astrophysical sources could, indirectly, see them, detecting the muon bundles crossing upward the apparatus.

Such a telescope can be either underground (underwater) or aboveground. For some typical detectors, like the Monopole, Astrophysics and Cosmic Ray Observatory (M.A.C.R.O.), already working at the Gran Sasso National Laboratory of the italian I.N.F.N. and the Deep Underwater Muon and Neutrino Detector (D.U.M.A.N.D. II), still under costruction in the deep ocean waters in front of the Hawaii islands shores, the detection rates have been recently estimated [2]: ≤ 0.1 events/year at MACRO (with its $\approx 10^3 m^2$ area MACRO is the largest underground detector ever built) and ≤ 5 events/year at DUMAND II (the predicted area for DUMAND would be $\approx 10^4 m^2$).

3. Neutralinos as possible dark matter candidates.

According to the introduction of Supersymmetry as a correct extension of the Standard Model a very natural dark matter candidate could be the Neutralino χ, which represents the lightest stable state of four neutral ones formed by the SUSY partner of the photon (photino), a neutral weak boson (B-ino) and two Higgs particles (Higgsinos) [12,13]. Unfortunately, so far all the investigations performed (directly) in colliders and (indirectly) in non-accelerator experiments about the neutralino's mass have been unsuccessful, implying that these particles, if they would exist, should be heavier than tens of GeV and lighter than 1TeV (Neutralinos with m_χ larger than TeV limit would, almost for sure, overclose the Universe).

In fact, searches done at LEP have placed lower limits on m_χ, i.e. $m_\chi > 45GeV$ [14], while measurements performed by the Kamiokande group have extended them in the range (30–80)GeV [15].

Therefore, recently the possibility that an indirect observation of neutralinos in neutrino telescopes could be successfully performed has been pointed out by many and that could be, obviously, lead to a clear confirmation of the validity of the supersymmetric theories [1,12,13]

In fact, according to a very attractive theoretical model [12,13], neutralinos embedding our Galaxy could be captured by the Sun, in its motion through the Galactic matter by means of a coherent scattering mechanism off the solar matter nuclei and be trapped inside of it, thanks to its intense gravitational field. Then, neutralinos, that could be thought to be mainly Higgsinos of 0.5 TeV mass, would, very probably, annihilate into energetic neutrinos ($E_\nu \geq 10GeV$), through W,Z decays or semi-leptonic heavy quark decays and, thus, providing neutrino telescopes some signals from the sun direction bearing a precise signature of an indirect observation of supersymmetric dark matter.

According to recent estimates [13] an ideal neutrino telescope with a very large area ($\geq 10^4 m^2$, at least) would be able to collect, very efficiently, signals of the order of $(10 - 10^3) events/year$ associated to high energetic neutrinos coming from the solar direction and produced by the annihilation of TeV neutralinos. As it is shown in Ref. [3] an apparatus of km-scale would enable us to investigate the full parameter space of the minimal supersymmetric model, even though, also with $10^3 m^2$ area detectors, like MACRO for example, it should be possible, in principle, to detect neutrinos from low energy SUSY dark matter particles.

4. A muon tracking RPC telescope for aboveground neutrino astronomy

The ambitious project of identifying astrophysical sources of High Energy neutrinos (like X-ray binaries in our Galaxy or Active Galactic Nuclei (A.G.N.)), christened "Neutrino Astronomy" is a new fascinating field of Astronomy growing up very fastly. In addition to the so-called atmospheric neutrinos produced by $\pi's$ and $\mu's$ decays, produced in air showers initiated by hadronic cosmic ray primaries in the atmosphere, in the high energy region ($E\nu \geq 10 GeV$), the detection of $\nu's$ coming from the above sources and the neutralino's solar annihilation can be performed by observing single upgoing muons produced in the rock underneath the experimental apparatus through Charged Current inclusive ν-interactions

$$\nu + N \to \mu \tag{2}$$

Such an experimental activity, certainly, requires very large area detectors (at least $\simeq 10^4 m^2$) that could be impossibly located inside an underground laboratory.

However, on the other hand, a surface experimental apparatus, no longer shielded by the cave hosting it, must, necessarily, cope with three main sources of background signals consisting in [16]:

(1) upgoing muons muons produced by atmospheric neutrinos;

(2) downgoing muons which are backscatterd in the ground around or under the detector;

(3) possible misidentification of downgoing muons, that is it should be able to distinguish between the upgoing muons, produced through eqn. (2) and the downgoing ones characterized by a much more intense flux.

The required up/down rejection power should be of the order of 10^{11}. Such a high rejection power can be achieved with a μ-tracking detector made of several planes of Resistive Plate Counters (RPC), able to provide, thanks to their typical performances (very good time ($\approx 2ns$) and space ($\approx 3cm$) resolutions), on each plane, a space-time correlation along the muon track.

Recently, the MINI collaboration has constructed an horizontal cosmic muon telescope instrumented with RPC's [17–19]. With this device the left/right rejection power for nearly horizontal cosmic muons has been studied by performing Time of Flight measurements. According to the preliminary results [16,19] obtained in this and other tests down with RPC's, it turns out that with a number of RPC planes ≥ 12 it is possible to reach the rejection power necessary to perform ν–Astronomy by a surface telescope covering a sensitive area $\geq 10^4 m^2$. Such a telescope would be effective in detecting single upward muons associated to the phenomenology described above, including neutralino's solar annihilation. Also, it could, in principle, be able to "observe" multiple upgoing muons at zenith angles $90^0 < \theta \leq 180^0$. In this case the physical background we expect would be essentially due to the downward muon bundles produced in standard cosmic–ray showers. On this subject work is still in progress.

5. Conclusions

We have shown that novel high energy phenomena at TeV scale can be, in principle, observable with Neutrino telescopes. In particular, muon tracking RPC telescopes can represent a very interesting way to realize aboveground $\nu - telescopes$.

Acknowledgements

We are deeply pleased to take the opportunity of this Celebration to acknowledge *Eduardo Caianiello*'s incommensurable teachings given to us both in scientific and human matters.

References

[1] P.Pistilli, Rapporteur Talk at the "22nd International Cosmic Ray Conference", Dublin Aug. 1991;

[2] D. A. Morris and R. Rosenfeld, Phys. Rev. **B44**, 3530 (1991)

[3] S. Barwick et al., Proc. of the "Third International Workshop on Neutrino Telescopes", Venice, 1991 (M. Baldo–Ceolin, ed.);

[4] SINGAO Coll.:(M. De Palma et al.), Nucl. Phys. B (Proc. Suppl.) **14B**, 69 (1990);

[5] M. P. Mattis, Lectures given at the "ICTP Summer School on High Energy Physics and Cosmology", Trieste 1991 (Preprint Los Alamos Laboratory, LA–UR–91–2926);

[6] A. Ringwald, Lectures given at the "International School of Subnuclear Physics, 28th Course: Physics up to 200 TeV", Erice 1990 (preprint CERN–TH–5949/90);

[7] A. Ringwald et al., Nucl. Phys. **B365**, 3 (1991);

[8] A. Ringwald and C. Wetterich, Nucl.Phys. **B353**, 303 (1991);

[9] J. Ellis et al., Proc. of the Large Hadron Collider Workshop, Aachen 4-9 October 1990, vol.II (G. Jarlskog and D. Rein (eds.)) CERN 90-10, ECFA 90-133;

[10] W. Buchmüller et al., DESY Preprint 91–053;

[11] G.R. Farrar and R. Meng, Phys. Rev. Lett. **65**, 3377 (1990);

[12] G.F. Giudice and E. Roulet, Preprint SISSA 103 EP (Aug. 1988);

[13] F. Halzen and T. Stelzer, Proc. of the "22nd International Cosmic Ray Conference", Dublin Aug. 1991 (preprint Univ. of Wisconsin–Madison MAD/PH/662 1991;

[14] ALEPH Coll.: (D. Decamp et al.), Phys. Lett. **B244**, 541 (1990);

[15] M. Mori et al., KEK preprint 91–62;

[16] MINI Coll.: (F. D'aquino et al.), HE-7, Proc. of the "22nd International Cosmic Ray Conference", Dublin Aug. 1991;

[17] MINI Coll.: (G. Iaselli et al.), Proc. of the "Vulcano Workshop 1990—Frontier Objects in Astrophysics and Particle Physics", Vulcano 1990

[18] MINI Coll.: (G. Iaselli et al.), Proc. of the "Second International Workshop on Neutrino telescopes", Venice 1990 (M. Baldo–Ceolin, ed.)

[19] MINI Coll.: (F. D'aquino et al.), Proc. of the "Third International Workshop on Neutrino telescopes", Venice 1991 (M. Baldo–Ceolin, ed.)

LATTICES: THEIR VORONOÍ CELLS,
SOME NEW GEOMETRIC INVARIANTS

L. Michel

Institut des Hautes,

Etudes Scientifique - 35, Raute de Chartres

F - 9111440 Bures sur Yvette - FRANCIA

It is a great pleasure to dedicate this paper to Eduardo Caianiello for his seventieth birthday. His remarkable scientific achievements made for the last forty five years should invite him to rest, but his tremenduous activity is rather increasing with age! Outside the field of physics, where many of his contributions have become classics, and besides his pionnering papers on models for the working of the brain, he has published about very general problems on "systems". The use of lattices for the theory of codes and the use of their Voronoí cells for the theory of communications (see e.g. J.H.Conway, N.J.A. Sloane, Sphere packings, lattices and groups, Springer 1988) might be an excuse to present some work on lattices at this international meeting in honour of Eduardo.

The work presented here has been made in collaboration with:
Peter Engel, Chemistry & Crystallography dpt, University of Bern, Switzerland
Marjorie Senechal, Mathematics dpt, Smith College, Northampton, Mass. USA

Although they have not seen the final form of this paper, their name should be added to mine if this work is quoted. Indeed we have worked together for the last five years and we plan to publish a small monography on the geometry of lattices with a minimal overlap with the book of Conway and Sloane quoted above. Some of our results will appear in the book "From number theory to physics" edited by Itzykson, Luck, Moussa and Walschmidt (Springer Verlag); chapter 10: "Introduction to lattice geometry" has been written by M.Senechal.

1. Voronoï cells of a lattice.

This first section is a new presentation of mostly known results on Voronoï cells of lattices in arbitrary dimension n; it is necessary as a background for the second and third parts which are entirely original. In a real vector space R^n, a lattice L is a closed subgroup $\sim Z^n$ generated by a basis $\{b_i\}$ of the vector space. For each basis of the lattice, we have a coordinate system in R^n; we note by ξ_i the coordinates of $x \in R^n$. For each basis we can define a fundamental domain of the lattice L by:

$$P_b : \{x, -\frac{1}{2} < \xi_i \leq \frac{1}{2}\}, \quad \text{volume}(P_b) = \det(b_i) \equiv \det L. \qquad 1(1)$$

Indeed this volume is independent from the choice of basis: the vectors of any other lattice basis $\{b_i'\}$ have integral components: $\{b_i'\} = \sum_j m_{ij} b_j$ with $|\det m_{ij}| = 1$, i.e. $m_{ij} \in GL_n(Z)$.

We consider only lattices in a Euclidean space. By choosing a point of the lattice as origin, we have an orthogonal space V_n whose scalar product is denoted by (x, y). For these lattices there is a natural way to define a fundamental domain independently of the choice of basis. This domain is known as the *Voronoï cell* of L: it is *the set of vectors $x \in V_n$ which are the shortest vectors in their L-coset $x + L$*; we denote its closure by $\mathcal{D}_0$. Thus $\mathcal{D}_0$ is the set

$$\mathcal{D}_0 = \{x \in V_n, N(x) \leq N(x - v) \ \forall v \in L\}. \qquad 1(2)$$

An equivalent interpretation of this equation is: *$\mathcal{D}_0$ is the set of points which are nearer to the origin than to any other lattice point*. Consequently, $\mathcal{D}_0$ is the intersection of the half-spaces:

$$\mathcal{D}_0 = \{x \in V_n, (x, v) \leq (v, v)/2 \ \forall v \in L\}. \qquad 1(3)$$

Thus the Voronoï cell of L is a convex body. Moreover, as the topological closure of a fundamental domain of the lattice:

$$\text{volume}(\mathcal{D}_0) = \det L. \qquad 1(4)$$

Since its definition respects the symmetry of the lattice, unlike the unit cell P_0, the Voronoï cell D_0 is invariant under the point group (crystallographers call it the holohedry) P of the lattice; in particular it has the origin as symmetry center. This property and 1(4) implies that a Voronoï cell is a bounded domain: indeed, if had a point at infinity, it would contain a whole straight line containing this point and passing by its symmetry center, and all its boundary hyperplanes would be parallele to this line: this is not compatible with a finite volume. The Voronoï cell[1] is also known as Dirichlet domain; in physics it is often called the Wigner-Seitz cell or, for the dual lattice of a crystal lattice, the Brillouin zone.

We can consider the Voronoï cell of any point of the lattice. All these cells form a regular paving of space. All points on the boundary ∂D_0 of D_0 are equidistant from 0 and at least a lattice point (let us denote it by c). We call these lattice points *corona vectors* and denote their set by C. We can also say that the corona vectors define those lattice points whose Voronoí cell touches D_0. Technically [2] :

$$C = \{c \in L, \ D_0 \cap D_c \neq \emptyset\} \Leftrightarrow C = L \cap \partial 2D_0. \qquad 1(5)$$

Geometrically, the union of the Voronoï cells centered on the corona vectors form a (n-dimensional) crown around D_0; this explains our choice of the expression: corona vectors.

Let c be a corona vector; since the middle of any pair of points of L is a symmetry center of the lattice, this is the case of the point $\frac{1}{2}c$; the corresponding symmetry exchanges D_0 and D_c: so

Lemma 1a. $\frac{1}{2}c$ *is symmetry center of the intersection* $D_0 \cap D_c$.

The last expression in 1(5) and the first definition of the Voronoï cell prove the lemma:

------

[1] We use the notation D in honor of Dirichlet which introduced first this object in dimension $n = 2$. It was also studied by Fedorov for $n = 3$ and by Minkowski for arbitrary n. But the first deep study of this geometrical object was made by Voronoï.

[2] The Voronoï cell at 0 of the lattice $2L$ is $2D_0$.

Lemma 1b. *The corona vectors are the lattice vectors which are shortest in their $L/2L$ coset.*

This result can also be expressed in formulas:

$$c \in C \;\leftrightarrow\; \forall 0 \neq v \in L,\; N(c+2v) - N(c) \geq 0 \;\leftrightarrow\; \forall 0 \neq v \in L,\; (c,v) + N(v) \geq 0,\; \leftrightarrow$$

$$\leftrightarrow\; \forall 0 \neq v \in L,\; (c+v,v) \geq 0. \tag{1(6)}$$

Using this last inequality we can generalize from $m = 2$ to every integer $m > 2$:

$$m^{-1}(N(c+mv) - N(c)) = 2((c,v) + N(v)) + (m-2)N(v) > 0 \;\Rightarrow$$

$$\Rightarrow\; (m > 2, \forall 0 \neq v \in L,\; N(c+mv) > N(c)). \tag{1(7)}$$

This is the proof of

Lemma 1c. *A corona vector is the shortest lattice vector in its coset L/mL for all integers $m \geq 3$.*

Hence

Corollary 1c. $|C| \leq 3^n - 1$ *and* $|C|$ *is even.*

The last statement comes from the invariance of L under the symmetry through the origin: so $c \in C \Leftrightarrow -c \in C$.

Since the number of corona vectors is finite, the number of bisector hyperplanes bounding the Voroni cell is finite; that and the boundedness of $\mathcal{D}_0$ that we have proven, shows that a Voronoï cell is a polytope. The boundary $\partial \mathcal{D}_0$ is made of $|F|$ faces, i.e. $n-1$ dimensional domains of the supporting hyperplanes. Each face is commun to two Voronoï cells. Using also lemma 1a, we have just proven the

Lemma 1d. *A Voronoï cell is a polytope which is centrosymmetric and has centrosymmetric faces.*

Remark that the $d < n-1$ dimensional faces are not in general centrosymmetric; this is only the case of those containing $\frac{1}{2}c$, the middle of a corona vector: they are the intersection $\mathcal{D}_0 \cap \mathcal{D}_c$.

The corona vector whose middle is a symmetry center of a face of $\mathcal{D}_0$ are called *face vectors*; we denote their set by F; we know that $f \in F \Leftrightarrow -f \in F$ and this implies $|F|$ is even. The equation of the plane bisector of a lattice vector v is

$$(v, x) = \frac{1}{2} N(v) \qquad 1(8)$$

and another definition of $\mathcal{D}_0$ is:

$$\mathcal{D}_0 = \{x, \forall f \in F, \ |(f, x)| \le \frac{1}{2} N(f)\}. \qquad 1(9)$$

Since a face contains the middle of one face vector only, we have the strict inequality:

$$f_1, f_2 \in F, \ f_1 \pm f_2 \ne 0, \quad |(f_1, f_2)| < \min(N(f_1), N(f_2)). \qquad 1(10)$$

We now prove a lemma which caracterizes the subset $F \subseteq C$.

Lemma 1e. *A lattice vector f is a face vector if and only if $\pm f$ are strictly shorter than the other vectors of their $L/2L$ coset.*

Proof of "if". Assume that in the same $L/2L$ coset there are other corona vectors $\pm c$, $N(c) = N(f)$; so $v = \frac{1}{2}(f + c) \in L$. Then one verifies that $\frac{1}{2} f$, the center of a face satisfies $1(8)$, the equation of the bisector hyperplane of $v \in L$; this is absurd. Proof of "only if". Assume that $\pm c$ are stricly shorter in their $L/2L$ coset and that the corona vector is not a face vector: $\frac{1}{2} c$ belongs to the boundary of a face of center $\frac{1}{2} f$; so $(f, c) = N(f)$. The middle of the vector $c' = 2f - c$ is in the same face (it is the symmetric of $\frac{1}{2} c$ through the face center); so c' is a corona vector in the same $L/2L$ coset as c. However $N(c') = N(c) + 4N(f) - 4(f, c) = N(c)$ which contradicts the hypothesis.

Lemma 1f. *The number $|F|$ of faces of a Voronoï cell satisfies $2n \le |F| \le 2(2^n - 1)$.*

The upper bound is an immediate consequence of lemma 1e. The lower bound means that at least n pairs of parallele hyperplanes are necessary to envelop a bounded domain; moreover if it is a Voronoï cell with $2n$ faces, each hyperplane of a parallele pair has to be perpendicular to the hyperplanes of the other pairs, so this is the Voronoï cell of an "orthorhombic P-lattice". As a particular case, for the cubic-P lattice, all faces are equal. For $n = 2$ the number of faces is therefore either 4 for orthorhombic-P or square lattices or 6 for the other lattices. These bounds on $|F|$ were found by Fedorov for $n = 3$ and by Minkovski for arbitrary dimension n. We remark that for the orthorhombic-P lattices (and in particular for the cubic-P ones) $|C| = 3^n - 1$: for these lattices the lower bound of $|F|$ and the upper bound of $|C|$ are both reached.

We denote by S the set of shortest vectors of a lattice and by $N(s)$ their commun norm.

Lemma 1g. *The lattice vectors of norm $< 2N(s)$ are face vectors and the vectors of norm $2N(s)$ are corona vectors; when they are in the same $L/2L$ cosets, they form a subset of a cross. All vectors of $L_{2N(s)}$ which are not face vectors, are sum of two orthogonal short vectors* [3] *.*

Let $N(v_1), N(v_2) \leq 2N(s)$; then condition ($\#$) $0 \neq v_1 \pm v_2 \in 2L$ requires:

$$0 \leq N(v_1 \pm v_2) - 4N(s) = (N(v_1) + N(v_2) - 4N(s)) \pm 2(v_1, v_2) \qquad 1(11)$$

This can be satisfied if and only if:

$$N(v_1) = N(v_2) = 2N(s) \quad \text{and} \quad (v_1, v_2) = 0. \qquad 1(12)$$

So if $N(v_1) < 2N(s)$ there are no linearly independent vector v_2 in the same $2L$ coset and norm $N(v_2) \leq 2N(s)$; that proves $N(v) < 2N(s) \Rightarrow v \in F$. We have also proven that no vectors of norm $2N(s)$ is in the same $L/2L$ coset than a shorter vector: hence it is a corona vector. If v is not a face vector, there are other corona vectors in the same coset, and condition ($\#$) shows that they form a cross or part of it. We label $2v = c_1 - c_2$, the difference of two corona vectors of norm $2N(s)$. The last expression in (6) shows that $(v + c_2, v) = 0$. So c_1 is the sum of the two orthogonal vectors and $N(c_1) = 2r = N(v + c_2) + N(v)$ shows that they are short vectors.

---

[3] *The last sentence has been proven by Venkov.*

Conversely if $c = x + y$ is the sum of two orthogonal short vectors: $x, y \in S$, $(x, y) = 0$, the four distinct vectors $\pm x \pm y \in L_{2N(s)}$ are in the same $L/2L$ coset, so they are corona vector but not face vectors. Combining this result with lemma 1e we obtain:

Corollary 1g. $S \subseteq F \subseteq C$.

Finally, we shall need some properties relating the vectors of L and those of its dual lattice. In the orthogonal space, the set of vectors whose scalar products with all vectors of L are integers form a lattice L^* which is called the dual lattice of L. Evidently $(L^*)^* = L$. If $\{b_i\}$ is a basis of L, the vectors $\{b_j^*\}$ defined by:

$$(b_i, b_j^*) = \delta_{ij} \qquad\qquad 1(13)$$

form a basis of L^* called the dual basis of $\{b_i\}$. Dual lattices have same holohedry P: if in the basis $\{b_i\}$ of L, the holohedry P acts by matrices $g \in GL_n(Z)$ on the vectors of L, in the dual basis P acts through the contragredient representation $g \mapsto (g^{-1})^\top$ on the vectors of L^*. We define a P-invariant non negative integer for each pair of vectors:

$$x \in L, \ y^* \in L^*, \quad |x, y^*| = \max_{g, g' \in P} (g.x, g'.y^*) \equiv \max_{g \in P} (g.x, y^*) = \max_{g \in P} (x, g.y^*).$$
$$1(14)$$

We will call $|x, y^*|$ the P-orbit scalar product of these two vectors. There are no simple relations between the Voronoï cells of two dual lattices.

2. Some geometric invariants of lattices.

In the previous section we have introduce two different types of fundamental domains for a lattice: P_b which depends on the choice of lattice basis and the Voronï cell D_0 which is basis independent. We denote by V the set of vertices of D_o. This Voronoï cell is the convex hull of V. For any basis of L, we can define δ_b as the smallest positive number such that

$$V \subset \delta_b \bar{P}_b \Leftrightarrow D_0 \subseteq \delta_b \bar{P}_b, \qquad\qquad 2(1)$$

where $\bar{P}_b$ is the topological closure of P_b. For each basis we can define δ_b as the smallest positive number such that $\mathcal{D}_0 \subseteq \delta_b P_b$. Let $\mathcal{B}(L)$ the set of bases of the lattice L. As we have seen it is a countable set, since the bases are tranformed from each other by elements of $GL_n(Z)$. So we can define the positive number:

$$\delta(L) = \min_{b \in \mathcal{B}(L)} \delta_b. \qquad 2(2)$$

This number is the same for any lattice obtained from L by an orthogonal transformation or a dilation. These properties are also shared by all integral geometric invariant we shall introduce in this section.

2.1 *The lattice invariants* σ, ϕ, γ.

Instead of the set V we consider here the three sets S, F, C; as we have seen, they are subsets of $2\partial\mathcal{D}_0$. In a similar fashion we define for each basis, $\sigma_b, \phi_b, \gamma_b$ as the smallest positive numbers which satify:

$$S \subset 2\sigma_b \bar{P}_o, \quad F \subset 2\phi_b \bar{P}_o, \quad C \subset 2\gamma_b \bar{P}_o. \qquad 2(3)$$

Remark that $\sigma_b, \phi_b, \gamma_b$ are positive integers since the vectors of S, F, C belong to the lattice. Moreover $S \subseteq F \subseteq C \subset 2\mathcal{D}_o$ implies:

$$\sigma_b \leq \phi_b \leq \gamma_b. \qquad 2(4)$$

We take the minima of these positive numbers over the set of bases $\mathcal{B}(L)$ of the lattice:

$$\sigma(L) = \min_{b \in \mathcal{B}(L)} \sigma_b, \quad \phi(L) = \min_{b \in \mathcal{B}(L)} \phi_b, \quad \gamma(L) = \min_{b \in \mathcal{B}(L)} \gamma_b. \qquad 2(5)$$

A simple proof per absurdo shows that the properties $2(4)$ are preserved:

$$\sigma(L), \phi(L), \gamma(L) \in Z; \quad \sigma(L) \leq \phi(L) \leq \gamma(L). \qquad 2(6)$$

In crystallography the *orthorhombic-P* lattices have an orthogonal basis which is not an "orthonormal" basis, except for the *cubic-P* lattices; in such basis $\bar{P}_o = \mathcal{D}_o$. Hence for orthorhombic-P lattices: $\sigma = \phi = \gamma = 1$

2.2 *The lattice invariants $\bar{\sigma}, \bar{\phi}, \bar{\gamma}$.*

In this section we introduce three new functions which also involve the dual lattice. In the basis $\tilde{b} = \{b_i\}$ of L, the coordinates of any $x = \sum_i \xi_i b_i \in R^n$ are $\xi_i = (b_i^*, x)$ where $\{b_i^*\}$ is the dual basis, i.e. $(b_i^*, b_j) = \delta_{ij}$. Remark that the $2n$ faces of P_o have for equation $(b_i^*, x) = \pm \frac{1}{2}$. With this explicit notation for the coordinates, the positive numbers defined in 2(3) can be defined equivalently:

$$\sigma_b = \max_{1 \leq i \leq n, s \in S}(b_i^*, s), \quad \phi_b = \max_{1 \leq i \leq n, f \in F}(b_i^*, f), \quad \gamma_b = \max_{1 \leq i \leq n, c \in C}(b_i^*, c). \qquad 2(7)$$

Since the sets S, F, C, V are invariant by the lattice holoedry P, we can replace the scalar products by the P-orbit scalar products defined in 1(14). Instead to take the min.max on the set of basis, we can defined three new invariants, by taking the min.max of these P-orbit scalar product, for *any* vector t^* of the dual lattice:

$$\bar{\sigma}(L) = \min_{t^* \in L^*} \max_{s \in S}|t^*, s|, \quad \bar{\phi}(L) = \min_{t^* \in L^*} \max_{f \in F}|t^*, f|, \quad \bar{\gamma}(L) = \min_{t^* \in L^*} \max_{c \in C}|t^*, c|. \qquad 2(8)$$

From the inclusions $S \subseteq F \subseteq C \subset L$, we have as in 2(6):

$$\bar{\sigma}(L), \bar{\phi}(L), \bar{\gamma}(L) \in Z; \quad \bar{\sigma}(L) \leq \bar{\phi}(L) \leq \bar{\gamma}(L). \qquad 2(9)$$

Similarly, we note that $\bar{\sigma}(L), \bar{\phi}(L), \bar{\gamma}(L)$ are positive integers. These invariants $\bar{\sigma}, \bar{\phi}, \bar{\gamma}$ are respectively equal to σ, ϕ, γ when we can extract a L^*-basis from the P-orbit of the minimizing dual vector t^*. More generally:

$$\bar{\sigma}(L) \leq \sigma(L), \quad \bar{\phi}(L) \leq \phi(L), \quad \bar{\gamma}(L) \leq \gamma(L). \qquad 2(10)$$

From these inequalities we obtain for the orhtorhombic-P lattices $\bar{\sigma} = \sigma = \bar{\phi} = \phi = \bar{\gamma} = \gamma = 1$.

We now give a geometrical interpretation of $\bar{\sigma}(L)$ and $\bar{\phi}(L)$. Every lattice defines a sphere packing: the spheres are centered at the lattice points and their radius r satisfies: $4r^2 = N(s)$, $s \in S$. These spheres are those inscribed in each Voronoï cell. The number of spheres tangent to a fixed one is $|S|$ and S is the set of centers of the sphere touching that centered at the origin.

So the minimal number of parallel hyperplanes containing the centers of all spheres tangent to the one centered in 0 is $2\bar{\sigma} + 1$; (this number has to be odd because the set of these spheres has 0 as symmetry center). We can also say:

Lemma 2.1 *Given a lattice L, the integer $\bar{\sigma}(L)$ is the minimal number of neighbour parallel hyperplanes on each side of a lattice hyperplane H which are necessary for containing all spheres touching those centered in H.*

Similarly: The minimal number of neighbour parallel hyperplanes containing the centers of all Voronoï polytopes having a commun face with a given one $\mathcal{D}_o$ is $2\bar{\phi}(L) + 1$. We can also say:

Lemma 2.2 *Given a lattice L, the integer $\bar{\phi}(L)$ is the minimal number of neighbour parallele hyperplanes such that the union of the Voronoï cells centered on these hyperplanes form a thick wall without holes ,i.e., the complement of the wall in the space R^n is the union of two connected components.*

Indeed, let t^* a vector of the dual lattice which yields in 2(8) the minimum corresponding to $\bar{\phi}$; consider the wall is made of the Voronoï polytopes centered in the hyperplanes H_m of equation $(t^*, x) = m$, $1 \leq m \leq \bar{\phi}(L)$. The hyperplane $H_{\bar{\phi}(L)}$ contains at least one face vector f such that $(t^*, f) = \bar{\phi}(L)$; if we remove from the wall the Voronoï cells centered in the hyperplane $H_{\bar{\phi}(L)}$, the face of $\mathcal{D}_o$ centered in $\frac{1}{2} f$ is not touched by a face of a Voronoï cell of the thinner wall $\cup_{m=1}^{\bar{\phi}-1} H_m$; so, following from o the lattice vector f we pass through a hole of the thinner wall.

By a long proof which relies partly on results obtained by computers, we have shown that for all lattices in dimension $n < 8$, $\bar{\sigma} = \bar{\phi} = 1$.

Using equation 1(10) and the obvious

$$\forall f \in F, \ |(f, v)| \leq N(f) \Leftrightarrow v \in C. \qquad 2(11)$$

we obtain for self dual lattices:

$$L = L^*, \ S = F, \quad \bar{\sigma}(L) = \bar{\phi}(L) = \bar{\gamma}(L) = N(s), \qquad 2(12)$$

318

or

$$L = L^*, \quad |S| < |F|, \quad \bar{\sigma}(L) < \bar{\phi}(L) = N(s) < \bar{\gamma}(L) = N_m(f), \qquad 2(13)$$

where $N_m(f)$ is the maximum of the norm of face vectors. As respective examples for these two equations:

$$L = E_8: \quad \bar{\sigma} = \sigma = \bar{\phi} = \phi = \bar{\gamma} = \gamma = 2; \qquad 2(14)$$

$$L = \text{Leech lattice} \equiv \Lambda_{24}: \quad \bar{\sigma} = 3 \le \sigma \le 4 = \bar{\phi} = \phi = \bar{\gamma} = \gamma; \qquad 2(15)$$

More generally, for integral lattices, since $L < L^*$, there are more integral linear forms, so the equality with the norm values have to be replaced by inequalities:

$$L < L^*, \quad \bar{\sigma}(L) \le \bar{\phi}(L) \le N(s), \quad \bar{\gamma}(L) \le N_m(f). \qquad 2(16)$$

3. The functions ρ on $\{AC\}_n$ and ρ' on lattices.

To use shorter expressions, we follow here the crystallography literature by calling an element of the set $\{CCF(GL_n(Z))\}$ of conjugation classes of finite subgroups of $GL_n(Z)$ an *arithmetic class*. Consequently we use two equivalent labels for this set:

$$\{AC\}_n \equiv \{CCF(GL_n(Z))\} \qquad 3(1)$$

This set carries a natural partial order by subgroup inclusion up to a conjugation. The function ρ tells which integers are absolutely necessary for writing the matrix elements of a finite subgroups $F < GL_n(Z)$ defined up to a conjugation.

Let F be a finite subgroup of $GL(n, Z)$ with $|F|$ elements. Let $g \in F$; we denote its matrix elements by g_{ij}. We define:

$$m(F) = \max_{g \in F} |g_{ij}| \qquad 3(2)$$

It is the the maximal absolute value of the $|F|.n^2$ elements of $g \in G$. We denote by $[F]$ the conjugacy class of F in $GL(n, Z)$. We define:

$$\rho([F]) = \inf_{F' \in [F]} m(F') \qquad 3(3)$$

This defines a function on $\{CCF(GL_n(Z))\}$, valued in the set N_+ of positive integers. If F_1 is a subgroups of the finite group $F_2 < GL_n(Z)$, then $[F_1] < [F_2]$. Let F_2' one group which minimizes $m(F_2)$ on the class $[F_2]$ and F_1' the conjugate of F_2 by the conjugation in $GL_n(Z)$ which transforms F_2 into F_2'. From 3(3) we have $\rho([F_1]) \leq m(F_1') \leq m(F_2') = \rho([F_2])$; this shows that ρ is a non decreasing function on $\{CCF(GL_n(Z))\}$.

C. Jordan has shown that $\{CCF(GL_n(Z))\} \equiv \{AC\}_n$ is a finite set for all dimensions. So we can define a function from N_+ to N_+ by:

$$\rho_n = \max_{[F] \in CCF(GL_n(Z))} \rho([F]) \qquad 3(4)$$

Since $\rho([F])$ is a non decreasing function on the arithmetic classes, to compute the value of ρ_n we need only to know $\rho([F])$ on $\mathcal{F}_n$, the set of the maximal elements of $\{CCF(GL_n(Z))\}$. Evidently all its elements are maximal Bravais classes. So an equivalent definition of ρ_n is:

$$\rho_n = \max_{[F] \in \mathcal{F}_n} \rho([F]) \qquad 3(4')$$

We denote by F^* the "contragredient group" of F, that is the group of matrices $g^{-1\top}$ with $g \in F$. Obviously:

$$\rho([F]) = \rho([F^*]). \qquad 3(5)$$

In crystallography, the arithmetic classes which are holohedry (i.e. symmetry of the lattice) are called Bravais classes. By definition, the set $\{BC\}_n$ of Bravais classes in dimension n, is a subset of $\{AC\}_n$. So we can define a function ρ' on the set of lattices in dimension n; for a lattice L, $\rho'(L)$ is the value of ρ on the Bravais class of L. Equation 3(4) implies that the function ρ' is invariant by lattice duality:

$$\rho'(L) = \rho'(L^*). \qquad 3(5')$$

Moreover the maximal arithmetic classes are Bravais classes so there is a maximum value ρ_n' of $\rho'(L)$ in dimension n; indeed $\rho_n' = \rho_n$. More generally, when $[F] \in \{CCF(GL_n(Z))\}$ is a Bravais class $[P]_{GL_n(Z)}$, the geometry of the lattices belonging to this Bravais class gives some information on the value of $\rho([P])$. This is what we shall see now by giving an equivalent definition of ρ'.

Let $\{b_j\}$ be a basis of a lattice L belonging to the Bravais class $[P]_{GL_n(Z)}$ and $\{b_i^*\}$ the dual basis. The matrix $g \in F$ has for elements:

$$g_{ij} = (b_i^*, g.b_j) \qquad 3(6)$$

where $g.v$ is the transformed of the vector v by g:

$$g.b_j = \sum_i b_i g_{ij} \qquad 3(7)$$

Equation 3(6) and the definition of the P-orbit scalar product in 1(14) yield a new definition of $m(P)$:

$$m(P) = \max_{1 \le i,j \le n} |b_i^*, b_j| \qquad 3(8)$$

We obtain $\rho([P]) = \rho'(L)$ by minimization over all bases:

$$\rho([P]) = \min_{b \in B} m(P). \qquad 3(9)$$

In plain words, $m(P)$ is the maximum of the absolute value of any coordinate of any vector of the orbits of the basis vectors. If we can extract a basis from F, the set of face vectors (respectively S, the set of shortest vectors), from the stability of F and S under the holohedry, 2(3) and 2(7) show that $m(P) = \phi_b$ (respectively $m(P) = \sigma_b$; this proves:

$$G < GL_n(Z), \ \ G.L = L, \quad \rho([G]) \le \begin{cases} \phi(L) & \text{when } \exists\{b_j\} \subset F, \\ \sigma(L) & \text{when } \exists\{b_j\} \subset S. \end{cases} \qquad 3(10)$$

An obvious lower bound for $\rho'(L)$ is $\bar{\sigma}(L)$. We also remark that the function ρ' is invariant by duality; this allows to improve (10) into

$$\max(\bar{\sigma}(L), \bar{\sigma}(L^*)) \le \rho([G] \le \begin{cases} \min(\phi(L), \phi(L^*)) & \text{when } \exists\{b_j\} \subset F, \\ \min(\sigma(L), \sigma(L^*)) & \text{when } \exists\{b_j\} \subset S. \end{cases} \qquad 3(10')$$

From the knowledge that E_8 is a root lattice and has therefore bases extracted from $S = F = \{\text{roots}\}$, from the equations 3(10') and 2(14) we deduce:

$$\rho([E_8]) \equiv \rho'(E_8) = 2. \qquad 3(11).$$

We have found a basis of short vectors (their norm is 4) for the Leech lattice. Its holohedry is $\tilde{C}_0 = C_0 \times Z_2(-I)$ where C_0 is the largest of the three Conway sporadic groups. From equation 2(15) we obtain:

$$3 \leq \rho(C_0) \leq 4. \qquad\qquad 3(12).$$

We have also proven $n \leq 7 \Rightarrow \rho_n = 1$ and that ρ is unbouded with n. Indeed we know that for $n \geq 4$, $\rho_n \geq E[n/4]$, the greatest integer $\leq n/4$.

The linearized polaron model treated by the diagonalization method and the Green function method.

N.N. Bogolubov Jr.
Moscow Steclov Mathematical Institute
Vavilova 42, Moscow, USSR

P. Vansant, F. Brosens and J.T. Devreese
Departement of Physics, University of Antwerp (UIA)
Universiteitsplein 1, B-2610 Antwerpen, Belgium

The Green function method for the linearized polaron [1] has been generalized to include a constant uniform magnetic field. The free energy for the linearized polaron in a magnetic field is obtained in closed form with the diagonalization of the hamiltonian by a canonical transformation [2] and with the above Green function method. We have proved that both theories lead to the same free energy concerning the part of the interaction between the electron and the phonons.

I. INTRODUCTION

The linearized polaron model in a magnetic field plays an important role as mathematical tool in the study of the Feynman polaron model. Devreese and Brosens [3] have proved that the linear polaron hamiltonian is equivalent to Feynman's quadratic action with a specific choice of the parameters. Therefore the expression, in closed form, of the partition function of the linear polaron model makes it possible to calculate the expectation values of the groundstate energy E_0, the free energy F, etc. [2,3,4,5]. For the Feynman polaron this calculation is done by a variational method which is based on the Feynman-Jensen inequality. This inequality is correct for the polaron in the abscense of a magnetic field. But applying a magnetic field, an imaginary term appears in the action. Therefore one cannot prove the correctness of the inequality in this case. Nevertheless, it will be used because it delivers an approximatif result wether the method is variational or not.

Also in the work of Bogolubov and Bogolubov Jr. [1],[6] the linear polaron model is important, for instance in the T-product operator treatment of the Fröhlich polaron. Nevertheless, the model has not been treated before in the presence of a magnetic field.

II. THE LINEARIZED POLARON MODEL BY
THE GREEN FUNCTION METHOD

A. The calculation of the expectation values.

The hamiltonian, studied by Bogolubov and Bogolubov Jr. [1] for an electron with bandmass m in a constant and uniform magnetic field $\vec{B}$ parallel to the z-axis, interacting with LO-phonons is:

$$H = \frac{1}{2m} \left[p_x^2 + (p_y + m\omega_c x)^2 + p_z^2 \right] + \frac{(K_0^2 + \eta^2)}{2} \mathbf{r}^2$$
$$+ \frac{i}{\sqrt{V}} \sum_{\mathbf{f}} S(f) \sqrt{\frac{\hbar}{2\nu(f)}} (\mathbf{f} \cdot \mathbf{r}) \left(b_f + b_{-f}^+ \right)$$
$$+ \sum_{\mathbf{f}} \hbar\nu(f) b_f^+ b_f + \frac{1}{2} \sum_{\mathbf{f}} \hbar\nu(f) \tag{1}$$

where $S(f) = S(|\mathbf{f}|)$ is real, $\nu(f) = \nu(|\mathbf{f}|) > 0$ with radial symmetry, $\mathbf{f} = (2\pi n_x/L,\ 2\pi n_y/L,\ 2\pi n_z/L)$ with n_α integers and $L^3 = V$ the volume of the system, b_f and b_f^+ are the phonon creation- and annihilation operators, η^2 is introduced for technical reasons, afterwards we take the limit $\eta \to 0$. To obtain the translation invariance of H we use:

$$K_0^2 = \frac{1}{3V} \sum_{\mathbf{f}} \frac{S^2(f) f^2}{\nu^2(f)} \tag{2}$$

The above Hamiltonian is exactly the same as the one used by Brosens and Devreese [2] except for the harmonic term. This means Brosens and Devreese use a free electron whereas Bogolubov, Bogolubov Jr. [1] and the present authors, use an harmonic oscillator with frequency $\eta \to 0$. The equations of motion are ($\alpha = 1, 2, 3$ or x, y, z):

$$m \frac{dr_\alpha}{dt} = p_\alpha + \delta(\alpha - 2) m\omega_c r_{\alpha-1} \tag{3}$$

$$\frac{dp_\alpha}{dt} = -\delta(\alpha - 1)\omega_c(p_{\alpha+1} + m\omega_c r_\alpha) - \left(K_0^2 + \eta^2 \right) r_\alpha$$
$$- \frac{i}{\sqrt{V}} \sum_f S(f) \left(\frac{\hbar}{2\nu(f)} \right)^{\frac{1}{2}} f_\alpha \left(b_f + b_{-f}^+ \right) \tag{4}$$

$$i\hbar \frac{db_f}{dt} = \hbar\nu(f) b_f - \frac{i}{\sqrt{V}} \left(\frac{\hbar}{2\nu(f)} \right)^{\frac{1}{2}} S(f) (\mathbf{f} \cdot \mathbf{r}) \tag{5}$$

$$-i\hbar \frac{db_f^+}{dt} = \hbar\nu(f) b_f + \frac{i}{\sqrt{V}} \left(\frac{\hbar}{2\nu(f)} \right)^{\frac{1}{2}} S(f) (\mathbf{f} \cdot \mathbf{r}) \tag{6}$$

324

The equilibrium correlation functions needed to calculate the free energy, are defined by using two-time Green functions:

$$\langle A(t), B(\tau)\rangle_{eq} = \int_{-\infty}^{\infty} I_{A,B}(\omega)e^{-i\omega(t-\tau)}d\omega \tag{7}$$

Those correlation functions can be calculated using the spectral intensity $I_{A,B}$. Therefore has been shown in [1] that:

$$\langle\langle A, B\rangle\rangle_{\Omega}\Big|_{\omega-i\epsilon}^{\omega+i\epsilon} = -2\pi i\frac{1 - e^{-\beta\hbar\omega}}{\hbar}I_{A,B}(\omega) \tag{8}$$

where:

$$\langle\langle A, B\rangle\rangle_{\Omega} = \frac{1}{\hbar}\int_{-\infty}^{\infty} I_{A,B}(\omega)\frac{1 - e^{-\beta\hbar\omega}}{\Omega - \omega}d\omega \tag{9}$$

The equation of motion for this Green function is:

$$-i\Omega\langle\langle A, B\rangle\rangle_{\Omega} = \frac{1}{i\hbar}\langle AB - BA\rangle + \langle\langle\frac{dA}{dt}, B\rangle\rangle_{\Omega} \tag{10}$$

If we put $A = r_\alpha, B = r_\beta$ in this formula and use the equation of motion (4), we obtain:

$$\langle\langle p_\alpha, r_\beta\rangle\rangle_{\Omega} = -im\Omega\langle\langle r_\alpha, r_\beta\rangle\rangle_{\Omega} - \delta(\alpha - 2)m\omega_c\langle\langle r_{\alpha-1}, r_\beta\rangle\rangle_{\Omega} \tag{11}$$

whereas $A = p_\alpha, B = r_\beta$ in formula (10) yields:
for $\alpha = 3$:

$$\langle\langle r_3, r_\beta\rangle\rangle_{\Omega} = \frac{\delta_{3\beta}}{[m\Omega^2 - \eta^2 + \Omega\Delta(\Omega)]} \tag{12}$$

for $\alpha = 1$:

$$\langle\langle r_1, r_\beta\rangle\rangle_{\Omega} = \frac{\delta_{1\beta}[m\Omega^2 - \eta^2 + \Omega\Delta(\Omega)] - im\omega_c\delta_{2\beta}}{[m\Omega^2 - \eta^2 + \Omega\Delta(\Omega)]^2 - m^2\omega_c^2\Omega^2} \tag{13}$$

and for $\alpha = 2$:

$$\langle\langle r_2, r_\beta\rangle\rangle_{\Omega} = \frac{\delta_{2\beta}[m\Omega^2 - \eta^2 + \Omega\Delta(\Omega)] + im\omega_c\delta_{1\beta}}{[m\Omega^2 - \eta^2 + \Omega\Delta(\Omega)]^2 - m^2\omega_c^2\Omega^2} \tag{14}$$

where:

$$\Delta(\Omega) = -\frac{1}{V}\sum_f \frac{S^2(f)}{6\nu^2(f)}f^2\left\{\frac{1}{\nu(f) + \Omega} + \frac{1}{\Omega - \nu(f)}\right\} \tag{15}$$

B. The free energy.

The free energy F is given by:

$$F = \frac{1}{\beta} \ln(Sp\ e^{-\beta H}) \tag{16}$$

In the following calculations we only derive the free energy corresponding to the interaction of the electron with the phonon field:

$$F_{int} = F - F_{part} - F_{\Sigma} \tag{17}$$

where F_{part} the free energy is corresponding to the particle not interacting with the phonon field:

$$F_{part} = -\frac{1}{\beta} \lim_{\eta \to 0} \ln\ Sp\ e^{-\beta H_{part}} \tag{18}$$

$$H_{part} = \frac{1}{2m} \left[p_x^2 + (p_y + m\omega_c x)^2 + p_z^2 \right] + \frac{\eta^2 \mathbf{r}^2}{2} \tag{19}$$

and F_{Σ} the free energy corresponding to the phonon field:

$$F_{\Sigma} = -\frac{1}{\beta} \ln Sp\ e^{-\beta H_{\Sigma}} \tag{20}$$

$$H_{\Sigma} = \frac{1}{2} \sum_f \left(p_f p_f^* + \nu^2(f) q_f q_f^* \right) \tag{21}$$

As in [1] we use the Hamiltonian for the linearized polaron system in a magnetic field and introduce the parameter λ:

$$H(\lambda) = \frac{1}{2m} \left[p_x^2 + (p_y + m\omega_c x)^2 + p_z^2 \right] + \frac{\eta^2 \mathbf{r}^2}{2} + \frac{\lambda^2 K_0^2 \mathbf{r}^2}{2}$$
$$\frac{i}{\sqrt{V}} \sum_f \lambda S_f q_f(\mathbf{f}.\mathbf{r}) + H_{\Sigma} \tag{22}$$

We see that:

$$H(0) = H_{part} + H_{\Sigma} \tag{23}$$
$$H(1) = H \tag{24}$$

Therefore we calculate F_{int} by:

$$F_{int} = \int_0^1 \frac{\partial F(\lambda)}{\partial \lambda} d\lambda = -\theta \int_0^1 \frac{\partial}{\partial \lambda} \ln\ Sp\ e^{-\beta H(\lambda)} d\lambda$$
$$= \int_0^1 \frac{Sp\ \frac{\partial H}{\partial \lambda}\ e^{-\beta H(\lambda)}}{Sp\ e^{-\beta H(\lambda)}} d\lambda = \int_0^1 \langle \frac{\partial H}{\partial \lambda} \rangle_{\lambda, eq} d\lambda \tag{25}$$

where the index λ in $\langle...\rangle_{\lambda,eq}$ indicates that averaging is performed with the Gibbs statistical operator, corresponding to $H(\lambda)$. Using the equation of motion (4), one obtains the following expression for F_{int} after taking the limits $\eta \to 0$, $V \to \infty$:

$$F_{int} = -\frac{i\hbar}{2\pi} \int_0^1 d\lambda \int_{-\infty}^{\infty} \frac{\Omega}{1 - e^{-\beta\hbar\Omega}} \frac{1}{\Omega}$$
$$\left. \frac{\lambda\Delta_\infty(\Omega)\{3[m\Omega + \lambda^2\Delta_\infty(\Omega)]^2 - m^2\omega_c^2\}}{[m\Omega + \lambda^2\Delta_\infty(\Omega)]\{[m\Omega^2 + \lambda^2\Delta_\infty(\Omega)]^2 - m^2\omega_c^2\}} \right|_{\omega-i\epsilon}^{\omega+i\epsilon} d\omega \tag{26}$$

where $\Delta_\infty(\Omega) = \lim_{V\to\infty} \Delta(\Omega)$.

Linearized polaron in the absence of a magnetic field. For the linearized polaron model in the absence of a magnetic field ($\omega_c \to 0$) we find:

$$F_{int}^{\omega_c\to0} = -\frac{3i\hbar}{2\pi} \int_0^1 d\lambda \int_{-\infty}^{\infty} \frac{\Omega}{1 - e^{-\beta\hbar\Omega}} \frac{1}{\Omega} \left. \frac{\lambda\Delta_\infty(\Omega)}{[m\Omega + \lambda^2\Delta_\infty(\Omega)]} \right|_{\omega-i\epsilon}^{\omega+i\epsilon} d\omega \tag{27}$$

Which is exactly the result obtained in [1]. In [1] the free energy has been evaluated also in closed form for a dispersionless phonon spectrum with frequency ν_0. They obtained:

$$F_{int}^{\omega_c\to0} = -3\theta \ln \sqrt{\frac{m + \frac{K_0^2}{\nu_0}}{m}} + \frac{3\hbar}{2}(\mu - \nu_0) - 3\theta \ln \left[\frac{1 - e^{-\beta\hbar\nu_0}}{1 - e^{-\beta\hbar\mu}}\right] \tag{28}$$

with:

$$\mu = \sqrt{\nu_0^2 + \frac{K_0^2}{m}} \tag{29}$$

Linearized polaron in a magnetic field. For the linearized polaron in the presence of a magnetic field we also consider F_{int} for dispersionless phonons with frequency ν_0, giving (see [1]):

$$\Delta_\infty(\Omega) = -\frac{K_0^2\Omega}{\Omega^2 - \nu_0^2} \tag{30}$$

Substituting (30) in the expression (26) for the free energy, and performing the λ-integral leads to:

$$F_{int} = -\frac{i\hbar}{4\pi} \int_{-\infty}^{\infty} \frac{d\omega}{1 - e^{-\beta\hbar\Omega}} \left\{ \ln\left(\Omega^2 - \mu^2\right) - 3\ln\left(\Omega^2 - \nu_0^2\right) - \ln\left(\Omega^2 - \omega_c^2\right) + \right.$$
$$\left. \ln\left(\left[\Omega^2 - \mu^2\right]^2 \Omega^2 - \omega_c^2\left(\Omega^2 - \nu_0^2\right)^2\right) \right\}\Bigg|_{\omega-i\epsilon}^{\omega+i\epsilon} \tag{31}$$

To solve the remaining integral, we use the complex analysis to take into account the poles and branches in 0, $x_{i=0,1,2,3}$, $-x_{i=0,1,2,3}$, ν_0, $-\nu_0$, ω_c, $-\omega_c$ with:

$$x_0 = \mu \tag{32}$$

$$x_{j=1,2,3} \text{ solutions of } x_j(x_j^2 - \mu^2) = (-1)^{j+1}\omega_c(x_j^2 - \nu_0^2) \tag{33}$$

We finally obtain:

$$F_{int} = -\theta \ln \sqrt{\frac{m + \frac{K_0^2}{\nu_0}}{m}} + \frac{\hbar}{2}(\mu - \nu_0) + \frac{\hbar}{2}\left(\sum_{j=1}^{3} x_j - 2\nu_0 - \omega_c\right)$$
$$-\theta \ln\left[\frac{\left(1 - e^{-\beta\hbar\nu_0}\right)^3}{\prod_{j=0}^{3}\left(1 - e^{-\beta\hbar x_j}\right)}\right] - \theta \ln\left(1 - e^{-\beta\hbar\omega_c}\right) \tag{34}$$

III. THE FREE ENERGY BY THE DIAGONALIZATION METHOD.

In [2] Brosens and Devreese derived the partition function for the linearized polaron system in a uniform magnetic field:

$$Z^{lin} = X \frac{m\omega_c V}{2\pi\hbar} \sqrt{\frac{m}{2\pi\hbar^2\beta}} \frac{\nu_0}{\omega} \frac{\left(2\sinh\frac{\beta\hbar\omega}{2}\right)^3}{\prod_{j=0}^{3}\left(2\sinh\frac{\beta\hbar\nu_j}{2}\right)} \tag{35}$$

where:

$$v_0^2 = \nu_0^2 + 2\gamma \tag{36}$$

$$v_{j=1,2,3} \text{ solutions of } v_j(v_j^2 - v_0^2) = (-1)^{j+1}\omega_c(v_j^2 - \omega^2) \tag{37}$$

and with X the partition function of the non-interacting phonons:

$$X = \prod_{\mathbf{k}}(n_{\mathbf{k}} + 1) \quad ; \quad n_{\mathbf{k}} = \frac{1}{e^{\beta\hbar\omega} - 1} \tag{38}$$

We take only into account the free energy corresponding to the particle and to the interaction between the particle and the phonon field. If one makes the substitutions $\omega \to \nu_0$ and $v_j \to x_j$ one obtains directly:

$$F_{int} + F_{part} = -\theta \ln \sqrt{\frac{m + \frac{K_0^2}{\nu_0^2}}{m}} + \frac{\hbar}{2}(\mu - \nu_0) + \frac{\hbar}{2}\left(\sum_{j=1}^{3} x_j - 2\nu_0\right)$$
$$-\theta \ln\left[\frac{\left(1 - e^{-\beta\hbar\nu_0}\right)^3}{\prod_{j=0}^{3}\left(1 - e^{\beta\hbar x_j}\right)}\right] - \theta \ln\left(\frac{m\omega_c V}{2\pi\hbar}\right) - \theta \ln\sqrt{\frac{m}{2\pi\hbar^2\beta}} \tag{39}$$

IV. COMPARISON OF THE TWO THEORIES

The free energy of an harmonic oscillator in a magnetic field for which we took the limit for the frequency $\omega \to 0$, is:

$$F_{part} = \frac{1}{2}\hbar\omega_c + \theta \ln\left(1 - e^{-\beta\hbar\omega_c}\right) \tag{40}$$

This free energy coincides with the free energy of the particle in our linearized polaron problem calculated by the Green function method. Using the same method as in [7] we obtain:

$$E_n = \left(n + \frac{1}{2}\right)\hbar\omega_c + \frac{\hbar^2 k_z^2}{2m} \tag{41}$$

with a total degeneracy of n:

$$\frac{m\omega_c V}{2\pi\hbar} \tag{42}$$

From the free energy for the particle (40), the degeneracy (42) and the term

$$-\theta \ln\sqrt{\frac{m}{2\pi\hbar^2\beta}} \tag{43}$$

originating from the free particle Hamiltonian (see [8]), we obtain in both cases for F_{int}:

$$F_{int} = -\theta \ln\sqrt{\frac{m + \frac{K_0^2}{\nu_0}}{m}} + \frac{\hbar}{2}(\mu - \nu_0) + \frac{\hbar}{2}\left(\sum_{j=1}^{3} x_j - 2\nu_0\right)$$

$$-\theta \ln\left[\frac{\left(1 - e^{-\beta\hbar\nu_0}\right)^3}{\Pi_{j=0}^3\left(1 - e^{-\beta\hbar x_j}\right)}\right] \tag{44}$$

This means we have proved that both theories obtain the same results for the free energy.

V. CONCLUSIONS.

We have studied the linearized polaron model in the presence of a constant and uniform magnetic field. We have proved that both theories, the diagonalization method [2] and the generalized Green function method [1] lead to the same results for the free energy concerning the part of the interaction between the electron and the phonons.

---

[1] N.N. Bogolubov Jr. N.N. Bogolubov. *Some aspects of polaron theory.* World Scientific, 1988.

[2] F. Brosens and J.T. Devreese. Rigorous groundstate energy for a linearized model of a polaron in a magnetic field. *Phys. Stat. Sol. (b) 145, 517*, 1988.

[3] F. Brosens J.T. Devreese. In path summation: achievements and goals. Eds. S. Lundqvist, A. Ranfagni, V. Sa-yakanit and L.S. Schulman, World Scientific, 1987.

[4] F. Brosens J.T. Devreese. Diagonalization of a linearized polaron model. in Proceedings of the Third International Conference on *'Path-Integrals from MeV to meV'*, Bangkok, 1989.

[5] F.M. Peeters and J.T. Devreese. Statistical properties of polarons in a magnetic field. i. analytic results. *Phys. Rev. **25**, 12, p7281-7301*, 1982.

[6] N.N. Bogolubov Jr. N.N. Bogolubov. Aspects of the polaron theory. *Comm. JINR,* P17-81-65, Dubna, 1981.

[7] C. Kittel. *Quantum theory of solids.* John Wiley & Sons, Inc., 1963. p. 219.

[8] R.P. Feynman. *Statistical Mechanics.* W.A. Benjamin, Inc., 1972.

COMPOSITE OPERATOR APPROACH
FOR
HIGHLY CORRELATED ELECTRON SYSTEMS

H. Matsumoto
Institute for Materials Research, Tohoku University
Katahira 2-1-1, Sendai 980, JAPAN

ABSTRACT

Strong coupling theory is one of the long lasting subjects in which no general method is established in quantum field theory. In recently discovered high Tc cuprate oxide superconductors, a highly correlated electron state near the Fermi level is one of the central interests to understand the mechanism of high Tc superconductivity. In this paper we present a method in which strong coupling interactions among electrons are treated by introducing composite electronic excitations describing simultaneous excitations of electrons and surrounding fluctuations. A complicated change of the electronic state in oxide superconductors from an insulator phase to a good metallic phase is explained by mixing and scattering processes among composite electronic excitations.

† Contributed to the Festschrift Volume *Structure: from Physics to General Systems* in honour of the 70th birthday of Prof. Eduardo R. Caianiello.

1. Introduction

How to treat strong correlation induced by a strong coupling is one of the difficult problems in quantum field theory. Pioneering works are found in a strong coupling theory by Wentzel[1] and an intermediate coupling theory by Tomonaga[2]. In those approaches, how particles dress clouds of other particles through the strong interaction is the interest of the argument. An approach where a strong interaction part is regarded as an unperturbed part and the kinetic part as a perturbation was proposed in a relativistic model by Caianiello[3]. In the area of solid state physics, one finds the approach by Hubbard[4] where the strong coupling at the equal site is treated to produce the level splitting in the atomic sense, which is now called the Hubbard splitting. Equations of motion and certain decomposition rules for multi-point Green functions are used to obtain a set of self-consistent equations among Green functions and the bridge between the band like behavior and the atomic like behavior was discussed in a narrow band of transition metals.

Recently there appears many materials in which many body effects and highly correlations among electrons are essential to determine physical properties of the system. Sometimes a system is largely modified by a small change of external conditions such as temperature, carrier density and applied magnetic field. Heavy fermion superconductors represented by $CeCu_2Si_2$, UBe_{13} and UPt_3[5] have attracted much interest of solid state physicists, in which electrons near the Fermi level are strongly renormalized with decreasing temperature and the conduction electron changes its mass about the order of 10^3 from room temperature to $1°K$. Recently discovered high Tc cuprate oxide superconductors[6] such as $La_{2-x}Sr_xCu O_4$ and $YBa_2Cu_3O_{7-x}$ change their phases from an antiferromagnetic insulator to a good metal by a small amount of carrier doping. The superconductivity appears only in a restricted metallic region close to the insulator[7]. This fact suggests that a certain electronic state suitable for the appearance of the superconductivity is realized in this intermediate region of the metal-insulator transition. In these phenomena, the problem is how one can describe so large change in renormalizations of electrons with a small change of parameters.

In the view point of quantum field theory, those systems have several new aspects. More or less electron levels with atomic character are involved. The transition energies among atomic levels depend on the occupation number, i.e. the creation and annihilation of electrons in these levels receive restriction according to the electron number. This necessitates a treatment of non-harmonic type canonical relations. Further, the above mentioned phenomena seem to occur in the intermediate region where the locality and itinerancy compete. There is so far no theory applicable in the intermediate region, although the local limit or itinerant limit are well studied. As for the treatment of non-harmonic oscillator type nature of quantum algebra, several methods have been proposed. In the slave boson method[8], additional operator degree of freedoms are introduced in such a way to allow application of Wick's theorem and restricted natures of the original operators are represented in forms of constraints among operators. In the non-crossing approximation(NCA)[9], modifications of local states from itinerant parts are firstly evaluated and afterwards an inter-site hopping is considered among modified local levels. These two methods have been shown successful in the analysis of the single impurity Kondo problem, while for the lattice case the former has difficulties in the treatment of constraints and the latter has difficulties in inclusion of inter-site effects. Attempts to generalize Wick's theorem for non-harmonic oscillator type quantum algebra are found in refs. 10 and 11, but there arise in this case difficulties in systematic summations of loop diagrams.

In this paper, we present our approach of the strong coupling theory developed recently in relation to the cuprate oxide superconductors[12−18]. This system is interesting not only for its high Tc superconductivity but also for the way of changing its phase from an insulator to a good metal by a small amount of carrier doping. The system is described by the p-d mixing model composed of p electrons at oxygen sites and d electrons at copper sites[19]. Experimentally[20], a large intra-site Coulomb repulsion U∼7eV at the Cu site has been concluded, which induces the Hubbard splitting in d levels. It is revealed by photoemission, X-ray absorption spectroscopy(XAS) and electron energy loss spectroscopy (EELS), that the insulator gap is the charge transfer gap formed between p and upper Hubbard levels. With carrier doping, there occurs two drastic changes in the electronic state experimentally[21−24]. The first is the rapid accumulation of the electron density of states at the Fermi level inside the charge transfer gap, which is not explained by the rigid shift of simple mixing bands. Then a transfer of the density of states from the upper Hubbard level to the low energy region is followed for further doping. The characteristic feature of this collapse of the upper Hubbard level is that it occurs without diminishing the level distance between the upper Hubbard and Fermi levels; only the intensity of levels interchanges. For a heavily doped region, the system shows an ordinary metallic phase.

In this paper, we show that the above mentioned complicated change of the density of states can be described by use of the notion of composite electronic excitations associated with Cu-O bonds. A simultaneous excitation of a p electron and neighboring fluctuations is described by composite operators. Due to strong covalency between p and d electrons and strong correlation at the Cu site, separate description of each electron freedom is no more true.

In the next section, we first review the result of a perturbation approach by starting from atomic levels, in which modification of states is described by dressing of clouds of fluctuations[12-14], and explain why we were lead to the description of composite electronic excitations. By identifying low lying electronic excitations of local character, a general scheme of the composite operator approach is developed to identify self-energies of composite operators induced by hopping interactions and decay processes among them. The non-applicability of Wick's theorem is covered by the notion of operator expansion by use of the orthogonality with respect to the norm defined through vacuum expectation values. Some of the numerical results are shown, which reproduces the doping dependence of the electronic density of states in oxide superconductors. Sect. 3 is devoted to concluding remarks.

2. Composite Operator Approach

The model is given by

$$H = \sum_i (\epsilon_d d^\dagger(i)d(i) + U n_\uparrow(i)n_\downarrow(i))$$
$$+ \sum_i \epsilon_p (p_x^\dagger(i + \frac{a_x}{2})p_x(i + \frac{a_x}{2}) + p_y^\dagger(i + \frac{a_y}{2})p_y(i + \frac{a_y}{2}))$$
$$+ \sum_i t_0 [d^\dagger(i)\{p_x(i + \frac{a_x}{2}) - p_x(i - \frac{a_x}{2}) - p_y(i + \frac{a_y}{2}) + p_y(i - \frac{a_y}{2})\} + h.c.],$$

$$(2.1)$$

where $d(i)$, $p_x(j)$ and $p_y(j)$ are annihilation operators for Cu $3dx^2 - y^2$ orbitals at the i-th site, O $2p_x$ and $2p_y$ orbitals at the j-th site, respectively. Here the spinor representation is used for electron field operators. In Eq. (2.1), $a_x(a_y)$ indicates the shift of the position in the $x(y)$ direction with Cu-Cu length a, $n_\uparrow(i) = d_\uparrow^\dagger(i)d_\uparrow(i)$ and $n_\downarrow(i) = d_\downarrow^\dagger(i)d_\downarrow(i)$. The Hamiltonian (2.1) represents the p-d mixing with the hopping energy t_0 on a CuO_2 plane with a strong Coulomb repulsion U at the Cu site.

The large on-site interaction U is handled by introducing $\eta_\sigma(i)$ and $\xi_\sigma(i)$ operators defined by[11]

$$\eta_\sigma(i) = d_\sigma(i)n_{-\sigma}(i) \tag{2.2a}$$

and

$$\xi_\sigma(i) = d_\sigma(i)(1 - n_{-\sigma}(i)) \quad . \tag{2.2b}$$

They have the following property with respect to H_0 which is the Hamiltonian H with $t_0 = 0$,

$$[\eta_\sigma(i), H_0] = \epsilon_\eta \eta_\sigma(i) \tag{2.3a}$$

and

$$[\xi_\sigma(i), H_0] = \epsilon_\xi \xi_\sigma(i) \tag{2.3b}$$

334

with $\epsilon_\eta = \epsilon_d + U$ and $\epsilon_\xi = \epsilon_d$. The operator $\eta_\sigma(i)$ represents the $n_d = 1 \leftrightarrow 2$ transition and $\xi_\sigma(i)$ does the $n_d = 0 \leftrightarrow 1$ transition, and they correspond to electronic excitations among atomic levels, forming the upper and lower Hubbard levels. We consider the situation where the p level is inside the Hubbard split levels closer to the upper Hubbard level and the p and upper Hubbard levels form the charge transfer gap in the insulator phase. Then we consider only the $n_d = 1 \leftrightarrow 2$ transition for the d electron freedom. The hopping interaction is treated as a perturbation. Although the mixing interaction is bilinear in field operators, it now induces several decay processes due to the restricted nature of $\eta_\sigma(i)$. In the following we use the notation $\eta_\sigma(x) = \eta_\sigma(i)$ etc.. Further since p electrons couple with d electrons in a definite combination, the bonding p electron is denoted as $p(x)$,

$$\gamma(-i\vec{\nabla}_x)p(x) \equiv \frac{1}{4}[p_x(x + \frac{a_x}{2}) - p_x(x - \frac{a_x}{2}) - p_y(x + \frac{a_y}{2}) + p_y(x - \frac{a_y}{2})] \quad (2.4)$$

with

$$\gamma(\vec{k})^2 = \sin^2 \frac{k_x a}{2} + \sin^2 \frac{k_y a}{2} \quad . \quad (2.5)$$

Let us use the renormalized upper Hubbard electron operator $r(x)$ as $r(x) = \sqrt{\frac{2}{n}}\eta(x)$ with the normalization $< \{r(\vec{x}), r^\dagger(\vec{y})\} > = \delta(\vec{x} - \vec{y})$. Equations of motion for $p(x)$ and $r(x)$ are given by

$$i\frac{\partial}{\partial t}p(x) = \epsilon_p p(x) + t_n r_\gamma(x) \quad (2.6a)$$

and

$$i\frac{\partial}{\partial t}r(x) = t_n p_\gamma(x) + \epsilon_\eta r(x) - \frac{t_n}{n}\sigma^\mu \delta n_\mu(x) p_\gamma(x) \quad , \quad (2.6b)$$

where $t_n = 2t_0\sqrt{\frac{n}{2}}$, $r_\gamma(x) = \gamma(-i\vec{\nabla})r(x)$, $p_\gamma(x) = \gamma(-i\vec{\nabla})p(x)$, $\sigma^\mu \delta n_\mu(x) = -(n(x)- < n >) + \vec{\sigma} \cdot \vec{n}(x)$ and $n(x) = d^\dagger(x) \cdot d(x)$, $\vec{n}(x) = d^\dagger(x)\vec{\sigma}d(x)$ with $\vec{\sigma}$ being the Pauli matrices. The equation (2.6) indicates that p and r levels mix by $t_n\gamma(\vec{k})$ and also in p-d hopping, spin and charge fluctuations are emitted and absorbed.

Through the interaction in Eq. (2.6b), electrons dress clouds of spin and charge fluctuations. Such effects are enhanced by formation of electron-hole pairs. By considering repetition of p electron-fluctuation loops, the density of states for p and d electrons is calculated in refs. 12 and 13. The result of ref. 13 shows that a certain state is formed above the bare p level even in the insulator phase and that, with carrier doping, such a new state develops at the Fermi level rapidly. Unfortunately, the approximation used there breaks down with increasing either carrier intensity, mixing strength or inverse temperature. The reason is that the inclusion from contributions of the newly developed state is insufficient and that there occurs violation of the unitarity. In this perturbation approach, fundamental electron excitations are p and d electron excitations at each site. The new state inside the bare charge transfer gap appears as a bound state of a p electron and spin and charge fluctuations. With the formation of the Fermi surface, involved particles for such formation of the bound state increase because of possible electron-hole pair creations. With the increase of available unoccupied states, electrons may gain more itinerancy by taking off clouds of fluctuations and may form an ordinary mixing band. However, to perform the above mentioned scenario in the actual calculation is quite difficult, since one needs to treat infinite order of loop corrections correctly without violating necessary requirements such as the unitarity and conservation of spin rotational symmetry. We will approach this problem in a slightly different view point in the following.

The appearance of the bound state mentioned above means that the p and d electrons are so strongly intertwined that the separate motion of each freedom is not good description. Instead of starting from the free motions of electrons without any restriction from Pauli principle, let us first consider how one can include effects of already existing electrons to the electron motion. Obviously, a p(d) electron cannot move due to the Pauli exclusion principle, when neighboring d(p) electron states with the same spin are occupied; that is, the way of the next neighbor hopping is very much affected by already existing electrons. This kind of effects are hard to include by use of freedoms only restricted at a single site; when one starts from states defined at single sites, much corrections are necessary. In order to express such configuration dependence, one needs to introduce composite operators composed of operators on several lattice points. Physically this indicates that the description starting from the independent motion of p and d electrons is not a good one but that electronic excitations are formed on a CuO_2 cluster as a whole; p electrons and charge and spin fluctuations of Cu ions are excited simultaneously. To represent this situation, we are lead to introduce composite electronic operators composed of a p electron and d spin and charge fluctuations[15-18].

Let us introduce a composite operator $p_\mu(x)$ defined by

$$p_\mu(x) = p_\gamma(x)\delta n_\mu(x) \quad , \tag{2.7}$$

where $p_\gamma(x) = \gamma(-i\vec{\nabla})p(x)$ with $\gamma^2(\vec{k})$ being given in Eq. (2.5). For the p electron with the Cu spin, we use

$$p_s(x) = \vec{\sigma}p_\gamma(x)\vec{n}(x) \tag{2.8a}$$

and for the p electron with the Cu charge, we use

$$p_0(x) = p_\gamma(x)\delta n(x) \quad . \tag{2.8b}$$

The equation of motion (2.6b) for $r(x)$ is rewritten as

$$i\frac{\partial}{\partial t}r(x) = t_n p_\gamma(x) + \epsilon_\eta r(x) - \frac{t_n}{n}(p_s(x) - p_0(x)) \tag{2.9}$$

and the equations for $p_{s,0}(x)$ are given as

$$i\frac{\partial}{\partial t}p_s(x) = \epsilon_p p_s(x) + t_n(h_s(x) - \psi_s(x) + \varphi_s(x)) \quad , \tag{2.10a}$$

and

$$i\frac{\partial}{\partial t}p_0(x) = \epsilon_p p_0(x) + t_n <2-n> r(x) + t_n(h_0(x) - \psi_0(x) + \varphi_0(x)) \quad , \tag{2.10b}$$

where $h_\mu(x) = r_{\gamma_1^2}(x)\delta n_\mu(x)$, $\psi_\mu(x) = p_\gamma(x)\delta n_{+\mu}(x)$ and $\varphi_\mu(x) = p_\gamma(x)\delta n_{-\mu}(x)$ with $h_{s,0}(x)$, $\psi_{s,0}(x)$ and $\varphi_{s,0}(x)$ being defined in a similar way as Eq. (2.8), and $\gamma_1^2(\vec{k}) = -\frac{1}{2}(\cos k_x a + \cos k_y a)$, $\delta n_{+\mu}(x) = p_\gamma^\dagger(x)\sigma_\mu r(x) - < p_\gamma^\dagger(x)\sigma_\mu r(x) >$ and $\delta n_{-\mu}(x) = r^\dagger(x)\sigma_\mu p_\gamma(x) - < r^\dagger(x)\sigma_\mu p_\gamma(x) >$.

We consider that the operators $\psi_n(x)$ $(= (p(x), r(x), p_s(x), p_0(x)))$ work as candidates to form quasi-stable electronic excitations. By a suitable combination of $\psi_n(x)$, one can represent a desired configurations of a p electron and neighboring spin and charge states of Cu ions. By taking the equations of motion for p electron and fluctuations as a whole, we can take into account of the situation where the presence of a hole, for example, modifies the surrounding fluctuations themselves.

Effects of mixing and decays among those electronic excitations $\psi_n(x)$ through the inter-site hopping are expressed as self-energies for those modes. Due to the composite nature of the operators, the notion of the self-energy needs to be generalized. Let us write Eqs. (2.6a), (2.9) and (2.10) in general in a form

$$i\frac{\partial}{\partial t}\psi_n(x) = j_n(x) \quad . \tag{2.11}$$

The retarded propagator $< R\psi_n(x)\psi_m^\dagger(y) >$ is denoted as follows,

$$< R\psi_n(x)\psi_m^\dagger(y) > = \frac{i\Omega}{(2\pi)^3}\int d\omega d^2 k\, e^{-i\omega(t_x - t_y)+i\vec{k}\cdot(\vec{x}-\vec{y})}S_{nm}(\omega,\vec{k}) \quad , \tag{2.12}$$

where the two dimensional case is considered and Ω is the volume of the unit cell. From the equation of motion

$$i\frac{\partial}{\partial t} < R\psi_n(x)\psi_m^\dagger(y) > = i\delta(t_x - t_y) < \{\psi_n(\vec{x}), \psi_m^\dagger(\vec{y})\} >$$
$$+ < Rj_n(x)\psi_m^\dagger(y) > \tag{2.13}$$

and defining a generalized self-energy as

$$< Rj_n(x)\psi_m^\dagger(y) > = \Sigma_{nn'}(-i\partial) < R\psi_{n'}(x)\psi_m^\dagger(y) > \quad , \qquad (2.14)$$

we have

$$S_{nm}(\omega,\vec{k}) = \left(\frac{1}{\omega - \Sigma(\omega,\vec{k})} I(\vec{k}) \right)_{nm} \qquad (2.15)$$

with $< \{\psi_n(\vec{x}), \psi_m^\dagger(\vec{y})\} > = \frac{\Omega}{(2\pi)^3} \int d^2 k e^{i\vec{k}\cdot(\vec{x}-\vec{y})} I_{nm}(\vec{k})$. We have

$$< \{j_n(\vec{x}), \psi_m^\dagger(\vec{y})\} > i\delta(t_x - t_y) + < Rj_n(x)\delta j_m^\dagger(y) >$$
$$= \frac{i\Omega}{(2\pi)^3} \int d\omega d^2 k e^{-i\omega(t_x - t_y) + i\vec{k}(\vec{x}-\vec{y})} \left(\Sigma(\omega,\vec{k}) I(\vec{k}) \right)_{nm} \quad , \qquad (2.16)$$

where $\delta j_m^\dagger(y) = j_m^\dagger(y) - \psi_{m'}^\dagger(y)\Sigma^\dagger(i\partial_y)_{m'm}$. Since the equal time vacuum expectation value should be time independent, we have in general $< i\frac{\partial}{\partial t} A(t,\vec{x}) B^\dagger(t,\vec{y}) > = < A(t,\vec{x})(-i\frac{\partial}{\partial t}) B^\dagger(t,\vec{y}) >$ and therefor we have

$$< \{j_n(\vec{x}), \psi_m^\dagger(\vec{y})\} > = < \{\psi_n(\vec{x}), j_m^\dagger(\vec{y})\} > \quad , \qquad (2.17)$$

which gives the static part of the self-energy in Eq. (2.16) and defines a generalized mean field $M_{0nm}(\vec{k})$ satisfying $M_0(\vec{k}) = M_0(\vec{k})^\dagger$. Further, since $< Rj_n(x)\delta j_m^\dagger(y) > = < R\delta j_n(x)\delta j_m^\dagger(y) >$ with $\delta j_n(x) = j_n(x) - \Sigma_{nn'}(-i\partial_x)\psi_{n'}(x)$, we see that $\delta M_{nm}(\omega,\vec{k})$ defined by $< Rj_n(x)\delta j_m^\dagger(y) > = \frac{i\Omega}{(2\pi)^3} \int d\omega d^2 k e^{-i\omega(t_x - t_y) + i\vec{k}(\vec{x}-\vec{y})}$ $\delta M_{nm}(\omega,\vec{k})$ satisfies the hermiticity relation $\delta M(\omega,\vec{k}) = \delta M(\omega^*,\vec{k})^\dagger$. We have

$$S(\omega,\vec{k}) = \left(I(\vec{k}) \frac{1}{\omega I(\vec{k}) - (M_0 + M(\omega,\vec{k}))} I(\vec{k}) \right) \quad . \qquad (2.18)$$

The matrix I determines the normalization of the electron operator $\psi_n(x)$. Since composite operators are included in $\psi_n(x)$, I is in general not the identity matrix. If one tries to identify a self-energy from Eq. (2.13) by use of a suitable decomposition rule, the self-energy so obtained is not easily seen to guarantee the unitarity condition, since both I and $\Sigma(\omega,\vec{k})$ participate in the total propagator. To handle M is much easier. As can be seen from Eqs. (2.17), the static part of the self-energy aside from the matrix I is determined by the anti-commutators between the source $j_n(x)$ and the electron field $\psi_m^\dagger$, that is, by taking the anti-commutator, one identifies the amplitude for a concerned component included in the source. The mean field approximation by use of a decomposition rule is replaced by the operator expansion by use of the norm defined by vacuum expectation values of anti-commutators,

$$Q = \sum_n c_n \psi_n + \delta Q \qquad (2.19)$$

with $c_n =< \{Q, \psi_m^\dagger\} > (I^{-1})_{mn}$. In the case where Wick's theorem is applied, the above definition reduces to a usual mean field approximation.

The left and right hand sides of Eq. (2.17) sometimes gives quite different expressions. The equality is valid only after taking the vacuum expectation values. A process that a mode represented by a higher order operator product changes to that of a lower order operator product should be same as its inverse process, and the equality represents a kind of the detailed valence. In an actual calculation, one must be careful not to use different approximations in the left and right sides of Eq. (2.17), otherwise the unitarity is easily violated. The explicit forms of $I(\vec{k})$, $M_0(\vec{k})$ and the dynamical correction $\delta M_(\omega, \vec{k})$ will be given elsewhere[18].

Roughly speaking, the electronic state in this approach is expressed by the mixing among p, r, p_s and p_0, where (p, p_s, p_0) are around ϵ_p and r is around ϵ_η. In Fig. 1, we show the results of numerical calculations. The bold solid line is for the d electron density of state, the fine solid line is for the p electron density of state. The parameter t_0 is chosen as $0.5eV$. The bare charge transfer gap $\Delta(\equiv \epsilon_\eta - \epsilon_p)$ is chosen as $2.5eV$. In this calculation, we evaluated the effect of fluctuation also self-consistently by considering single site decay effects. The detailed formulation will be presented elsewhere[18]. In Fig. 1, we observe the effect of the composite excitations as a small shoulder of the main p band in the insulator phase and the peak structure is enhances when the Fermi surface is formed. The analysis shows that the electron state near the Fermi level is dominated by p_s component, and for a smaller Δ, the peak structure is more broadened in the insulator phase and the shift of the intensity to the Fermi level is much enhanced. In Fig. 2, we show the t_0 dependence of the density of states in the metallic region. We choose $\Delta = 1.2eV$ and $t_0 = 0.5eV$ for the case (a) and $t_0 = 0.7eV$ for the case (b). One can see drastic change in the spectral distribution. The accumulation of the intensity at the Fermi level is more drastic in (b), so that it seems to form one band around the Fermi level. In order to understand the origin of the intensity transfer without shrinking the level distances, we present the imaginary parts of self-energies defined by

$$S_{pp\uparrow}(\omega, \vec{k}) = \frac{1}{\omega - \epsilon_p - 4t_0^2\gamma^2(\vec{k})\Sigma_p(\omega, \vec{k})} \tag{2.20}$$

and

$$\Sigma_p(\omega, \vec{k}) = \frac{1}{\omega - \tilde{\epsilon}_\eta - \Sigma_\eta(\omega, \vec{k})} \quad , \tag{2.21}$$

where $\tilde{\epsilon}_\eta = \epsilon_\eta - 2b\frac{t_n}{n}$ is the renormalized η level. The dashed line indicate the position of $\tilde{\epsilon}_\eta$. One can see the crossover of the renormalized upper Hubbard level and the composite excitations. This crossover induces the drastic decrease of intensity at the upper Hubbard level without collapsing the level distances. This may be interpreted that, due to the increase of unocuupied states, electrons hop more freely among p and d electron sites, and the trapped states (i.e. composite excitations) becomes higher energy states.

3. Concluding Remarks

In this paper, we present our recent approach for highly correlated electron system in which an on-site strong coupling is treated by restricted electron operators and configuration dependence is described by introducing composite electronic operators composed of electron and fluctuations. A method to define self-energies for such composite electron operator is presented. A complicated change of the electronic state in cuprate oxide superconductors can be explained by use of the mixing and decay processes among so introduced composite excitations. Specially a change from a highly correlated electron state to a band like state is explained by the crossover from the composite excitation dominate region to the upper Hubbard dominated region in the p-d mixing scheme.

Acknowledgments

The author would like to express his sincere thanks to Prof. E. Caianiello, in this occasion of the celebration of his 70th birthday, for his heartily friendship and continuous encouragement on the author's research, specially during the period of the author's stay as a member of the department of theoretical physics at University of Salerno. The author also would like to thank Prof. M. Tachiki, Prof. F. Mancini, Dr. T. Koyama, Dr. S. Takahashi, Mr. M. Sasaki and Mr. S. Ishihara for valuable discussions. This work was supported by Grant-in-Aid from the Ministry of Education, Science and Culture, Japan.

References

1. G. Wentzel, Helv.Phys. Acta. **13**, 269 (1940).

2. S. Tomonaga, Prog.Theor. Phys. **1**, 83 (1946).

3. E. R. Caianiello and G. Scarpetta, Nuovo Cimento **22**, 448 (1974).

4. J. Hubbard, Proc. Roy. Soc. **A276**, 238 (1963); **A277**, 237 (1964); **A281**, 401 (1964).

5. See, for example, G. R. Stewart, Rev. Mod. Phys. **56**, 755 (1984).

6. J. G. Bednorz and K. A. Müller, Z. Phys. **64**, 189 (1986).

7. J. B. Torance, A. Bezinge, A. I. Nazzal, T. C. Huang, S. S. P. Parkin, D. T. Keane, S. J. LaPlaca, P. M. Horn and G. A. Held, Phys. Rev. **40**, 8872 (1989).

8 S. E. Bars, J. Phys. **F7**, 2637 (1977); P. Coleman, Phys. Rev. **B29**, 3035 (1984).

9. Y. Kuramoto, Z. Phys. **B53**, 37 (1983); H. Kojima, Y. Kuramoto and M. Tachiki, Z. Phys. **B54**, 293 (1984); C. -I. Kim, Y. Kuramoto and T. Kasuya, Solid State Commun. **62**, 627 (1987); N. Grewe, T. Pruschke and H. Keiter, Z. Phys. **B71**, 75 (1988).

10. E. V. Anda, J. Phys. **C14**, L1037 (1981).

11. H. Matsumoto and H. Umezawa, Phys. Rev. **B31**, 4433 (1985).

12. H. Matsumoto, M. Sasaki and M. Tachiki, Phys. Rev. **B43**, 10247 (1991).

13. H. Matsumoto, M. Sasaki and M. Tachiki, Phys. Rev. **B43**, 10264 (1991).

14. S. Ishihara, H. Matsumoto and M. Tachiki, Phys. Rev. **B42**, 10041 (1990).

15. M. Sasaki, S. Ishihara, H. Matsumoto and M. Tachiki, to be published in Physica C. (1991).

16. M. Tachiki, S. Ishihara, M. Sasaki and H. Matsumoto, in Toshiba summer school at Kyoto, Japan (1991), to be published.

17. H. Matsumoto and M. Tachiki, to be published in Suppl. Prog. Theor. Phys.

18. H. Matsumoto, M. Sasaki, S. Ishihara and M. Tachiki, *Composite Operator Approach in the p-d Mixing Model of Oxide Superconductors*, (IMR, Tohoku University, preprint 1991).

19. V. J. Emery, Phys. Rev. Lett. **58**, 2794 (1987).

20. See, for example, references in refs. 14–16.

21. T. Takahashi, H. Matsuyama, K. Kamiya, T. Watanabe, K. Seki, H. Katayama-Yoshida, S. Sato and H. Inokuchi, Physica **B165 & B166**, 1221 (1990).

22. S. Uchida, T. Ido, H. Takagi, T. Arima, Y. Tokura and S. Tajima, Phys. Rev. **B43**, 7942 (1991).

23. H. Romberg, M. Alexander, N. Nücker, P. Adelmann and J. Fisk, Phys. Rev. **B42**, 8768 (1990).

24. C. T. Chen, F. Sette, Y. Ma, M. S. Hyberstsen, E. B. Stechel, W. M. C. Foulkes, M. Schluter, S-W. Cheong, A. S. Cooper, L. W. Rupp, Jr., B. Batlogg, Y. L. Soo, Z. H. Ming, A. Krol and Y. H. Kao, Phys. Rev. Lett. **66**, 104 (1991).

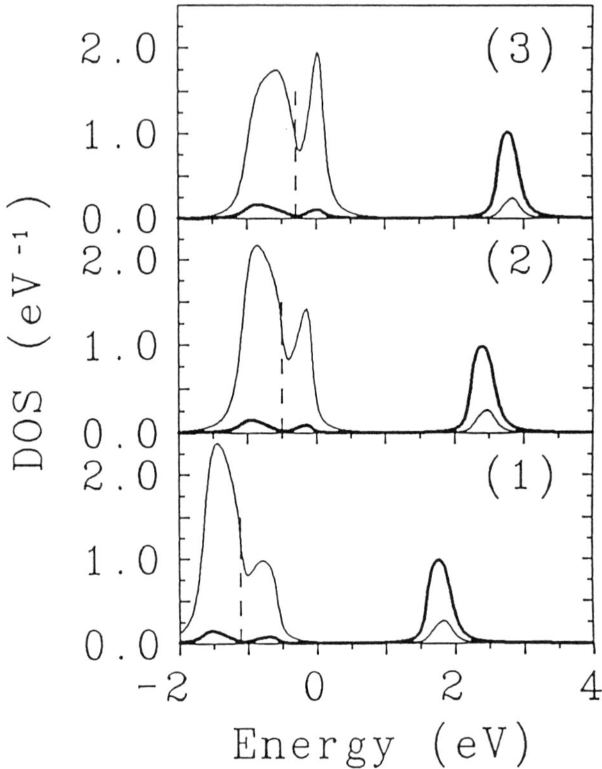

Fig. 1 Electron density of states for p(fine solid lines) and d(bold solid lines) electrons calculated from a composite operator approach[19]. Parameters are $t_0 = 0.5eV$ and $\Delta = 2.5eV$. (1) is the insulator phase, the hole number is (2)0.078 and (3)0.353.

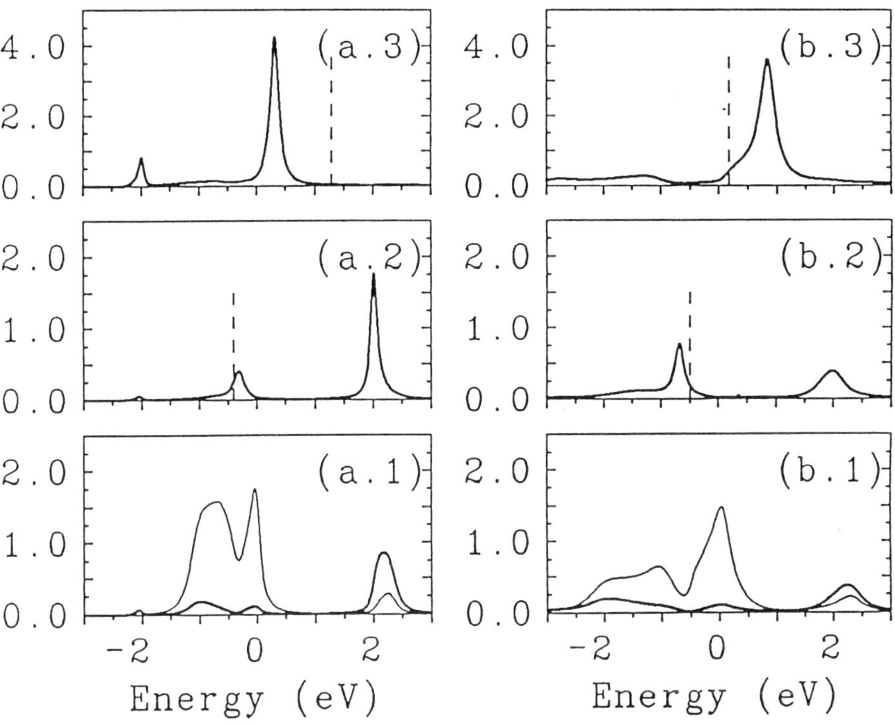

Fig. 2 Comparison of the electron density of states for p(fine solid lines) and d(bold solid lines) electrons. Parameters are $t_0 = 0.5eV(a)$ and $t_0 = 0.7eV(a)$ and $\Delta = 1.2eV$. The electron number $n_p + n_d$ is (a)2.78, (b)2.68. (1) is for full spectral functions, (2) is for $(-\frac{1}{\pi})Im\Sigma_p$ and (3)is for $(-\frac{1}{\pi})Im\Sigma_\eta$.

ON THE QUANTUM SPIN-GLASS PROBLEM

by

L. De Cesare, I. Rabuffo

Dipartimento di Fisica Teorica e S.M.S.A.
Università di Salerno, 84100 Salerno, Italy

and

K. Lukierska - Walasek, K. Walasek

Institute of Physics, Pedagogical University
of Zielona Góra, Plac Slowiański 6,
65-069 Zielona Góra, Poland

Abstract

We review some recent results concerning the phase diagrams of the quantum infinite range transverse Ising and XY spin-glass models as obtained from the "cavity field approach" via a Trotter - Suzuki transformation. This approach, which avoids the use of the replica trick and can be applied to different spin-glass models on the same footing, appears to be a promising tool towards a satisfying theory of spin-glass physics.

1. Introduction

The theory of spin glasses [1,2] has acquired, in the last twenty years, a central role in condensed matter physics, starting from the initial objective [3-5] to understand the strange behavior of certain magnetic alloys. New techniques and algorithms have been developed that can be successfully applied in other fields. Optimization [2,6] and biological [2,7] applications are examples of problems involving a wide range of complex systems, that otherwise are very difficult to handle.

The theory, although developed at the mean field level [1,5] shows a very rich and complex structure and can be now considered sufficiently well established for classical spin glass (SG) systems. The same progress is lacking at the present for quantum SG's. Only in the recent years [8-31], a growing interest has been devoted to understand the role of quantum fluctuations. In particular, the quantum infinite range transverse Ising SG model, which is a model for the proton glasses [32] as the mixed hydrogen-bonded ferroelectric $Rb_{1-x}(NH_4)_xH_2PO_4$ system [33], has been studied very extensively [10-27] by means different approaches with and without the use of the replica trick [3]. In contrast, little attention has been dedicated to the quantum XY SG model [18, 28-31], though it seems to be relevant for magnetic spin glasses with very strong anisotropy, as $Cd\,Mn$ alloys [34].

In any case, mathematical difficulties arising from the use of the replica method [2] and uncontrollable approximations [16-18,26,28-30] have forced to find alternative methods for acquiring more reliable results.

The cluster variation method, originally formulated for pure spin systems [35], has been recently extended both to classical [36] and to quantum [14,19,28] SG's in the two cluster appoximation. While in the quantum transverse Ising case [19] it seems to work sufficiently well, unfortunately, in the XY case [28] it supplies physical results for the SG phase diagram, only in the low temperature limit along the critical line in the temperature-transverse field plane, as pointed out in ref. [29].

A new method, named "cavity field approach", has been developed [2,37,38] for the classical Sherrington-Kirkpatrick (SK) model with infinite-ranged two spin interactions [4]. This seems more promising towards a proper description of SG systems. It relies on well known mathematics (i. e. probability theory) and deals with quantities having transparent physical meaning such as local magnetizations, random magnetic fields, etc., without any use of the replica trick. Quite recently [39,40], the cavity method has been successfully extended to quantum transverse Ising [39] and XY [40] versions of the SK model via a Trotter-Suzuki transformation [41] for taking into account the noncommuting nature of the spin operators.

The main objective of this paper is to review the essential aspects of the cavity method results for the above mentioned quantum infinite-range SG models, in which the quantum degrees of freedom are correctly taken into account beyond the usual static approximation [10,12,13].

The review is organized as follows. In section 2, the two quantum SG models of interest for us are presented together with their Trotter-Suzuki representation. The quantum extension of the "cavity fields approach" is outlined in section 3 and the related SG phase diagram equations are exibited in section 4. Finally in section 5 some concluding remarks are drawn.

2. Quantum infinite range SG models and Trotter-Suzuki representation.

The Hamiltonians of the quantum Ising and XY SG models consisting of N interacting spins in a transverse field Γ read:

$$H_I = -\frac{1}{2} \sum_{i,j=1}^{N} J_{ij} \hat{\sigma}_i^z \hat{\sigma}_j^z - \Gamma \sum_{i=1}^{N} \hat{\sigma}_i^x \tag{2.1}$$

$$H_{XY} = -\frac{1}{2} \sum_{i,j=1}^{N} J_{ij} \left(\hat{\sigma}_i^x \hat{\sigma}_j^x + \hat{\sigma}_i^y \hat{\sigma}_j^y \right) - \Gamma \sum_{i=1}^{N} \hat{\sigma}_i^z \tag{2.2}$$

Where $\hat{\sigma}_i^\mu (\mu = x, y, z)$ are the Pauli spin operators for the ith spin in the μ-direction. The exchange couplings J_{ij} are assumed to be quenched random variables with two possible values $J_{ij} = \pm J/\sqrt{N}$ (infinite range spin models) with equal probability for $\pm$ signs. (The same results can be also obtained for Gaussian distributions [2] with $[J_{ij}]_{av} = 0$ and $\left[J_{ij}^2 \right]_{av} = J^2/N$, where $[\ldots]_{av}$ denotes the quenched average over the J's.)

For the quantum SG models (2.1)-(2.2), it is not clear at the present how to apply directly the basic ideas of the original classical version [2] of the cavity fields approach, owing to the presence of noncommuting spin operators. An elegant way to ovecome this difficulty is to use the Trotter-Suzuki transformation [41] based on the relationship for operator A and B:

$$e^{A+B} = \lim_{M\to\infty} \left(e^{A/M} e^{B/M} \right)^M \tag{2.3}$$

Equation (2.3) can be applied to the partition function of an arbitrary d-dimensional quantum model

$$Z = Tr e^{-\beta H} \tag{2.4}$$

by decomposing the Hamiltonian H into sums of convenient terms $H = H_1 + H_2$. This gives:

$$Z = \lim_{M\to\infty} Tr \left(e^{-\beta H_1/M} e^{-\beta H_2/M} \right)^M \equiv \lim_{M\to\infty} Tr e^{-\beta \mathcal{H}_M} \tag{2.5}$$

where β is the inverse of temperature T and $\mathcal{H}_M$ denotes the effective Hamiltonian of a classical $(d+1)$-dimensional model for the Mth approximant. So, the original quantum model in d dimensions is exactly mapped into a $(d+1)$-dimensional classical one with effective Hamiltonian $\mathcal{H}_M$ consisting of $M(M\to\infty)$ d-dimensional hyperplanes in the extra Trotter-Suzuki dimension. In this way, the quantum fluctuations may be taken correctly into account, especially in computer simulations.

If the previous transformation is used for quantum models (2.1)-(2.2) in d-dimensions, they are exactly mapped in $(d+1)$-dimensional classical spin models with effective Hamiltonian for the Mth approximants [18,30,39,40]:

$$\mathcal{H}_M^{(I)} = -\frac{1}{2} \sum_{i,j=1}^N \frac{J_{ij}}{M} \sigma_{i,k}\sigma_{j,k} - \beta^{-1} K_M \sum_{i=1}^N \sum_{k=1}^M \sigma_{i,k}\sigma_{i,k+1} - N\beta^{-1} A_M \tag{2.6}$$

and

$$\mathcal{H}_M^{(XY)} = -\frac{1}{2} \sum_{i,j=1}^N \sum_{k=1}^M \frac{J_{ij}}{M} \left(\sigma_{i,k}^x \sigma_{j,k}^x + \sigma_{i,k}^y \sigma_{j,k}^y \right) +$$
$$- \beta^{-1} \sum_{i=1}^N \sum_{k=1}^M \sigma_{i,k}^y \left(I_M^{(1)} \sigma_{i,k}^x + I_M^{(2)} \sigma_{i,k+1}^x \right) - N\beta^{-1} B_M \tag{2.7}$$

348

with

$$A_M = \frac{1}{2}\ln\left(\frac{1}{2}\sinh\frac{2\beta\Gamma}{M}\right) \quad ; \quad K_M = \frac{1}{2}\ln\left(\coth\frac{\beta\Gamma}{M}\right) \tag{2.8}$$

$$B_M = \ln\left(\frac{1}{2}\cosh^{1/2}\frac{2\beta\Gamma}{M}\right) \quad ; \quad I_M^{(1)} = i\frac{\pi}{4} \quad ; \quad I_M^{(2)} = -i\arctan\left(e^{-\frac{2\beta\Gamma}{M}}\right). \tag{2.9}$$

In eqs. (2.6)-(2.9), $\sigma_{i,k} = \pm 1$ is an Ising spin variable and $\sigma_{i,k}^{\mu}(\mu = x,y)$ are XY classical spin variables for the ith site in k^{th} hyperplane, and $k(= 1,2,\ldots,M)$ denotes a site in the Trotter-Suzuki direction. Ultimately, the limit $M \to \infty$ must be taken for obtaining the exact thermodynamic properties.

3. Quantum cavity field approach.

The basic idea of the original "cavity fields approach" [2] is to compare the behavior of the classical SK model moving from N to $N+1$ spins. The system reacts at the inclusion of the new spin with a reshuffling of the various configuration levels in the ultrametricity topology of equilibrium states. The comparison appears quite meaningful since the reaction effects are very pronounced due to the existence of infinitely many valleys in the free energy of a SG [2].

With this procedure, the system of N classical spins is mapped in a $(N+1)$-spin one in the presence of a random field which is the field originally measured in the new site once the newcomer has been removed (hence the name "cavity field").

For the quantum SG models of interest, the direct extension of the previous idea to noncommuting spin operators is not clear. However, one can overcome the difficulty by applying properly the classical point of view to the corresponding Trotter-Suzuki effective models (2.6)-(2.7) where only classical spin variables are involved and the quantum degrees of freedom are described by additional spin-couplings in the extra dimension.

For clarity reasons we shall discuss separately the two quantum SG models by stressing, when necessary, the relevant differences connected with the different role played by quantum fluctuations.

A) *The Transverse Ising SG Model* [39].

The cavity field approach to quantum SK Ising model relies on a comparison of the behavior of the classical spin model (2.6) moving the total number of spins from NM to $(N+1)M$, when a *chain* of spins, located convenctionally at sites $\{(0,k); k = 1,\ldots,M\}$, is added to the original effective spin model. The reaction will be a reshuffling of the configurations inside a pure state (a thermodynamic phase). Then one can define [2,39,40] the cavity fields as the magnetic fields $\{h_k; k = 1,\ldots,M\}$ produced by a given configuration of the NM original spins and acting on the new sites $\{(0,k); k = 1,\ldots,M\}$ once the corresponding spins $\{\sigma_{0,k}\}$ have been removed:

$$h_k = \sum_{i=1}^{N} J_{0i}\sigma_{i,k} \qquad (k = 1,\ldots,M) \tag{3.1}$$

Here, the new couplings are choosed as random variables assuming the values $\pm J/\sqrt{N}$ with equal probability for $\pm$ signs.

By construction, the random variables $\{J_{0i}\}$ are uncorrelated with the spins $\{\sigma_{i,k}\}$ which are in equilibrium among themselves but not with newcomers. With the assumption of validity of the "clustering property" [2,37,38] inside a pure state in the thermodinamic limit ($N \to \infty$), the fields $\{h_k\}$ are random variables controllable through the central limit theorem and hence obeying the M-dimensional Gaussian propability distribution [39]:

$$P\left(\{h_k\}\right) = \left(\frac{\beta}{2\pi J^2}\right)^{M/2} det^{-1/2}\|\chi_{k,k'}\| \times$$
$$\times \exp\left[-\frac{\beta}{2J^2}\sum_{k,k'=1}^{M}\left(\chi^{-1}\right)_{k,k'}\left(h_k - h_{\alpha(N)}\right)\left(h_{k'} - h_{\alpha(N)}\right)\right]. \tag{3.2}$$

In eq. (3.2),

$$h_{\alpha(N)} = \langle h_k\rangle_{\alpha(N)} = \sum_{i=1}^{N} J_{0i}m^{(i)}_{\alpha(N)}\,, \tag{3.3}$$

$m^{(i)}_{\alpha(N)} = \langle\sigma_{i,k}\rangle_{\alpha(N)}$ defines the local magnetization and $\langle\cdots\rangle_{\alpha(N)}$ denotes an ensemble averaging over configurations inside the pure state $\alpha(N)$ of the NM-spin system. The parameters $\chi_{k,k'}$ are defined by the equations:

$$\chi_{k,k'} = \beta\left(\frac{1}{N}\sum_{i=1}^{N}\langle\sigma_{i,k}\sigma_{i,k'}\rangle_{\alpha(N)} - q\right) \tag{3.4}$$

$$\sum_{k''=1}^{M} \left(\chi^{-1}\right)_{k,k''} \chi_{k'',k'} = \delta_{k,k'} \tag{3.5}$$

where

$$q = \frac{1}{N} \sum_{i=1}^{N} \left(m_{\alpha(N)}^{(i)}\right)^2 \tag{3.6}$$

is the Edwards-Anderson SG order parameter [3]. Of course, the quantities $m_{\alpha(N)}^{(i)}$ and q do not depend on k because of the translational invariance along the Trotter-Suzuki dimension.

Since the probability distribution (3.2) is statistically independent of energy of the old spin system, it can be shown [2,39] that the difference between the free energies of the two systems with $(N+1)M$ and NM spins in the pure states $\alpha(N+1)$ and $\alpha(N)$ for fixed M, respectively, is given by:

$$F_{\alpha(N+1)} - F_{\alpha(N)} = -\beta^{-1} \ln Tr e^{-\beta \mathcal{H}_I^{(0)}} - \beta^{-1} A_M \tag{3.7}$$

where

$$\mathcal{H}_I^{(0)} = -\frac{J^2}{2M^2} \sum_{k,k'=1}^{M} \chi_{k,k'}\sigma_{0,k}\sigma_{0,k'} - \beta^{-1} K_M \sum_{k=1}^{M} \sigma_{0,k}\sigma_{0,k+1} - \frac{h_{\alpha(N)}}{M} \sum_{k=1}^{M} \sigma_{0,k} \tag{3.8}$$

In the presence of the new spins $\{\sigma_{0,k}\}$, the probability distribution for the random fields $\{h_k\}$ is no more Gaussian. It is correlated with the values of $\{\sigma_{0,k}\}$ and reads:

$$\wp\left(\{h_k\}, \{\sigma_{0,k}\}\right) = e^{\beta\left(F_{\alpha(N+1)} - F_{\alpha(N)}\right)} P\left(\{h_k\}\right) \times$$
$$\times \exp\left[A_M + K_M \sum_{k=1}^{M} \sigma_{0,k}\sigma_{0,k+1} + \frac{\beta}{M} \sum_{k=1}^{M} \sigma_{0,k}h_k\right]. \tag{3.9}$$

So, using eqs. (3.9) and (3.1) and the traslational symmetry along the extra dimension, it is easy to check that the average magnetization in an 0th site of the $(N+1)M$-spin system

$$m_{\alpha(N+1)}^{(0)} = \frac{1}{M} \sum_{k=1}^{M} \langle\sigma_{0,k}\rangle_{\alpha(N+1)} = \langle\sigma_{0,k}\rangle_0 \tag{3.10}$$

satisfies the equation:

$$\sum_{i=1}^{N} J_{0i} m_{\alpha(N+1)}^{(i)} = J^2 \chi m_{\alpha(N+1)}^{(0)} + h_{\alpha(N)} \tag{3.11}$$

where

$$\chi = \frac{1}{M^2} \sum_{k,k'=1}^{M} \chi_{k,k'} \tag{3.12}$$

with $\chi_{k,k'} = \chi(|k - k'|)$.

The parameters $\chi_{k,k'}$ are to be determined self-consistently by demanding that [2,39]

$$\left[\langle \sigma_{0,k}\sigma_{0,k'}\rangle_0\right]_{av} = \left[\langle \sigma_{i,k}\sigma_{i,k'}\rangle_{\alpha(N)}\right]_{av} \tag{3.13}$$

and in the thermodynamic limit no distinction is to make between $\left[\langle \sigma_{i,k}\sigma_{i,k'}\rangle_{\alpha(N)}\right]_{av}$ and $\frac{1}{N}\sum_{i=1}^{N}\langle \sigma_{i,k}\sigma_{i,k'}\rangle_{\alpha(N)}$. Then from eq. (3.4) it is immediate to see that the self consistency equations for $\chi_{k,k'}$ are :

$$\left[\langle \sigma_{0,k}\sigma_{0,k'}\rangle_0\right]_{av} - q = \beta^{-1}\chi_{k,k'} \quad . \tag{3.14}$$

Eq. (3.11), together with (3.12) and (3.14), is the quantum version of the TAP equation [2,5] for the transverse SK Ising model, from which the main SG properties can be derived at mean field level. Of course, at $\Gamma = 0$, it reduces to that for the classical SK Ising SG [2,5].

B) *The Transverse XY SG model* [40].

The cavity results for the quantum SG model (2.2) can be obtained with little modifications of the previous arguments starting from the Trotter-Suzuki representation (2.7). The only delicate point is that now the energies of the equivalent classical $2NM$-spin model are complex numbers. The difficulty can be overcome [40] assuming in (2.7) real values for the parameters $I_M^{(j)}(j = 1, 2)$ and making at the end of calculation a proper analytic continuation to the imaginary values (2.9). With this in mind, we proceed to a comparison of the behavior of our model when a chain of spins, localized at new sites $\{i = 0, k = 1, \ldots, M\}$, is added to the original $2NM$-spin system. In the present case, the cavity fields $\left\{\vec{h}_k \equiv (h_k^x, h_k^y)\right\}$ produced by a given configuration of $2NM$ original spins and acting on the new sites in absence of the new spins $\left\{\sigma_{0,k}^{\nu}; \nu = x, y\right\}$ are

$$h_k^{\nu} = \sum_{i=1}^{N} J_{0i}\sigma_{i,k}^{\nu} \qquad (\nu = x, y; k = 1, \ldots, M) \tag{3.15}$$

beying the $2M$-dimensional Gaussian probability distribution

$$P\left(\{h_k\}\right) = \left(\frac{\beta}{2\pi J^2}\right)^M det^{-1/2}\left\|\chi_{k,k'}^{\nu,\nu'}\right\| \exp\left[-\frac{\beta}{2J^2} \sum_{k,k'=1}^{M} \sum_{\nu,\nu'}^{(x,y)} \left(\chi^{-1}\right)_{k,k'}^{\nu,\nu'} \times\right.$$

$$\left.\times \left(h_k^\nu - h_{\alpha(N)}^\nu\right)\left(h_{k'}^{\nu'} - h_{\alpha(N)}^{\nu'}\right)\right]$$

$$(3.16)$$

with

$$h_{\alpha(N)}^\nu = \langle h_k^\nu \rangle_{\alpha(N)} = \sum_{i=1}^{N} J_{0i} m_{\alpha(N)}^{(i)\nu} \tag{3.17}$$

$$m_{\alpha(N)}^{(i)\nu} = \langle \sigma_{i,k}^\nu \rangle_{\alpha(N)} \tag{3.18}$$

and $\chi_{k,k'}^{\nu,\nu'}$ defined by the equations:

$$\chi_{k,k'}^{\nu,\nu'} = \beta \left(\frac{1}{N}\sum_{i=1}^{N} \langle \sigma_{i,k}^\nu \sigma_{i,k'}^{\nu'} \rangle_{\alpha(N)} - q^{\nu,\nu'}\right) \tag{3.19}$$

$$\sum_{\nu''=(x,y)} \sum_{k''=1}^{M} \left(\chi^{-1}\right)_{k,k''}^{\nu,\nu''} \chi_{k'',k'}^{\nu'',\nu'} = \delta_{\nu,\nu'}\delta_{k,k'}. \tag{3.20}$$

Here, $\langle\ldots\rangle_{\alpha(N)}$ denotes averaging over configurations inside the pure state $\alpha(N)$ of the $2NM$-spin system and

$$q^{\nu,\nu'} = \frac{1}{N}\sum_{i=1}^{N} m_{\alpha(N)}^{(i)\nu} m_{\alpha(N)}^{(i)\nu'} \tag{3.21}$$

define an anisotropic SG order parameter.

The difference between the free energies of the two systems with $2(N+1)M$ and $2NM$ spins in the two pure states $\alpha(N+1)$ and $\alpha(N)$, respectively, now reads:

$$F_{\alpha(N+1)} - F_{\alpha(N)} = -\beta^{-1}\ln T_r e^{-\beta \mathcal{H}_{XY}^{(0)}} - \beta^{-1}B_M \tag{3.22}$$

where

$$\mathcal{H}_{XY}^{(0)} = -\frac{J^2}{2M^2}\sum_{k,k'=1}^{M}\sum_{\nu,\nu'}^{(x,y)} \chi_{k,k'}^{\nu,\nu'}\sigma_{0,k}^\nu \sigma_{0,k'}^{\nu'}$$

$$-\beta^{-1}\sum_{k=1}^{M} \sigma_{0,k}^y \left(I_M^{(1)}\sigma_{0,k}^x + I_M^{(2)}\sigma_{0,k+1}^x\right) + \tag{3.23}$$

$$-\frac{1}{M}\sum_{k=1}^{M}\sum_{\nu=(x,y)} h_{\alpha(N)}^\nu \sigma_{0,k}^\nu \quad .$$

Then, the new nongaussian probability distribution of the random fields $\{h_k^\nu\}$ in the presence of the spins $\left\{\sigma_{0,k}^\nu\right\}$ will be :

$$\wp\left(\{h_k^\nu\},\{\sigma_{0,k}^\nu\}\right) = e^{\beta\left(F_{\alpha(N+1)}-F_{\alpha(N)}\right)}P\left(\{h_k^\nu\}\right) \times$$

$$\times \exp\left[B_M + \sum_{k=1}^{M}\sigma_{0,k}^y\left(I_M^{(1)}\sigma_{0,k}^x + I_M^{(2)}\sigma_{0,k+1}^x\right) + \frac{\beta}{M}\sum_{k=1}^{M}\sum_{\nu=(x,y)}\sigma_{0,k}^\nu h_k^\nu\right].$$

$$(3.24)$$

As last step, one finds that the average anisotropic magnetization in a 0th site of the $2(N+1)M$-spin system:

$$m_{\alpha(N+1)}^{(0)\nu} = \frac{1}{M}\sum_{k=1}^{M}\langle\sigma_{0,k}^\nu\rangle_{\alpha(N)} = \langle\sigma_{0,k}^\nu\rangle_0 \qquad (3.25)$$

satisfies the equations:

$$\sum_{i=1}^{N}J_{0i}m_{\alpha(N+1)}^{(i)\nu} = J^2\sum_{\nu'(x,y)}\chi^{\nu,\nu'}m_{\alpha(N+1)}^{(0)\nu'} + h_{\alpha(N)}^\nu \qquad (3.26)$$

with

$$\chi^{\nu,\nu'} = \frac{1}{M^2}\sum_{k,k'=1}^{M}\chi_{k,k'}^{\nu,\nu'} \qquad (3.27)$$

determined by the self-consistency equations:

$$\left[\langle\sigma_{0,k}^\nu\sigma_{0,k'}^{\nu'}\rangle_0\right]_{av} - q^{\nu,\nu'} = \beta^{-1}\chi_{k,k'}^{\nu,\nu'} \ . \qquad (3.28)$$

At this stage we can consider the parameters $I_M^{(j)}(j=1,2)$ to have the original imaginary values (2.9) since the averaging $\langle\ldots\rangle_0$ with $\mathcal{H}_{XY}^{(0)}$ complex is well defined and the correlation functions are real.

Eqs. (3.26)-(3.28) are the generalization of the TAP equation for the quantum XY SG model as obtained within the cavity fields approach.

4. Quantum SG-phase Diagrams.

From eqs. (3.11) and (3.26) one can extract the equations for the SG phase transition temperature as function of the transverse field Γ for our quantum SG

354

models. Indeed, assuming a continuous SG transition, linearization of eqs. (3.11) and (3.26) near the transition line gives:

$$\sum_{i=1}^{N} J_{0i} m^{(i)}_{\alpha(N+1)} = \chi_0^{-1} \left(1 + J^2 \chi_0^2\right) m^{(0)}_{\alpha(N+1)} \qquad (4.1)$$

and

$$\sum_{i=1}^{N} J_{0i} m^{(i)\nu}_{\alpha(N+1)} = \chi_0^{-1} \left(1 + J^2 \chi_0^2\right) m^{(0)\nu}_{\alpha(N+1)} \qquad (\nu = x, y) \qquad (4.2)$$

for the Ising and XY models, respectively. The parameter χ_0 has to be calculated self-consistently from eqs. (3.12) and (3.27) at $h_{\alpha(N)} = h^\nu_{\alpha(N)} = 0$ and $q = q^{\nu,\nu'} = 0$, with $\chi^{\nu,\nu'}_{k,k'} = \delta_{\nu,\nu'} \chi^{(0)}_{k,k'}$ for the XY model.

Then, the transition line $T_c(\Gamma)$ (or $\Gamma_c(T)$) in the (Γ, T)-plane can be determined from the equation [2,5]

$$det \left\| \left(1 + J^2 \chi_0^2\right) \delta_{ij} - \chi_0 J_{ij} \right\| = 0 \qquad (4.3)$$

taking into consideration only the largest eigenvalue $2J$ of the random matrix $\|J_{ij}\|$ for $N \to \infty$. Thus, the SG transition point condition becomes:

$$\chi_0 J = 1 \qquad (4.4)$$

whih, after a self-consistent calculation of χ_0 , gives in principle the freezing temperature as a function of Γ for the infinite-range Ising and XY SG models in a transverse field. A numerical investigation of the related phase diagrams for different values of M was given in refs. [39,40] (see also refs. [18,26]). In particular, for $M = 10$, one finds $\Gamma_c(0) \sim 1.5J$ and $T_c(0) = J$ for the Ising model and $\Gamma_c(0) \sim 1.44J$ and $T_c(0) \sim 0.77J$ for the XY one. The results are in good agreement with those obtained [18,26] for the same quantum models via the replica trick.

5. Concluding Remarks.

Although many progresses have been made in the last years, the SG problem is far from being solved.

For classical SG's, most of the available results, supported by numerical simulations, are still at mean field level and even this first "simple" step has required up to now a lot of effort.

In this review we have shown that, as for the classical case, also the quantum SG problem may be successfully attacked using the cavity method via a Trotter-Suzuki transformation. This approach avoids any reference to the replica trick that, althogh it remains a powerful tool in investigating frustrated systems, appears still lacking of a precise mathematical ground. We were able to obtain self-consistent TAP-like equations which are more complicated than their classical counterpart due to the proper inclusion of the quantum degrees of freedom but still manageable for numerical calculations. Even if our results are limited to a pure state and do not take into account fluctuations, there are indications [2,37,38] that significant progress can be made along this line, also in the quantum context, by means of an accurate study of the rearrangement at each level of the ultrametric state tree [2].

References

[1] K. Binder and A.P. Young, Rev. Mod. Phys. 58, 801 (1986); D. Chowdhury, *Spin Glasses and Other Frustrated Systems*, World Scientific, Singapore (1986).

[2] M. Mezard, G. Parisi, M.A. Virasoro, *Spin Glass Theory and Beyond*, World Scientific Lecture Notes in Physics, Vol. 9, (1987).

[3] F. Edwards and P.W. Anderson, J. Phys. F 5, 965 (1975)

[4] D. Sherrington and S. Kirkpatrick, Phys. Rev. Lett. 35, 104 (1975)

[5] D.J. Thouless, P.W. Anderson and R.G. Palmer, Phil. Mag. 35, 593 (1977)

[6] S. Kirkpatrik, C.D. Gelatt Jr. M.P. Vecchi, Science 220, 671 (1983)

[7] E.R. Caianiello, J. Theor. Biol. 2, 204 (1961); E.R. Caianiello and A. de Luca, Kybernetic 3, 33 (1966); W.A. Little, Math. Biosci. 19, 101 (1974); T. Kohonen, *Self-organization and Associative Memory*, Springer Verlag, Berlin (1984); reprints contained in Ref. [2].

[8] R.A. Klem J. Phys. C 12, L735 (1979); A.J. Bray and M.A. Moore, J. Phys. C 13, L655 (1980)

[9] H.-J. Sommers, J. Magn. Mater. 22, 267 (1981); H. J. Sommers and K.D. Usadel, Z. Phys. B 47, 63 (1982); K.D. Usadel, K. Bien, and H. J. Sommers, Phys. Rev B 27, 6957 (1983).

[10] H. Ishii and T. Yakamoto, J. Phys. C 18, 6225 (1985).

[11] Y.V. Federov and E.F. Shender, Pis'ma Zh. Eksp. Teor. Fiz. 43, 526 (1986) [JETP Lett. 43, 681 (1986)].

[12] K.D. Usadel, *Solid State Comm.* 58, 629 (1986).

[13] K. Walasek and K. Lukierska-Walasek, Phys.Rev. B 34, 4962 (1986).

[14] T. Yokota, Phys. Lett.A 125, 482 (1987); Phys. Rev. B 40, 9321(1989)

[15] T. Yamamoto and H. Ishii, J. Phys. C 20, 6053 (1987).

[16] K.D. Usadel and B. Schmidtz, Solid State Commun. 64, 975 (1981).

[17] T.K. Kopeć, J. Phys. C 21, 297 (1988).

[18] K.D. Usadel, Nucl. Phys. B (Proc. Suppl.)5A, 91 (1988).

[19] K. Walasek and K. Lukierska-Walasek, Phys. Rev. B38, 725 (1988).

[20] D. Thirumalai, Q. Li and T.R. Kirkpatrick, J. Phys. A 22, 3339 (1989).

[21] P. Ray, B. Chakrabarti, and A. Chakrabarti, Phys. Rev. B 39, 11828 (1989).

[22] T.K. Kopeć, K.D. Usadel, and G. Büttner, Phys. Rev. B39 12418 (1989); T. Kopeć, G. Büttner, and K.D. Usadel, Phys. Lett. A 150, 70 (1990).

[23] Q. Jiang and Z.Y. Li, Phys. Rev. B 40, 11264 (1989).

[24] Y.Q. Ma and Z.Y. Li, Phys. Lett. A 145, 19 (1990; 148, 134 (1990)

[25] G. Büttner and K.D. Usadel, Phys. Rev. B 41, 428 (1990).

[26] Y.Y. Goldshmidt and P.Y. Lai, Phys. Rev. Lett. 64, 2467 (1990).

[27] L. De Cesare, K. Lukierka-Walasek, and K. Walasek (unpublished).

[28] L. De Cesare, K. Lukierska-Walasek, I. Rabuffo, and K. Walasek, Phys. Lett A 145, 291 (1990)

[29] G. Büttner, T.K. Kopeć, and K.D. Usadel, Phys. Lett. A 149, 248 (1990).

[30] G. Büttner and K.D. Usadel, Z. Phys. B 8, 131 (1991).

[31] L. De Cesare, K. Lukierska-Walasek, I. Rabuffo and K. Walasek, (unpublished)

[32] V. Dobrosavljevic and R.M. Stratt, Phys. Rev. B 36, 8484 (1987), and references cited therein.

[33] R. Pirc, B. Tadic and R. Blinc, Z. Phys. B 61, 69 (1959); P. de Gennes, Solid State Comm.1, 132 (1963); R.B. Stinchcombe, J. Phys. C 6, 2549 (1973)

[34] H. Albrecht, E.F. Wasserman, F.T. Hedgecock and P. Monod, Phys. Rev. Lett. 48, 819 (1982); K. Baberscheke, P. Pureur, A. Fert, R. Wendel and S. Senoussi, Phys. Rev B 29, 4999 (1984)

[35] T. Morita and T. Tanaka, Phys Rev. 145, 288 (1966)

[36] K. Nakanishi, Phys. Rev. B 23, 3514 (1981)

[37] M. Mézard, G. Parisi and M.A. Virasoro, Europhys. Lett. 1,77 (1986)

[38] R. Brunetti and G. Parisi, Europhys. Lett. 11, 281 (1990)

[39] L. De Cesare, K. Lukierska-Walasek and K. Walasek, Phys. Rev. B (1992), in press.

[40] L. De Cesare, K. Lukierska - Walasek, I.Rabuffo and K. Walasek, Phys. Rev. B 45, 1041, (1992)

[41] M. Suzuki, Progr. Theor. Phys. 56, 1454 (1976); M. Suzuki, Phys. Rev. B 31, 2957 (1986) and in *Quantum Monte Carlo Methods* (Springer Series in Solid State Sciences, M. Cardona, P. Fulde, K. Klitzing, H.-J. Von Qeisser (eds.)), vol. 74, Berlin, Heidelberg, New York (1986)

POSITION SPACE RENORMALIZATION GROUP
FOR ISING QUANTUM SYSTEM

F. Esposito and U. Esposito

Dipartimento di Scienze Fisiche
Università di Napoli "Federico II"
Piazzale V. Tecchio 80 - 80125 Napoli, Italy

1. Quantum Ising chain at zero temperature

It is a great honour for us to have been invited to contribute to the festschrift of Prof. Eduardo R. Caianiello.

We began our scientific activity in quantum field theory under Caianiello's supervision applying *his* renormalization procedure to some model of quantum electrodynamic. It seems appropriate to contribute with our more recent works on renormalization group for quantum spin systems.

Macroscopic properties of many systems can be investigated by using the real-space renormalization group method. This method introduced for Ising-line classical systems by Niemeijer and van Leuween [1] was extended to quantum systems at finite temperature by Rogiers and Dekeyser [2]. Here we present a method which treats in a unified way the zero and finite temperature formalism; it can be applied to quantum Ising model for any value of the spin and can be used for any dimensionality provided that the lattice can be divided into cells which present a central site. Our method [3] can be also applied to Heisenberg [4] systems and to diluted [3], bond mixed [5] and anisotropic [6] systems.

Very often, to the first order of perturbation expansion, it does not generate proliferation terms.

We illustrate the method on a spin 1/2 linear chain in a transverse field with an Ising interaction described by the Hamiltonian

$$H = -\Gamma \sum_i \sigma_i^x - J \sum_i \sigma_i^z \sigma_{i+1}^z \qquad (1.1)$$

where J is the coupling parameter, Γ is the magnetic field and $\sigma_i^{x,z}$ are the i-site spin 1/2 Pauli matrices. This model has the property of self duality [7]. Preservation of this property at every stage of the renormalization procedure [8] gives us the possibility to locate the critical point exactly and to obtain the exact value for the thermal exponent.

We group the spins into cells, as shown in figure 1. We denote by i_0 the central site and by i_p ,$(p = \pm 1)$ the site p of the i-cell.

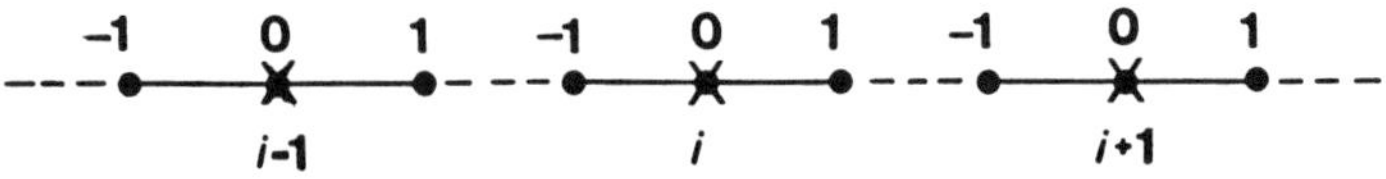

Fig. 1 -*Each cell consists of $n_s = 3$ sites. The full and broken lines represent the intra-cell and intercell interaction, respectively.*

Using the trick of Fernandez-Pachico [8] we divide the Hamiltonian into an intra-cell term H^0 and an inter-cell V, we remove from H^0 the central site field-dependent term and put it into perturbation term V

$$H^0 = \sum_i H_i^0 \qquad H_i^0 = \sum_{p=\pm 1} \left(-\Gamma \sigma_{ip}^x - J \sigma_{ip}^z \sigma_{i0}^z\right) \qquad (1.2)$$

$$V = -\Gamma \sum \sigma_{i0}^x - J \sum_i \sigma_{i1}^z \sigma_{i+1,-1}^z$$

The first step is to diagonalise H_i^0 exactly. This diagonalisation can be simplified introducing the operators L^1 and L^{-1} defined by

$$L^\alpha |\beta\rangle = \delta_{\alpha\beta} |\alpha\rangle \qquad \alpha, \beta = \pm 1 \qquad (1.3)$$

where $|1\rangle$ and $|-1\rangle$ are eigen-states of σ^z. Now the free Hamiltonian H_i^0 can be written

$$H_i^0 = L_{i0}^1 \sum_{p=\pm 1} M_{ip}^+ + L_{i0}^{-1} \sum_{p=\pm 1} M_{ip}^- \tag{1.4}$$

where

$$M_{ip}^+ = -\Gamma \sigma_{ip}^x - J\sigma_{ip}^z \quad ; \quad M_{ip}^- = -\Gamma \sigma_{ip}^x + J\sigma_{ip}^z \tag{1.5}$$

Due to the shift of the central site, the diagonalisation of H_i^0 can be reduced to a single site problem with the aid of $L^{\pm 1}$ standard basis operators. The eigenstates and eigen-values of $M_{ip}^{\pm}$ can be easily founded and then it is straightforward to obtain the eigenvalues and the eigen-states of H_i^0. If $|m\rangle_i$ is the ground state of the i-cell, the ground state of the chain $|m\rangle$ can be written

$$|m\rangle = \prod_{i=1}^{N/3} |m\rangle_i \tag{1.6}$$

where N is the total number of sites.

Guided from the value of the central-site spin we attribute to the cell the same value of spin of central-site. At zero temperature only the lowest levels of H_i^0 are important and we retain only the two lowest states. These states have the same energy and opposite spin; therefore can be easily interpreted as eigen-states of the cell spin operator s_i^x.

The first order of perturbation expansion can be obtained for the original Hamiltonian (1.1) and as result the new (renormalized) Hamiltonian, apart from a constant term, preserves the original form

$$H' = -c - \Gamma' \sum_i s_i^x - J' \sum_i s_i^z s_{i+1}^z \tag{1.7}$$

The renormalized parameters are given by the recursion relations

$$c = \frac{2}{3} N \left(\Gamma^2 + J^2 \right)^{\frac{1}{2}}$$

$$\Gamma' = \frac{\Gamma^3}{\Gamma^2 + J^2} \tag{1.8}$$

$$J' = \frac{J^3}{\Gamma^2 + J^2}$$

The recursion relations (1.8) give the exact critical field $(\Gamma/J)_c = 1$, exact critical exponent $\nu = 1$ and an approximate dynamical exponent $z = 0.63$ which coincide with results of Uzelac et al [9]. It is possible to extend calculations at second-order in V but the renormalization transformation does not change. We note that at first and second order expansion does not appear proliferation terms. In order to find the magnetic exponent β we introduce a small longitudinal field h and we add to the Hamiltonian small term

$$V_h = -h \sum_i \sigma_i^z$$

and we consider its effect on the renormalization transformation.

In table 1 we make comparison between our and others results.

Table 1 - *Comparison of our results with exact and previous renormalization-group results; here n_s=number of sites in the cell.*

	$(\Gamma/J)_c$	ν	β
exact value	1	1	0.125
Hirsch e Marenko [10]	1.09	0.88	0.170
Uzelac et al [9] $(n_s = 3)$	1	1	0.188
Uzelac et al [9] $(n_s = 17)$	1	1	0.161
Plascak [12]	1.28	1.47	0.50
Present	1	1	0.162

2. Finite-temperature extension

The finite-temperature extension of our zero-temperature technique is simplified with respect to that of other authors.

All the energy levels and corresponding eigenstates can be divided into two groups. All the states of the first group can be transformed into those of the second group, reversing all the site-spins. We associate the cell-spin variable $\sigma_i = 1$ to the first group states and the value $\sigma = -1$ to reversed-spin states of the second group.

There is a lot of arbitrariness in this associations and we choose the conventions, referred to as "central spin" and "majority rule". As can be seen diagonalising the free Hamiltonian H^0, the cell states are eigen-states of the central site spin operator σ_{i0}^z so that we can associate the cell-spin variable σ_i with its eigenvalues ± 1. This is the "central spin" convention.

If the transverse fild Γ goes to zero the cells-state are reduced to products of eigen-states of σ_i^z. In this case the cell-spin variable σ_i can be chosen according to the usual "majority rule" [1]. Now, a free-energy preserving mapping from the reduced Hamiltonian βH to an effective cell Hamiltonian $\beta' H'$ ($\beta = 1/k_B T$) can be defined according to the equation:

$$\langle \{\sigma_i\} | e^{-\beta' H'} | \{\sigma_i'\}\rangle = Tr_{\{\tau_i\}} \langle \{\sigma_i\}, \{\tau_i\} | e^{-\beta H} | \{\sigma_i'\}, \{\tau_i\}\rangle \qquad (2.1)$$

Here τ is an additional quantum number which specifies completely the states with a given value of σ_i and

$$| \{\sigma_i\} \{\tau_i\}\rangle = \prod_{i=1}^{N/3} |\sigma_i, \tau_i\rangle \qquad (2.2)$$

N is the number of sites, $N/3$ is the number of the cells and $\{\sigma_i\}, \{\sigma_i'\}$ are interpreted on the left-hand side of (2.1) as z components of new cell spin variables.

If we assume that the new Hamiltonian H' can be written in the form

$$H' = E^{0'} + H_1' + H_2' + \cdots \qquad (2.3)$$

where $H^{0'}$ is a constant which does not depend of the cell spin operators, the recursion relation between parameters of site and cell Hamiltonian can be written

$$y' = y^3 \quad , \quad K' = \frac{th^2 K \left(1 + y^2\right)^{1/2}}{1 + y^2} K$$

$$y' = th^2 \left[2K(1 + y^2)\right] y^3 \quad ; \quad K' = \frac{1}{1 + y^2} \frac{1}{4} \left\{1 + th^2 \left[K(1 + y^2)^{1/2}\right]\right\}^2 \qquad (2.4)$$

for the "central-spin" and "majority-rule" convention, respectively. Here we denoted by $K = J/k_B T$ $\Gamma/k_B T = yK$. We readily see that non-trivial fixed points K^* does not exist as the ratio K'/K in both cases is less than 1 as expected. Moreover, if we perform the $T = 0$ limit of the recursion relations (2.4) we reduced to (1.8) obtained at $T = 0$.

364

3. Extension to higher dimension

We apply our method to three types of planar lattices: honeycomb, triangular and square described by the Hamiltonian

$$H = -\Gamma \sum_j \sigma_j^x - \frac{1}{2} J \sum_{jd} \sigma_j^z \sigma_{j+d}^z \qquad (3.1)$$

where j is a site index and $j + d$ is one of the first nearest neighbours: we split the lattice into cells (fig. 2) each consisting of a central site and $n_s - 1$ surrounding sites. For honeycomb and triangular lattice $n_s = 4$ and for the square one $n_s = 5$. We denote by i a cell index, by i_0 the central site of the i cell and i_p or i_q the other sites of the i cell; p and q take the values 1,2,3 for the honeycomb and triangular lattices and the values 1,2,3,4 for the square lattice.

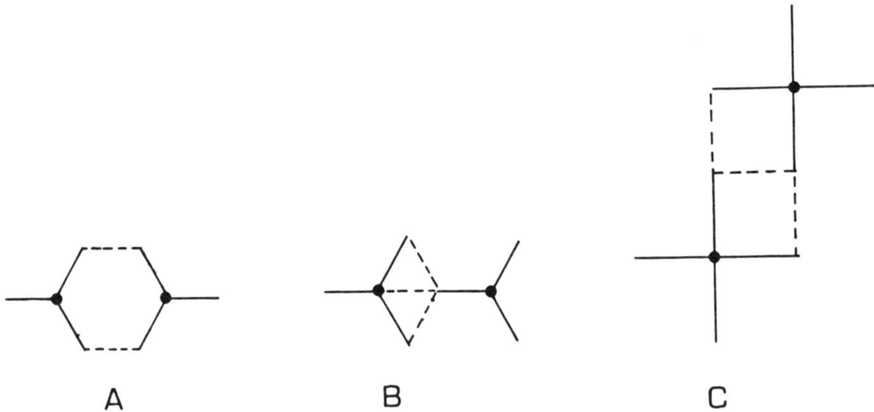

Fig. 2 - *Spin blocks and their couplings. A hexagonal lattice; B triangular lattice; C square lattice.*

The intra-cell term H^0 can be written in the form

$$H^0 = \sum_i H_i^0 \qquad H_i^0 = \sum_p \left(-\Gamma \sigma_{ip}^z - J \sigma_{ip}^z \sigma_{i0}^z \right) \qquad (3.2)$$

and the inter-cell V perturbation term can be written

$$V = -\Gamma \sigma_{i0}^x - \frac{1}{2} J \sum_{i\delta pq} \varepsilon_{p,q}^{i,i+\delta} \sigma_{ip}^z \sigma_{i+\delta,q}^z \qquad (3.3)$$

for honeycomb and square lattice. For triangular lattice, as i_p an i_q denotes sites different from the central one, we must add to equation (3.3) a term of the type

$$V_1 = -\frac{1}{2} \sum_{i,\delta q} \left(\varepsilon_{0,q}^{i,i+\delta} \sigma_{i0}^z \sigma_{i+\delta,q}^z + \varepsilon_{q,0}^{i,i+\delta} \sigma_{iq}^z \sigma_{i+\delta,0}^z \right)$$

In equation (3.3), $i + \delta$ denotes the nearest-neighbour cell of the i cell. The factor $\varepsilon_{p,q}^{i,i+\delta}$ is equal to 1 if the sites i_p and $i + \delta, q$ are the nearest neighbours and is zero otherwise.

The diagonalisation procedure of H_i^0 is performed as for the linear chain. The zero-temperature renormalization scheme proceeds as follows: we keep only the two lowest energy states and consider them as the eigen-states of a new cell spin operator s_i^z . The renormalized Hamiltonian, which preserves the energy of the low-lying energy states, is given, up to the second order in the perturbation V [10] by

$$H' = \sum_{m,n} |m\rangle \left[H_{m,n}^0 + V_{m,n} + \sum_{\alpha} V_{m\alpha} V_{\alpha n} \frac{1}{2} \left(\frac{1}{E_m - E_\alpha} + \frac{1}{E_n - E_\alpha} \right) \right] \quad (3.4)$$

Here $|m\rangle$ and $|n\rangle$ denote the ground state of the whole lattice.

Retaining only the linear terms in the perturbation V, we obtain the firs order approximation and the recursion relations give us the non-trivial fixed point $y^* = (\Gamma/J)^*$ and the corresponding critical exponent ν. In table 2 we reported the first order fixed point y^* and critical exponent ν.

Table 2

	y^*	ν^*
honeycomb	1.544	0.581
triangular	2.20	0.69
square	1.95	0.65

The second-order correction can be calculated and gives essentially a prolifera tion term which involve the next nearest neighbours of the cell i. The recursion relations can be closed if we include a term of the same form in the site Hamiltonian and consider it as a second-order correction.

In table 3 we report the second order critical field and critical exponent.

366

Table 3 - *Second order critical field and critical exponent ν.*

	$(\Gamma/J)_c$	ν	β
honeycomb	1.67	0.637	0.42
triangular	2.93	0.673	
square	2.09	0.717	0.33
series result [14]		0.638	0.32

We calculate the arithmetic means of the critical exponent y_c and we report in table 4 making comparison with other results:

Table 4

Present	Subbarao [15]	Hirsch [10]	series
2.23			
0.68	0.92	0.65	0.63
0.37	0.51	0.28	0.32

Our results turn out be comparable with those of Hirsch but improved with respect to those of the projection method (Sultaras). The finite temperature extention, now is performed as for the linear chain.

The free energy preserving map from the reduced Hamiltonian $\beta H(\beta = 1/K_B T)$ to an effective cell Hamiltonian, as for the linear chain, gives us the recursion relations between the Hamiltonian parameters. We see that no fixed point exists when Γ and β are finite. Two fixed points exist at $T = \Gamma = 0$.

In fig. 3 we report the critical lines joining the two fixed points for the three types of lattice.

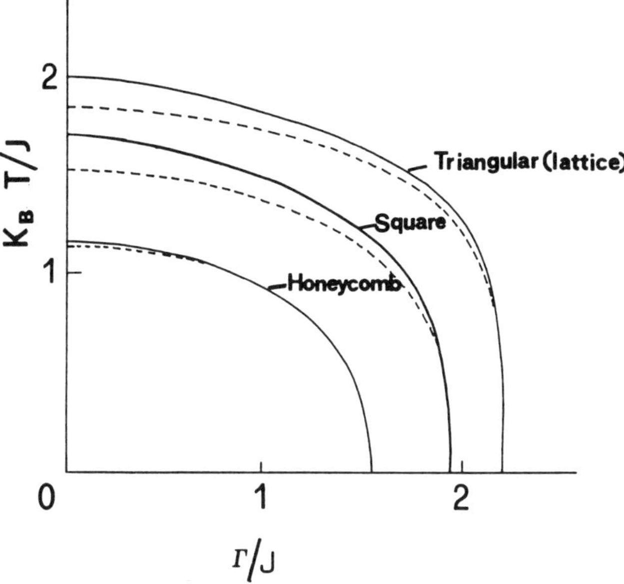

Fig. 3 -*Phase diagram in presence of transverse field; full line and dashed line represent majority and central spin convention.*

In table 5 we report the values of the dimensionless critical T_c at first and second order for both conventions.

A comparison is made with the exact results [16]

Table 5

	Majority rule		Central spin		
	first ord.	second ord.	first ord.	second ord.	Exact
honeycomb	1.152	1.378	1.135	1.344	1.519
triangular	1.966	2.996	1.820	2.715	3.641
square	1.686	1.882	1.519	1.733	2.269

In table 6 we report the magnetic exponent y_h at first and second order by using "central site" and "majority rule" convention.

Table 6

	Majority rule		Central spin		
	first ord.	second ord.	first ord.	second ord.	Exact
honeycomb	2.286	2.147	2.319	2.198	1.875
triangular	2.451	1.943	2.634	2.134	1.875
square	2.086	1.970	2.297	2.165	1.875

The real space renormalization group method allows to investigate thermodynamics quantities as specific heat, magnetization and susceptibility.

We observe that the renormalization transformation can be iterated and after n steps we get:

$$H^{(n)} = E_1 + E_2 + \ldots E_n + H_n \tag{3.5}$$

In (3.5) H_n retains the original form with the old parameters Γ, J, N, substituted by the new ones, Γ_n, J_n, N_n. $E_i (i = 1 \ldots n)$ is a constant term which does not depend of the spin operators, appearing at the i-th step of the renormalization procedure.

The free energy per site can be obtained in the standard way [1] summing up all the constant term

$$f(J, \Gamma) = \sum_{p=0}^{\infty} \frac{E_{p+1}}{N_p} \frac{1}{n_s^p} \tag{3.6}$$

where n_s is the number of sites in a given cell. Performing the first and second derivative respect to temperature we easily obtain from (3.6) internal energy and specific heat respectively.

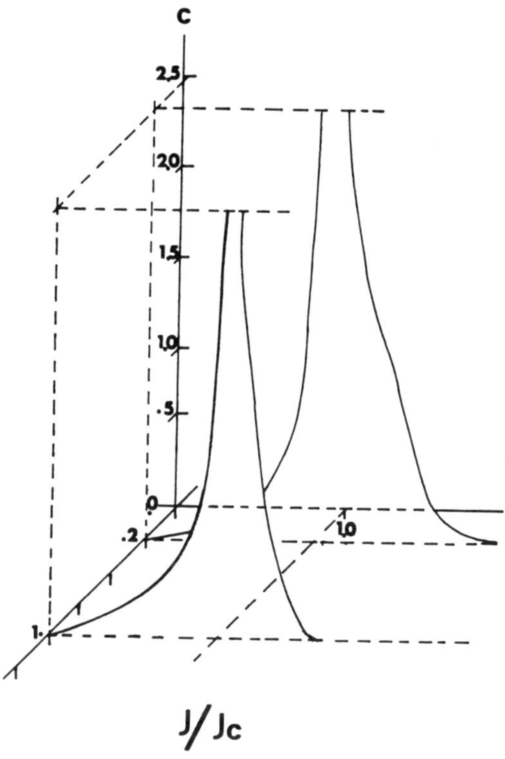

Fig. 4 -*Specific heat versus* J/J_c $\gamma - 0.2$ *and* $\gamma = 1$.

In fig. 4 we present our result of specific heat as a function of J/J_c for different values of transverse field $\gamma = \Gamma/J$. In fig. 5 we compare our approximation (square lattice; $n_s = 5$) with exact results reported by Onseger [17]. We note that there is a fairly good agreement for $J < J_c$, whereas for $J > J_c$ the agreement is not so good. This is a general feature of our results, due to the fact that we have used recursion relation obtained in the first order in J

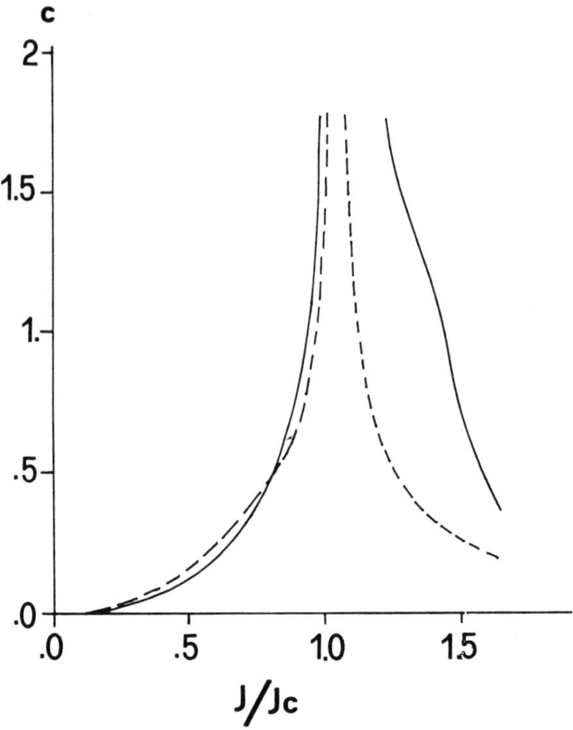

Fig. 5 -*Specific heat versus* J/J_c $\gamma = 0$. *Full line, present result, dashed line Onsager exact results.*

4. Spin-1 quantum Ising chain

The extension to higher spin system [18] is not trivial, because the appearance of proliferation terms generated in the renormalization transformation obliges us to consider critically different model, as for instance the Blume-Emery-Griffiths model [19]. Previous authors [18] have only considered the $T = 0$ case; as far as we know, nobody has considered a position space renormalization group approach to higher spin quantum systems at finite temperature. Here we present a method which is the natural extension to spin 1 system presented in §1. It can be applied to quantum Ising model for any value of spin and for any dimensionality provided the cell present a central site. Our method avoids proliferation terms if one includes in the Hamiltonian a single-site anisitropy term. The Hamiltonian can be written in the form

$$H = \Gamma \sum_i \sigma_i^x - J \sum_i \sigma_i^z \sigma_{i+1}^z + D \sum_i \left(\sigma_i^z\right)^2 \tag{4.1}$$

where D is the anisotropy parameter and Γ^i the external field. $\sigma^{x,z}$ are the spin-1 matrices.

As before we divide the Hamiltonian (4.1) into intra-cell term H^0 and an inter-cell term V and denote by i_2 the central site and by $p = 1, 3$ the site p of the i-th cell.

$$H^0 = \sum_i H_i^0 \qquad H_i^0 = \sum_{p=1,3} H_{ip}^0$$

$$H_{ip}^0 = \Gamma \sigma_{ip}^x - J \sigma_{ip}^z \sigma_{i2}^z + D \left(\sigma_{ip}^z\right)^2 + \frac{1}{2} D \left(\sigma_{i2}^z\right)^2 \tag{4.2}$$

$$V = \Gamma \sum_i \sigma_{i2}^x - J \sum_i \sigma_{i3}^z \sigma_{i+1,1}^z$$

Introducing the operators L^1, L^0, L^{-1} defined by $L^\alpha |\beta\rangle = \delta_{\alpha\beta}|\alpha\rangle$ with $\alpha, \beta = 1, 0, -1$ H_i^0 can be written as

$$H_{ip}^0 = M_{ip} L_{i2}^1 + N_{ip} L_{i2}^{-1} + T_{ip} L_{i2}^0 \tag{4.3}$$

where

$$M_{ip} = \Gamma \sigma_{ip}^x - J \sigma_{ip}^z + D \left(\sigma_{ip}^z\right)^2 + \frac{1}{2} D$$

$$N_{ip} = \Gamma \sigma_{ip}^x + J \sigma_{ip}^z + D \left(\sigma_{ip}^z\right)^2 + \frac{1}{2} D \tag{4.4}$$

$$T_{ip} = \Gamma \sigma_{ip}^x + D \left(\sigma_{i\,ip}^z\right)^2$$

and easily diagonalized. We divide the eigenstates of H_i^0 into three groups, each consisting of nine states, according to the value of the z-component of the central site spin and we assign to each group a cell-spin value consistent with the central site value.

For the ground state $(T = 0)$ only the lowest levels are important and we retain only the three lowest states belonging to the first, second and third group, respectively. They can be interpreted as eigenstates of the cell spin operator s_i^z.

The renormalized Hamiltonian in the first order of the perturbation expansion furnishes the recursion relations. A detailed numerical study of this relations for the parameter $x = \Gamma/J$ and $y = D/J$ gives us the flow-line picture as shown in fig. 4. We find a tricritical point T^* at $(0.2505; 0.8318)$ and a discontinuity fixed point B^* at $(0,1)$. Using appropriate recursion relations we find also the fixed ferromagnetic point $F^*(0,-2)$ found by Humber [20]. We report in tables 7 and 8, tricritical fixed point and exponents and a comparison with previous results directly obtained for quantum spin-1 models. Our results are consistent with those found by other authors.

Table 7 - *Tricritical point*

present	$x^* = 0.2505$	$y^* = 0.8318$	$z^* = 0$
Hu [18]	$x^* = 0.3452$	$y^* = 0.8689$	$z^* = 0.0866$
Humber [20]	$x^* = 0.2912$	$y^* = 0.8945$	$z^* = 0.0686$
Boyanovsky and Masperi [18]	$x^* = 0.35$	$y^* = 0.74$	$z^* = 0$

Table 8 - *Tricritical exponents.*

	y_1	y_2
present	1.8479	0.6425
Hu [18]	1.7562	0.6017
Humber [20]	1.8168	0.9321
Burkardt [21]	1.797	0.7987
Berker and Wortis [22]	1.92	0.7192
	1.8373	0.9181
Adler [23]	3.1	0.6
	2.0	0.4
-expansion	1.968	1.2

In fig. 6 we report the phase diagram in the $x - y$ plane.

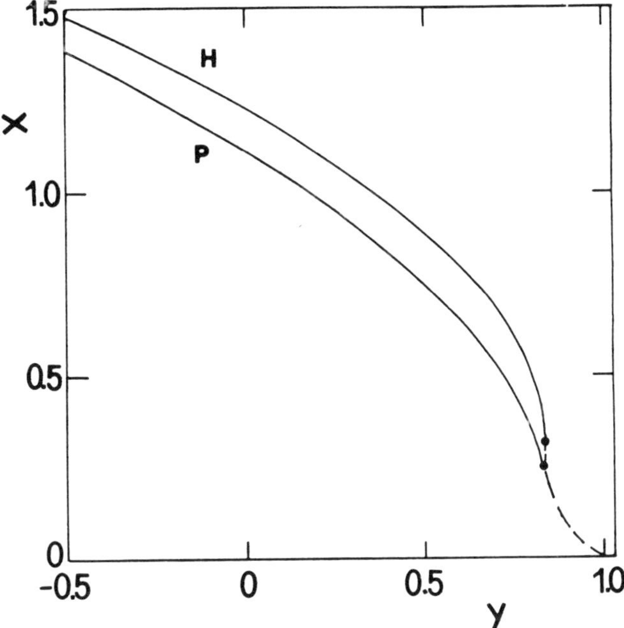

Fig. 6 - *Phase diagram the $x = \Gamma/J$ and $y = D/J$ plane. The tricritical point (solid circle) separate the first-order line (dashed) from the second-order critical line. The curve P are present results, the curve H are Humber [20] results.*

We denote the tricritical point by a solide circle. This tricritical point separate the first-order line (dashed) from the second-order critical line. The curve a shows present results, the curve b the Humber results [20].

The finite renormalization transformation is performed by summing up, in the matrix representation of the partition function, all the degrees of freedom different from the cell spins. The renormalization which preserves the free energy is

$$\langle \{s_i\} | e^{-\beta' H'} | \{s_i'\} \rangle = Tr \langle \{s_i\}, \{\tau_i\} | e^{-\beta H} | \{s_i\} \{\tau_i\} \rangle \tag{4.5}$$

By ispection of recursion relations we don't find any nontrivial fixed point as expected. Moreover, if we perform the $T = 0$ limit, the $T \neq 0$ recursion relations reduces to ones obtained a $T = 0$.

Acknowledgements

We wish to thank Mr. F. Belardo for his technical aid in making the figures.

References

[1] Niemeijr Th and van Leuween J.M.J. 1976 *Phase Transitions and Critical Phenomena* 6, ed. C. Domb and M.S. Green (London: Academic) p.425

[2] Rogiers J. and Dekeyser R. 1976 Phys. Rev. B 13, 4886

[3] G. Kamieniarz, L.S. Campana, A. Caramico D'Auria and U. Esposito J. Phys. C. Solid State Phys. 20 (1987) 1337

L.S. Campana, A. Caramico, D'Auria, U. Esposito and G. Kamieniarz J. Phys. C. Solid State Phys. 20 (1987) 1841

L.S. Campana, A. Caramico D'Auria, U. Esposito and G. Kamieniarz J. Phys. C. Solid State Phys. 20 (1987) 5161

[4] L.S. Campana, A. Caramico D'Auria, U. Esposito and G. Kamieniarz Phys. Rev. B 39 (1989) 9224

L.S. Campana, A. Caramico D'Auria, U. Esposito and G. Kamieniarz Phys. Rev. B 41 (1990) 6733

L.S. Campana, A. Caramico D'Auria and U. Esposito Phys. Stat Sol. (b) 165, 561 (1991)

L.S. Campana, A. Caramico D'Auria, F. Esposito and G. Kamieniarz Phys. Stat. Sol. (in press)

[5] L.S. Campana, A. Caramico D'Auria and U. Esposito Phys. Stat. Sol. (b) 151, K177 (1989)

[6] L.S. Campana, A. Caramico D'Auria, F. Esposito and G. Kamieniarz J. Phys. C Solid State Phys. 21 (1988) 589

[7] Fradkin E. and Susskind L. (1978) Phys. Rev. D 17, 2637

[8] Fernandez-Pacheco A(1979) Phys. Rev. D 19, 3173

[9] Uzelac K, Jullien R. and Pfeuty P. (1980) J. Phys. A: Math. Gen. 13, 3735

[10] Hirsch J.E. Phys. Rev. B 20, 3907 (1979)

Hirsch J.E. Nakeuzo G.F. Phys. Rev. B 19, 2656 (1979)

[11] Indezen J.O., Maritain A. and Stella A.L. (1982) J. Phys. A: Math. Gen. 17, L697(1984)

[12] Plascak J.A. J.Phys. A: Math. Gen 17, L697 (1984)

[13] Jullien R., Pfeuty P., Fields J.M., Domiach S. Phys. Rev. B 18, 3568 (1978)

[14] Yanase A., Takeshige Y. and Suzuki H. J. Phys. Soc. Japan 41, 1108 (1976)

[15] Subbarao K. Phys. Rev. Lett. 37, 1712 (1976)

[16] Domb C. Adv. Phys. 9, 149 (1960)

[17] l. Onsager Phys. Rev. 65, 117 (1944)

[18] B. Hy Phys. Lett A 75, 372, (1980)

H. Humber Phys. Rev. B 21, 3899 (1980)

D. Boyanovsky and L. Masperi Phys. Rev. D 21, 1550 (1980)

Y. Gefen, Y. Imry and D. Mukamel Phys. Rev. B 23, 6099 (1981)

[19] M. Blume, Y.J. Emery and R.B. Griffiths Phys. Rev. A 4, 1071 (1971)

[20] H. Humber Phys. Rev. B 21, 3899 (1980)

[21] T.W. Burkhardt Phys. Rev. B 5, 1106 (1972)

[22] A.M. Berker and M. Wortis Phys. Rev. B 14, 4946 (1976)

[23] J. Adler, A. Aharony and J. Oitmaa, J. Phys. A 11, 963 (1978).

INTRINSIC AND EXTRINSIC JOSEPHSON EFFECT – TWO INTRIGUING ASPECTS OF HIGH Tc SUPERCONDUCTORS

ANTONIO BARONE

Dipartimento di Scienze Fisiche , Facolta' di Ingegneria , Universita' di Napoli Federico II , Napoli , Italy

RUGGERO VAGLIO

Dipartimento di Fisica , Universita' di Salerno , I–84081 Baronissi (Sa), Italy

ABSTRACT

The role of the Josephson effect in high-T_C superconductors is discussed in connection with the properties arising from intrinsic effects of low dimensionality of these materials. The superconducting weak coupling which stems from the granular nature of sinterized samples are also discussed showing how d.c. and r.f. properties can be explained in the light of the Josephson effect. Finally, the occurrence of Josephson effect in actual junction devices will be described and discussed.

The constant aim and success of Eduardo Caianiello in the search of a general view of different aspects of Science is worldwide recognized. Our attempt of synthesis in the present paper is, in this sense, further stimulated by the circumstance of the 70th birthday of Eduardo to whom this work is dedicated.

INTRODUCTION

High T_C Superconductors, five years after the discovery by Bednorz and Muller [1], are still in the limelight of the world's scientifc community. Advances in this topic have been achieved both in terms of underlaying physics as well as for the potential, in some case near term, practical applications. However, as stated by the most serious among the numerous projection made in the early stages of this new chapter of condensed matter, a lot of the puzzling aspects remain open to both theoretical and experimental investigations.

In dealing with problems of high complexity, such as the physics of high-Tc superconductivity, there is a justified tendency to look for the most elementary phenomenological aspects, common features, possible general ideas and realiable tools which could provide a higher degree of confidence in the approach. This paper does not represent an exception to this trend. Indeed the Josephson effect is adopted as a key for understanding the two different aspects of high-T_C superconductors, one concerned with the intrinsic bulk properties of the material, the other referring to the realization of Josephson junctions both as a tool for providing basic physics information as well as for the possibility of realizing a variety of device applications.

LAYERED STRUCTURE

It is widely recognized that high T_C materials like $Y_1Ba_2Cu_3O_7$ (YBCO) or $Bi_2Sr_2CaCu_2O_8$ (BSCCO) are layered materials where the CuO planes play the role of separate layers. Due to this circumstance the electron spectrum is strongly anisotropic, implying a significantly large ratio of effective masses m_c/m_{ab}. Here m_{ab} and m_c indicate the effective mass of the electron moving on the CuO planes and across them respectively. The experimental data of conductivities allow an evaluation of the mass ratio as $\sigma_{ab}/\sigma_c \approx m_c/m_{ab} \approx$ 100-200 for YBaCuO samples at 100 K [2]. For BSCCO materials this ratio is higher.

Such type of supercondutors can be therefore reported as a multilayered system with Josephson interaction occurring between adjacent layers. Accordingly, these anisotropic superconductors are also characterized by two coherence lengths ξ_{ab} , in ab-plane, and ξ_c in the perpendicular direction. The ratio of these lengths is given by $\xi_{ab}/\xi_c (m_c/m_{ab})^{1/2} \gg 1$.

The ratio between ξ_c and the interlayer distance d plays a very important role in the understanding of the physics of layered superconductors. Indeed if ξ_c is larger than d, then we will have three dimensional anisotropic superconductors whereas in the opposite case the superconductivity will have the two-dimensional character, with a weak Josephson coupling between the layers. The theory [3,4] gives the value $\xi_c = d/\sqrt{2}$ which separates the two limiting cases.

Experimental measurements of the upper critical field Hc_2 give the information on the actual dimensionality of the anisotropic superconductor. In Ref. 5 measurements of Hc_2 have been performed for both YBCO and BSCCO compounds. The results show that for the former compound the three dimensional approximation is valid.

For BSCCO material deviation of the results from the 3D theory for magnetic field almost parallel to the ab plane indicates a two dimensional character of the superconductivity. Using the relations :

$$Hc_2 \ (//c) = \frac{\Phi_0}{2\pi\,\xi_{ab}} \qquad\qquad Hc_2 \ (\mid c) = \frac{\Phi_0}{2\pi\,\xi_{ab}\,\xi_c} \qquad\qquad (1)$$

in Ref. 6 the coherence lengths are estimated.

For BSCCO material $\xi_{ab}(0) \approx 30$ Å and $\xi_c(0) \approx 1.6$ Å, while interlayered spacing between CuO is 12 Å and c-axis lattice spacing 30.6 Å. Therefore one can conclude that the BSCCO compound is a strongly 2D layered system with Josephson interaction between the layers. Near the critical temperature $\xi_c(T) \sim \xi_c(0) \ (1-T/T_c)^{-1/2}$ can be larger tha $d/\sqrt{2}$, and accordingly, the three dimensional regime realizes. Estimation show that the Josephson regime for BSCCO ($\xi_c, d/\sqrt{2}$) takes place at all temperatures, T for (T_c-T) larger than a few Kelvin. For YBCO a two dimensional regime can probably be achieved only at a quite low temperature.

In high-T_c materials it has been observed a temperature dependence of the upper critical field Hc_2 exhibiting the positive sign of the second order derivative $(d^2Hc_2)/(dT^2)$ near T_c in contrast with conventional superconductors where such a derivative is negative. Once again for a possible explanation one can resort to the role of the Josephson effect in the interaction between layers. The Ginzburg-Landau equation governing the order parameter of the nth layer $\Psi_n(r)$ in such a system can be expressed in the form :

$$ (2) $$

$$\left[-\alpha + \frac{1}{2m} \left(\nabla - \frac{2ie\,\vec{A}}{c} \right) + \beta|\psi_n|^2 \right]\psi_n(r) + \frac{1}{2md^2} \left(2\psi_n - \psi_{n+1} - \psi_{n-1} \right) = 0$$

which accounts for the negative values of $d^2Hc_2/(dT^2)$ at T not too far from T_c. Moreover, the Josephson-like behavior occurs also in the description of the vortex structure and vortex dynamics in the layered high-T_c superconductors. Indeed coreless vortices as those occurring in large Josephson junctions (junctions dimensions exceeding the Josephson penetration) do exist also in this case (see Fig.1a). In addition, with respect to the Josephson junction, if the vortex lies within the layers, a current I along the layers and perpendicular to the vortex will produce a Lorentz force F_L acting on the vortex in a direction perpendicular to the layers(see Fig.1b). If this force is not large enough, namely for small current values the votex will remain "pinned" and superconductivity at the layer will be destroyed. Such intrinsic pinning effect occurring in highly layered Josephson structures has bee subject to numerous investigations [7].

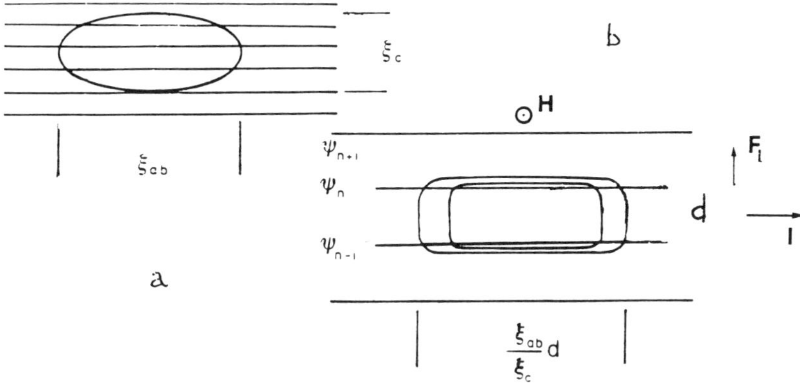

Fig.1 : Abrikosov vortex (a) and coreless Josephson vortex (b) (see text).

Thermal fluctuations can supply the necessary additional energy to allow vortex motion across the layers. Such "flux creep" induced by the thermal activation process has been widely investigated since many years. The action of the Lorentz force can bend the vortex and push it acrosa the layer producing a shift of the magnetic flux Φ_0 by the layered structure period. As a consequence of this flux change in time a measurable voltage will arise.

It is interesting to observe that as a further modeling in terms of Josephson junction, also macroscopic quantum tunneling can occur at relatively low temperature. Indeed a crossover temperature T_0 has been evaluated below which MQT starts to dominate over thermal activation [8].

NATURAL GRAIN BOUNDARIES

In the previous section, although qualitatively , we have seen how heawily Josephson effect is present in the intrinsic bulk properties of high T_c

superconductors. In a rigorous sense the word "intrinsic" should be confined to the interlayer Josephson coupling just discussed. However the weak super-conductivity effects occurring at grain boundaries of a granular sample are still occasionally referred as "intrinsic" though they attain more properly to the morphology of the sample.

Bulk samples of the high Tc cuprate superconductors are obtained by a typical ceramic sintering process , consisting of grinding and mixing powders of the constituent oxides, followed by an high pressure compaction and high temperature ($\geq 850\,^{0}C$) reaction in oxygen .

Due to this process the material has a characteristic granular structure , that strongly influences the transport properties.

The first results reported at the beginning of 1987 on the critical currents of high Tc superconductors were indeed rather discouraging, with values of J_c of the order of 10^2 A/cm^2 and a very strong magnetic field dependence of the critical current. It was already clear in those days that such low J_c values were not an intrinsic property of the material and in fact the magnetic field data could easily be interpreted in terms of a simple model in which the material is schematically treated as a series-parallel random array of Josephson junctions .

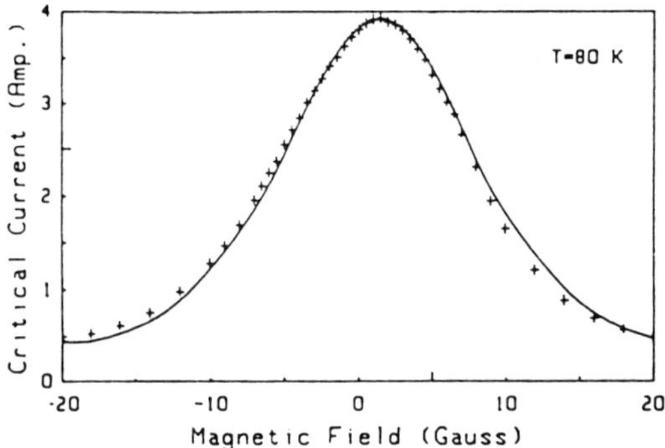

Fig. 2. Low magnetic field dependence of the critical current for a $YBa_2Cu_2O_7$ sintered pellet . The continuous line is computed by eq. ! assuming H_0 = 13 gauss , N = 26 and Fi = 0.50 + 0.05 i (i = 0,1, ... , 25).

A good fit of the J_c-H data (Fig. 2) could be obtained using the simple formula [9] :

$$Jc(H) = [Jc(0)/N] \sum_{1}^{N} \frac{\sin(\pi H / H_0 Fi)}{\pi H / H_0 Fi} \qquad (3)$$

The low values of the Josephson critical currents associated to the grain boundaries can be related to the tendency of the high Tc cuprates to deplete oxygen at the free surfaces. This effect creates a layer with strongly reduced superconducting properties at the grain surfaces that, together with the low values of the superconducting coherence lenght ξ ($\xi \approx .18 \hbar v_f / KTc \approx 10 \text{Å}$), induces a strongly depressed superconducting order parameter Δ at the interfaces ($J_c \sim \Delta$, see next section)

As soon as single crystals and epitaxial thin films of the high Tc cuprates become available , it was shown that the intrinsic critical currents (for a current flowing in the a-b plane) were indeed fairly high , reaching 10^7 A/cm^2 in epitaxial films , a value that compares well with the critical current values of metallic superconductors.

A key experiment was conducted on YBCO thin films by Mannhart et al. [10] at IBM , Yorktown Heights . These authors were able to measure both the intrinsic critical current and the Josephson grain boundary critical current in the same sample . They found J_c(grain) $\approx 10^7$ A/cm^2 and J_c (grain-boundaries) $\approx 10^5$ A/cm^2 , with a temperature dependence indicating a flux-creep limited current density in the first case and a Josephson junction limited current density in the second .

In any case , even if the grain boundary problem is , in principle , avoidable , for most applications single crystals are not suitable , and the epitaxial growth technology can only be applied using very special substrates and in very restrictive conditions. This means that in most present and potential applications sintered bulk materials or polycrystalline (or textured) thin films have to be used and this explains why Josephson models are still very popular in the physics and technology of high Tc superconductors.

As an example sophysticated Josephson array models have been also used to explain the low field magnetization and the microwave adsorption properties of high Tc pellets [11,12].

Recently the Josephson effect between grains has been also introduced to account for the r.f. properties of high Tc polycrystalline film surfaces [13].

The Superconductivity Groups of the Universities of Napoli and Salerno are presently carrying a common research project in this field, so we will describe this last topic in more detail.

In the model the film is described as a network of superconducting grains coupled via Josephson junctions. The basic element of such network and the equivalent circuit for the admittances are represented in Fig. 3 .

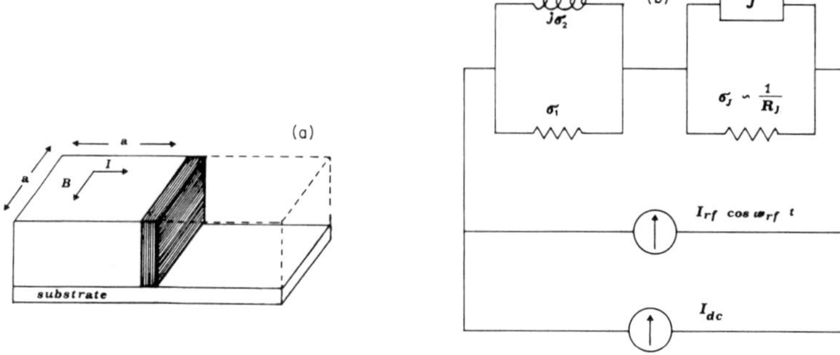

Fig. 3. Basic element of the Josephson junction network simulating a granular thin film and relative equivalent circuit for the admittances.

The element "J" is described by the Josephson current-phase relation :

$$I(t) = I_c \sin \left[(2e/\hbar) \int_o^t V(t')\,dt' \right] \qquad (4)$$

The shunt resistance of the junction R_j is assumed to be constant.
For small r.f. signal amplitudes $(I_{rf} \ll I_c)$ the Josephson element can
be well approximated by an inductance of value [13] :

$$L_j = \hbar / 2e I_c F(I_{dc}) \qquad (5)$$

where :

$$F(I_{dc}) = [1 - (I_{dc} / I_c)^2]^{1/2} \qquad (6)$$

If we further assume that the junction conductivity is dominated by the
Josephson element $(\hbar\omega / 2e I_c R_j F \ll 1)$, we are led to an effective conductivity σ_{jeff} of the grain boundary that , averaged over the grain dimension,
can be written as :

$$\sigma_{j\,eff} = \{\mu_o \omega \lambda_j^2 F^{-1} [(\hbar\omega/2e I_c R_j F) + j]\}^{-1} \qquad (7)$$

where $\lambda_j = (\hbar/a2e J_c \mu_o)^{1/2}$, a is the average grain dimension and J_c is the critical Josephson current density.

The surface impedance Z_{sj} of the polycrystalline film, is then obtained by replacing, in the standard formulas, the normal conductivity with the overall conductivity of the basic network element :

$$\sigma_{eff} = [\sigma_{j\ eff}^{-1} + (\sigma_1 + j\sigma_2)^{-1}]^{-1} \qquad (8)$$

σ_1 and σ_2 ($\sigma_2 = \pi\Delta\sigma_n/\hbar\omega$) are the real and immaginary parts of the intragrain conductivity. At low temperatures and frequencies we get:

$$R_{sj} = \frac{1}{2\lambda_{eff}} \ \frac{\sigma_1 + \sigma_r}{\sigma_2^2} \qquad (9)$$

$$X_{sj} = \mu_o \omega \lambda_{eff} \qquad (10)$$

where we have introduced :

$$\lambda_{eff} = (\lambda_1^2 + \lambda_j^2 F^{-1})^{1/2} \qquad (11)$$

$$\sigma_r = \left(\frac{\lambda_j}{\lambda_1}\right)^2 \frac{\pi\Delta}{2e I_c R_j} F^{-2} \sigma_n \qquad (12)$$

here σ_r is frequency and temperature independent (as long as we assume J_c and $I_c R_j$ to be temperature independent as well).

Eqs. (9),(11),(12) give the temperature and magnetic field dependence of the surface resistance Rs and predict the presence of an anomalously large residual term (related to σ_r).

A good agreement with the data on the surface resistance of high Tc superconductors can be generally obtained by this formula using J_ca and $I_c R_j$ as fitting parameters. Along this line, to sistematically check the model, new experiments are presently carried out at the Universities of Napoli and Salerno, using a ring resonator ,on both YBCO and BSCCO films of different morfologies [14].

ARTIFICIAL JOSEPHSON STRUCTURES

The development of superconducting electronic devices is tightly connected with the capability of producing reliable Josephson junctions.

Since the first high Tc samples produced were bulk pellets, early measurements of the Josephson effect in high Tc cuprates where performed using point contact or "break-junction" arrangements. Though point contacts are intrinsically unstable and therefore not suitable for practical applications, very high Tc cuprates. As an example , by microwave irradiation of a point contact YBCO junction. Tsai and cooworkers [15] were able to carefully determine the value of the flux quantum Φ_0 from the a.c. Josephson effect voltage-frequency relation ($V = n\Phi_0 f$) . Since the flux quantum Φ_0 is related to the value of the charge q of the supercurrent carriers ($\Phi_0 = h/q$) , the experiment could clearly prove that electrons pairs were responsible of the superconducting properties also for the new superconductors.

Experiments on YBCO-Nb point contacts also showed Josephson behavior indicating the s-wave character of the electron pairing. Point contacts however fail in producing neat quasiparticle tunneling characteristics as observed in conventional superconductors, even if, in the best cases, evidence of a gap-like structure was present [16].

When high quality single crystals and epitaxial films become available , both pair (Josephson) and quasiparticle tunneling experiments were performed in more reliable structures.

Josephson weak-link type junctions require to realize a constriction of the dimension of ξ ($\approx 10 \overset{o}{A}$) , which is .of course , impossible. The first bridge-type junctions were then realized isolating single grain boundaries in a polycrystalline (textured) thin film using constrictions of the order of $1\mu m$. This technique was producing reasonable . though scarcely reproducible. "resistively shunted junction" (RSJ) I-V curves . but $I_c R_j$ products of the order of few μV only .

More recently , a variety of sophisticated techniques have been adopted to produce weak-link type junctions. The most successfull was probably the bi-epitaxial grain boundary junction method developed at the IBM Laboratories in Yorktown Heights [17]. In this case a narrow line ($\approx 1\mu m$) is patterned across an artificially grown bicrystal of known angle. Typically $SrTiO_3$ oriented bicrystals are used as substrates in an epitaxial growth process (laser ablation or other techniques) of an YBCO film. Quite reproducible results have been obtained by this method and almost ideal RSJ I-V characteristics. The $I_c R_j$ products have been found to depend on the misorientation angle with a clear scaling behavior in respect to the critical current density value [18] [Fig.4a]. An interpretation of the $I_c R_j$ v.s. Jc dependence can be given in terms of a simple model [19] . In fact , if the order parameter is a function of the position at the grain boundary interface ($\Delta = \Delta (x)$), the usual BCS relation $I_c R_j = \pi \Delta_0 / 2e$ is substituted by the integral relation :

$$I_c = (\pi / \Delta_0 e) \int_0^\infty \frac{\Delta(x) \Delta'(x)}{R_j(x)} dx \tag{13}$$

$$\left(R_j = \Delta_0 \left(\int_0^\infty \frac{\Delta'(x) dx}{R_j(x)} \right)^{-1} \right)$$

and the $I_c R_j$ product is now a function of R_j (Jc).

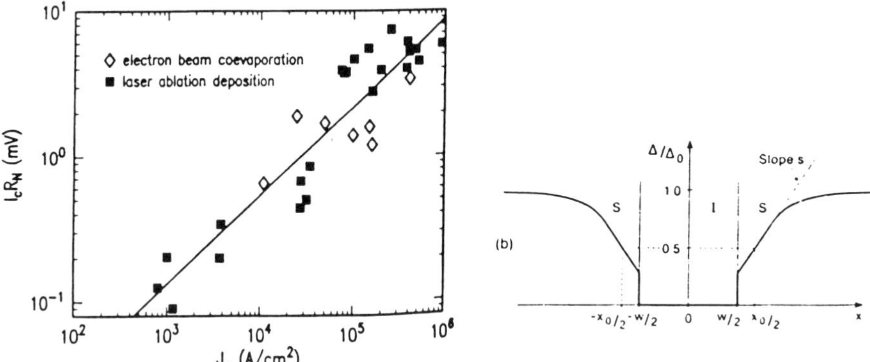

Fig. 4 (a): $I_c R_j$ product as a function of Jc for a series of YBCO artificial grain boundary junctions with different misorientation angles.
(b): Order parameter profile $\Delta(x)$ assumed to fit the data reported in a.

Assuming the order parameter profile $\Delta(x)$ shown in Fig. 4b, one can fit closely the data reported in Fig. 4a [19]. Moreover the $I_c R_j$ v.s. J_C relation is rather "universal" since it has been proved to hold also for natural grain boundaries [14].

Different approaches have been also succesfully adopted to obtain weak-link Josephson behavior in high Tc films, including artificial grain boundaries induced by substrate steps, superconductivity weakning by ion implantation and many others.

The "tunnel junction" approach has also been deeply investigated but this last approach has proved to be more difficult. As an example junctions based on natural YBCO surface oxide barriers generally do not show Josephson currents and the quasiparticle characteristics, though much more reproducible with respect to the point contact results, are far from being ideal (BCS) and are difficult to interpret [20].

Proper RSJ Josephson behavior has been recently observed in YBCO-PBCO-YBCO junctions but also in this case quasiparticle tunneling was far from ideal [21].

CONCLUSIONS

The role of the Josephson effect in high-Tc materials is essential. For the material in bulk form Josephson weak- coupling occurs in the structure of a pellet at intergrain boundaries whereas, more intrinsically , it is responsible of the actual coupling between layers in anisotropic crystals. In addition it is of paramount importance the possibility of realizing reliable Josephson junctions for a variety of important appplications.

Indeed these topics would certainly deserve a more deep and extensive analysis. Here we have just outlined the importance of the Joesephson effect as a common feature to the various aspects of the fascinating topic of high Tc oxide materials.

REFERENCES

1) J.G. Bednortz and K.A. Muller, Z. Phys. B 64, 189 (1986).

2) K.S. Bedell, D. Coffrey, D.E. Meltzer, D. Pines and R. Shrieffer " High Temperature superconductivity", Addison - Wesley Eds. (1989)

3) W.E. Lawrence and S. Doniach, Proc. 12th Int. Conference on Low Temperature Physics, Kyoto, Japan (1970).

4) A. Barone, A.I. Larkin, Yu. N. Ovchinnikov , J. Superconductivity 3 , 155 (1990).

5) M.J. Naughton et al. Phys. Rev. B 38 , 9280 (1988).

6) T.M. Plastra , B. Battlog , L.E. Schneemayer, R.B. Van Dover and J.V. Waszak, Phys. Rev. B 38 , 5102 (1988).

7) Yu. N. Ovchinnikov and B.I. Ivlev, Phys. Rev. B 43 , 8024 (1991), M. Tashika and S. Takahashi, Solid State Comm. 70 , 291 (1989).

8) G. Blatter and V. Geshkenbein, , to be published .

9) G. Paterno' , C. Alvani , S. Casadio , U. Gambardella and L. Maritato , Appl. Phys. Letts. 53 , 609 (1988).

10) J. Mannhart, P. Chaudhari , D. Dimos , C.C. Tsuei, J. Chi, and T.R. Mc Guire, Phys. Rev. Letts. 61, 2476 (1988).

11) R. De Luca, S. Pace and B. Savo , Phys. Letts. A, 154 A, 185 (1991).

12) Giura , Fastampa , Marcon , Phys. Rev. B 40 , 4437 (1989).

13) C. Attanasio, L. Maritato and R. Vaglio , Phys. Rev. B 43 , 6128 (1991).

14) A. Andreone , C. Attanasio, G. Balestrino, A. Di Chiara, L. Maritato, S. Marra, E. Milani, G. Peluso, and R. Vaglio , to be published .

15) J.S. Tsai , Y. Kubo and J. Tabuchi, Phys. Rev. Letts. 58, 1979 (1987).

16) A. Barone , A. Di Chiara, G. Peluso, U. Scotti di Uccio, A.M. Cucolo, R. Vaglio, F.C. Matacotta and E. Olzi , Phys. Rev. B 36 , 7121 (1987).

17) P. Chaudhari , J. Mannhart, D. Dimos , C.C. Tsuei, J. Chi, M.M. Oprysko and M. Scheuerman , Phys. Rev. Letts. 60, 1653 (1988).

18) R. Gross, P. Chaudhari , M. Kawasaki and A. Gupta, Phys. Rev. B 42 , 10735 (1990).

19) J. Mannhart and P. Martinoli, Appl. Phys. Letts. 58 , 643 (1991).

20) A.M. Cucolo, R. Di Leo , P. Romano, L.F. Schneemeyer and J.V. Waszczak , Phys. Rev. B 44 , 2857 (1991).

21) J. Gao, W.A.M. Aarnink, G.J. Gerritsma and H. Rogalla, Physica C 171 , 126 (1990).

PHASE TRANSITIONS IN JOSEPHSON ARRAYS AND SUPERCONDUCTING GRANULAR FILMS

Gaetano Giaquinta, Giuseppe Falci and Rosario Fazio

Istituto di Fisica, Facoltà di Ingegneria
viale A. Doria 6 , 95125 Catania
ITALY

ABSTRACT: We will briefly review the low temperature properties of Josephson junction arrays. We will discuss the phase transitions which take place in this systems,   emphasizing the T=0 superconductor-insulator transition driven by quantum fluctuations. The phase diagram is discussed employing a variational method due to Giachetti-Tognetti and Feynmann-Kleinert.

388

Introduction

It is a quite remarkable fact that the birth of most we are going to summarize in what follows has to be traced back to the seminal lectures delivered by P.W. Anderson at the second Ravello Summer School on The Many Body Problem directed by Professor E. Caianiello and edited by him in the Proceedings at the very early sixties. Insofar these well celebrated books have been of the outmost relevance in the educational training of the theoreticians and subsequent generations - and we belong to this class - we feel that we have a further motivation to thank Prof. E. Caianiello as a Physicist, as a master and as a man of culture.

It is by now a well established fact that Macroscopic Coherence Effects (MCE) have to be neatly distinguished from truly Macroscopic Quantum Effects (MQE). The first ones refer to the feasibility of the experimental evidence, on a truly macroscopic scale, of cooperative effects due to the coherent superposition of an extremely large number of degrees of freedom, each governed by a quantum dynamical equation; on the other we refer to the actual possibility of being able to single out truly quantum mechanical effects (essentially Quantum fluctuations) within the dynamics of a single degree of freedom. Superconductors in themselves are considered the best candidates where both MCE and MQE can be essentially observed, partially after the astonishing progress recently gained with lithographyc techniques: we mean essentially Josephson Junction Arrays (JJA) and Granular Superconductors (GS) (we will not talk about single junctions and we refer to the review of G. Schön and A.D. Zaikin [1]).

In a single Josephson junction the relevant dynamical degree of freedom is the phase difference field $\varphi(\mathbf{r},t)$ between the macrowave functions $\Delta_{1,2}(\mathbf{r},t) = |\Delta_{1,2}|\exp\{i\varphi_{1,2}(\mathbf{r},t)\}$ of the two weakly coupled superconductors which realize the Josephson contact. In this context the field Δ_i has to be understood as the anomalous Gor'kov pair field corresponding to the coherent superposition of Cooper pairs in the restricted ensemble, or alternatively as the (microscopic) Ginzburg-Landau order parameter; this point of view, more general than the gap function in the simple BCS model is best suited to both physical and computational reasons insofar as it allows us to develop the theory by the methods of Quantum Statistical Field Theory. The variable φ has to be considered as a MACROSCOPIC VARIABLE subjected to the Josephson dynamical equation:

$$\dot{\varphi} = (4\pi e/h)V$$

V being the externally applied bias. As you the Josephson current due to the tunneling of Cooper pairs is calculated to be:

$$I_S = I_O \sin \varphi$$

whereas the dissipative current due to the thermally excited quasi-particle tunneling (in the model of the Resistance Shunted Junction) contributes with term

$$I_{QP} = V/R_N \; (h/4\pi e R_N)\dot{\varphi}$$

R_N being the normal state resistance of the junction. To these terms a displacement current

$$I_C = (Ch/4\pi e)\ddot{\varphi}$$

has to be added, C being the junction capacitance. Physically it corresponds to charging effects due to the charge imbalance by Cooper pairs transfer from one electrode to the other of the junction (and viceversa) viewed as a capacitor.

When the dissipative term is neglected, current conservation implies a dynamical equation for the phase difference field which is formally the same of the CLASSICAL HAMILTONIAN EQUATIONS OF MOTION for a system with one degree of freedom φ governed by the Hamiltonian function:

$$H_{class} = \frac{4e^2}{2C} n^2 + E_J \,[1 - \cos\varphi]$$

It is quite a remarkable fact that although a *classical limit* exists for the dynamics of φ (despite the fact that the quantum of action h explicitly appears in the coefficients of the evolutive equation) so that is legitimate our previous assumption of φ as a macroscopic variable, its physical significance lies intrinsically only at the quantum level.

Insofar as the charge conservation is a strict implication of global gauge invariance (and Superconductivity is nothing else that the occurrence of spontaneously gauge symmetry) when we transcribe the problem quantum mechanically via the procedure of canonical quantization:

$$H_{class}(A,B) \;\rightarrow\; H^{op}(A^{op}, B^{op})$$

$$\{A,B\}_{P.B.} \;\rightarrow\; [A^{op}, B^{op}]_-$$

the occurrence of quantum fluctuations in the dynamics of the macroscopic variable φ is driven by the operatorial correspondence

$$n^{op} = -\,i\partial_\varphi \qquad\qquad \varphi^{op} = i\partial_n \qquad\qquad\qquad (6a)$$

that implies the uncertainty relation

$$[\varphi^{op}, n^{op}]_{-} = -i \qquad\qquad (6b)$$

Then the observability of Quantum Fluctuations, as distinct from thermal ones, should be possible in principle when the ground energy satisfies $\hbar\Omega \geq kT, \alpha E$ (with $\alpha \approx 1$). If thermal dissipation due to quasi-particle excitation is accounted for, the classical oscillator equation for φ acquires a damping term, with $(\hbar/2e)^2 R_N^{-1}$ as the damping coefficient, so that the condition for the quantum fluctuation regime becomes

$$\min[\hbar\Omega_c, \hbar\Omega] \geq kT, \alpha E$$

Ω_c being the critical frequency introduced by the friction coefficient.

From what we said so far it emerges that:

1) below the superconductive critical current, and field, of a single Josephson junction, MQC can be inhibited by Quantum Fluctuations driving the system into a resistive state, despite the fact that we are still in the presence of two weakly coupled superconducting systems, and that

2) dissipative effects play a relevant role in selecting the actual dynamical regime.

II. Phase transitions

The effect of quantum fluctuation is very important to understand the physics of Josephson arrays (JJA) and granular films (GF). In the flowing we will try to briefly review the *status of art* of the field. Due to the limitations of space we cannot go into details; we will try, instead, to stress the main theoretical approaches to this problem. We will not discuss about the physics ultrathin films; although in some respects can be compared with the JJA and GF, there are important difference that we do not have the time to discuss in the paper.

Both in JJA and GF are formed by a certain number of metal islands which are coupled by the tunneling of Cooper pairs (CP) and Quasi-Particles (QP). The islands can be at the sites of a regular array (as in the JJA) or at disordered positions (as in GF). In these systems a distinction is in order between local and global superconductivity. At a certain critical temperature T_{CO} the islands undergo a transition to a superconducting state and develop a non zero order parameter $\Delta\exp\{i\varphi_r\}$. The neighbour islands are coupled through the Josephson term $-E_J \cos(\varphi_i - \varphi_j)$, where $\mathbf{r}$ and $\mathbf{r'}$ denote the position of the islands in the array. If the phase configurations are disordered the system is still in a resistive phase. This is what is called local superconducting state. At a different temperature T_C smaller than T_{CO} the phases order and the global ordering sets in.

The global superconducting state is destroyed by thermal and/or quantum fluctuations. One of the most striking observation in these systems is the existence of a T=0 superconductor-insulator transition when the value of the normal state resistance is of the order of $h/4e^2$ (see Ref.2). Depending on the various mechanisms of dissipation an extra resistive phase can be present at zero temperature. A recent suggestion that at the T=0 transition the system is metallic with a value of the resistance which is universal [3] seems supported by the experiments on ultrathin film[4] ; we will not discuss further this point.

The phase transition to the phase coherent state can be studied considering the analogy of the JJA with an XY Model. We do not go into the details of the mapping; we refer, for the interested reader, to the volume of Physica Ref.5 where a number of articles on the subject are presented and further references can be found.

The model which takes into account the general features of the low temperature properties [6] of the JJA is described by the following action [7]:

$$S[\varphi] = S_{ch}[\varphi] + S_J[\varphi] + S_D[\varphi] \tag{1}$$

where the three terms describe the charging energy

$$S_{ch}[\varphi] = \frac{1}{16E_0} \sum_i \int_0^\beta d\tau \left(\frac{d\varphi_i}{d\tau}\right)^2 + \frac{1}{16E_C} \sum_{<i,j>} \int_0^\beta d\tau \left(\frac{d\varphi_{ij}}{d\tau}\right)^2 \tag{1.a}$$

the Josephson energy

$$S_J[\varphi] = - E_J \sum_{<i,j>} \int_0^\beta d\tau \cos \varphi_{ij}(\tau) \tag{1.b}$$

and the dissipative contribution $S_D[\varphi]$ which is non local in the (imaginary) time. This last can arise from the tunneling of quasi-particles

$$S_D[\varphi] = \sum_{<i,j>} \int_0^\beta d\tau \int_0^\beta d\tau' \alpha(\tau - \tau') \left\{ 1 - \cos\left[\frac{\varphi_{ij}(\tau) - \varphi_{ij}(\tau')}{2}\right]\right\} \tag{1.c}$$

or normal electrons due to an Ohmic shunt

$$S_D[\varphi] = \frac{1}{2} \sum_{<i,j>} \int_0^\beta d\tau \int_0^\beta d\tau' \alpha(\tau - \tau') \left[\frac{\varphi_{ij}(\tau) - \varphi_{ij}(\tau')}{2}\right]^2 \tag{1.d}$$

The two terms in S_{ch} represent the charging energy due to the self-capacitance $E_0 = e^2/2C_0$ and the nearest-neighbour capacitance $E_C = e^2/2C$, respectively. The action S_J describes the Josephson coupling between the islands due to the tunneling of Cooper pairs. The different form of the dissipative part arises from the nature itself of the tunneling; for the (1.c) the charge at the electrodes changes in units of e due to the quasiparticles while in the second case (1.d) a continuous flow of charge is assumed. The subscripts i,j label the islands on the lattice, and $\varphi_{ij} = \varphi_i - \varphi_j$ refers to nearest-neighbours. The dissipative kernel in the case of gapless superconductors is

$$\alpha(\tau) = \alpha \frac{(1/\beta)^2}{\sin^2(\pi\tau/\beta)} \qquad (2)$$

where the dimensionless tunneling conductance is defined as

$$\alpha = \frac{h}{4e^2 R_t} \approx 6.45 \ k\Omega/R_t$$

In the case of quasiparticle tunneling an additional summation over the winding numbers should be included to calculate the partition function.

The system described by the action (1) has been studied by many authors, disregarding dissipation [8] and considering the effect of the (1.c) [9] or (1.d) [10] on the phase diagram. Most of the works concern with the case of a self-capacitance much larger than the mutual capacitance between the islands. In the opposite case an additional phase transition occurs an the phase diagram has new phases [11,12].

In the rest of the paper we will briefly review the application of a variational method developed by Giachetti-Tognetti and Feynmann-Kleinert [13] (GTFK)for the study of the phase diagram [14]. We will concentrate on the case in which the dissipation is due entirely to Ohmic shunts and on the self-charging model ($C=0$ and $C_0 \neq 0$). Like in some previous works [15] we will assume that the relevant configuration in driving the transition are the vortices of the field φ_i. We calculate the effective interaction energy renormalized by the quantum fluctuations and as a consequence we can calculate the phase boundary. The exact evaluation of the partition function as a functional integral from the action (1) is impossible. The aim is to express the partition function as an integral over the "classical " phases

$$Z = \int d\varphi^o e^{-\beta W(\varphi^o)} \qquad (3)$$

and

$$\varphi^0 = \frac{1}{\beta} \int_0^\beta d\tau \varphi(\tau) \tag{4}$$

The variational method consists in the variational calculation of the effective potential $W(\varphi^0)$. In the present case the variational action S_{tr} can be obtained introducing the matrix $K_{ij}(\varphi^0)$ and replacing the Josephson term

$$-\int_0^\beta d\tau \sum_{<i,j>} E_J \cos \varphi_{ij}(\tau) \rightarrow \int_0^\beta d\tau \sum_{ij} (\varphi_i(\tau) - \varphi_j^0) \frac{1}{2} K_{ij}(\varphi_i(\tau) - \varphi_j^0) + \beta L \tag{5}$$

The coefficients $K_{ij}(\underset{\sim}{\phi}^0)$ and the function L are determined variationally with the use of the Jensen-Bogoliubov inequality

$$Z \leq Z_{tr} e^{\langle S_{tr} - S \rangle_{tr}} \tag{6}$$

In the considered case $W(\varphi^0)$ describes a classical XY model where the renormalized coupling constant depends on the various parameters defined above $E_J \rightarrow \check{K}(T,C,E_J,R_N)$. It is therefore possible to borrow al the results form the classical Kosterlitz-Thouless-Berezinskkii theory of vortex unbinding and in particular to study the phase boundary and the flux flow resistance above the critical point.

Without going into details[15] it is possible to show that the renormalized Josephson coupling is given by

$$K = 2 \frac{\partial V_{sm}}{\partial \sigma^2} = E_J e^{-\frac{1}{2}\sigma^2(\varphi)^0} \tag{7}$$

where

$$\sigma^2(\varphi^0) = \frac{2}{z\beta N} \sum_{\omega k}{}' z_k \left\{ \frac{1}{8E_0}\omega^2 + \frac{\alpha}{2\pi} z_k |\omega| + K z_k \right\}^{-1}$$

and the classical effective potential is

$$W(\underset{\sim}{\phi}^0) = (1 + \frac{1}{2}\sigma^2) V_{sm} - \frac{1}{2\beta} \sum_{\omega k}{}' \ln \frac{D_{\omega k}}{D_{\omega k}(K=0)} \tag{5}$$

where we introduce the smeared potential

$$V_{sm}(\varphi^0) = \sum_{<i,j>} \int \frac{d\xi}{(2\pi\sigma^2)^{1/2}} \, V(\xi) \, \exp\{-\frac{(\xi-\varphi_{ij}^0)^2}{2\sigma^2(\varphi^0)}\}$$

and the propagator

$$D_{\omega k}(\varphi^0) = \{\frac{1}{8E_0}\omega^2 + \frac{\alpha}{2\pi}z_k |\omega| + K\, z_k\}^{-1}$$

and the quantity $z_k = \sum_{i=1}^{z} (1 - \cos(\mathbf{k}\mathbf{a_i}))$.

In the region where the quantum effects are most relevant the self-consistent equation for K can approximated to the form

$$\frac{8\alpha^2 E_0}{\pi E_J} = \left(\frac{K}{E_J}\right)^{1-\frac{z\alpha}{2}} + \frac{\alpha}{2\beta E_J}\left(\frac{K}{E_J}\right)^{-\frac{z\alpha}{2}}$$

When $\alpha < \frac{2}{z}$ the solution is

$$K \approx E_J \left(\frac{\pi}{16\beta\alpha E_0}\right)^{\frac{2}{z\alpha}} \tag{6}$$

whereas per $\alpha > \frac{2}{z}$ and at very low temperatures

$$K \approx E_J \left(\frac{8\alpha^2 E_0}{\pi E_J}\right)^{\frac{2}{2-z\alpha}} \tag{7}$$

A $T = 0$ the equation for K e coincides with that obtained with the Self Consistent Harmonic Approximation. We stress that at T=0 the coherent state is achieved only if

$$\alpha > \frac{2}{z}. \tag{8}$$

At $T \neq 0$ a numerical solution of the self consistent equations leads to the phase diagram plotted in Figure 1.

Acknowledgments

We want to thank G. Schön and V. Tognetti for the constant support and many useful discussions. We also acknowledge many discussions with A. van Otterlo, V. Scalia, A. Priolo, H. Van der Zant.

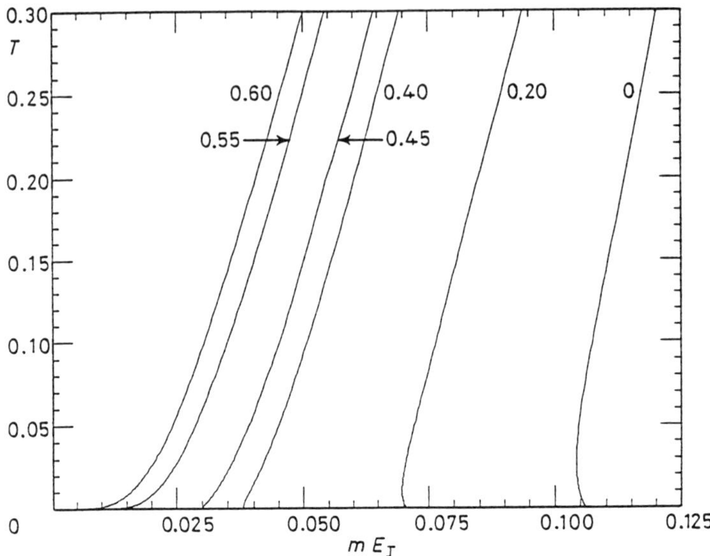

Fig. 1 The phase diagram in the (mE_J - T) plane ($m=1/8E_0$) is shown for various values of the dissipation α.

References.

[1]   G. Schön and A. D. Zaikin, Phys. Rep.**198** (1990) 237

[2]   H.M. Jaeger et al., Phys. Rev. B **34** (1986) 4920; Phys. Rev. B **40** (1989) 182; L.G. Geerligs and J.E. Mooij, Physica B **152** (1988) 212; L.J. Geerligs, M. Peters, L.E.M. de Groot, A. Verbruggen, and J.E. Mooij, Phys. Rev. Lett. **63**, 326 (1989)

[3]   M.N. Cha, M.P.A. Fisher, S.M. Girvin, M. Wallin and A.P. Young , Phys. Rev. B **44** (1991) 6883

[4]   Y. Liu, K.A. McGreer, B. Nease, D.B. Haviland, G. Martinez, J.W. Halley and A.M. Goldman, Phys. Rev. Lett. **67**, 2068 (1991)

[5]   For a number of articles see e.g. Proc. NATO Advanced Research Workshop *Coherence in Superconducting Networks* , eds. J.E. Mooij and G. Schön, Physica B **152** (1988).

[6]   in this paper we do not consider the case in which also the fluctuations of the modulus of the order papermeter are important.

[7]   V. Ambegaokar,U. Eckern and G. Schön,Phys. Rev. Lett. **48** (1982)1745; U. Eckern, V. Ambegaokar and G. Schön, Phys. Rev. **B 30** (1984) 6419.

[8]   E. Simanek, Phys. Rev. **B 22** (1980) 495; K. B. Efetov, Zh.Eksp.Teor.Fiz. **78** (1980) 2017 (Sov. Phys. JETP **51** (1980) 1015); P. Fazekas, B. Mühlschlegel and M. Schröter, Z. Phys. **B 57** (1984) 193; R. Fazio and G. Giaquinta, Phys. Rev. **B 34** (1986) 4909; R.S. Fishman, Phys. Rev. Lett. **63** (1989) 89. L. Jacobs, J.V. José and M.A. Novotny, Phys. Rev. Lett. **53**, 2177 (1984)

[9]   S. Chakravarty, G.L. Ingold, S. Kivelson and A. Luther, Phys. Rev. Lett. **56** (1986) 2303; A. Kampf and G. Schön, Phys. Rev. B **36** (1987) 3651; R. Fazio, G. Falci and G. Giaquinta, Sol. St. Comm. **71** (1989) 275; W. Zwerger, J. Low Temp. Phys. **72**, 291 (1988); M.P.A. Fisher, Phys. Rev. B **36**, 1917 (1987).

[10]   J. Choi and J. V. José, Phys. Rev. Lett. **62** (1989) 1904 ; G. Falci R. Fazio, V. Scalia and G. Giaquinta, Phys. Rev. **B43** (1991) 13053 ; B. Mirhashem and R.A. Ferrell, Physica **152 C**, 361 (1988); Phys. Rev. B **37**, 649 (1988)

[11] J.E. Mooij, B.J. van Wees, L.J. Geerligs , M. Peters, R. Fazio, and G. Schön, Phys. Rev. Lett. **65**, 645 (1990)

[12] R. Fazio and G. Schön,Phys. Rev. **B43** (1991) 5307.

[13] R. Giachetti and V. Tognetti, Phys. Rev. Lett. **55**, 912 (1985); R.P. Feynmann and H. Kleinert, Phys. Rev. **A34** (1986) 5080

[14] G. Falci R. Fazio and G. Giaquinta, Europhys. Lett. **14**, 145 (1991)

[15] E. Simanek and S. Stein, Physica **A129** (1984) 40

[16] For the details see also G. Falci, PhD thesis, University of Catania, unpublished.

THE PHYSICAL NATURE OF DRESSED ATOMS
IN QED AND IN QUANTUM OPTICS

G. Compagno, L. Lo Cascio

Istituto di Fisica dell'Università, Via Archirafi 36,

I - 90123 Palermo, Italy

R. Passante

Istituto per le Applicazioni Interdisciplinari della Fisica,

Consiglio Nazionale delle Ricerche, Via Archirafi 36,

I - 90123 Palermo, Italy

F. Persico

Istituto di Fisica dell'Università and Istituto per le

Applicazioni Interdisciplinari della Fisica, Via Archirafi 36,

I - 90123 Palermo, Italy

Abstract

We summarize the results of recent investigations on the theory of dressed atoms, which is of general use in the nonrelativistic theory of the atom-photon interaction. To this aim, a modified version of the usual quantum theory of measurement is presented, which takes into account the finite duration of measurement process. This theory is first applied to investigate the nature of atoms dressed by the zero-point fluctuations of the vacuum electromagnetic field. In this way it is shown that in a general measurement the atom is perceived as dressed by the high-frequency photons only. Successively the theory is used to investigate the case of an atom dressed by a strong monochromatic radiation field, and it is shown that the experimental apparatus perceives the atom as dressed or as bare depending on the parameters characterizing the measurement, with special reference to the duration of the measurement process. Conceptual and practical consequences of these results are discussed.

1. What is a dressed atom?

The expression "dressed atom" is of current use in QED and in quantum optics. It is referred, however, to two quite different physical configurations.

In the first configuration a ground state neutral atom is considered in interaction with the vacuum fluctuations of the e.m. field. In the ground state of the total atom + field system, these vacuum fluctuations, which are of course of a quantum nature, induce virtual absorption and reemission processes of photons by the atom. These processes are virtual because the bare energy of the atom + field system is not conserved. Thus the corresponding fluctuations δE in the bare energy must have a finite duration δt, according to the Heisenberg uncertainty relation

$$\delta E \sim \hbar/\delta t \tag{1.1}$$

During one of these fluctuations, photons are emitted and reabsorbed after δt. Because of their limited lifetime these photons are called virtual photons. Since the fluctuations take place continuously, a steady-state cloud of virtual photons is formed around the neutral atom. The complex object (bare atom + cloud of virtual photons) is called a dressed atom [1]. In this sense the concept described by this expression is in the same class as that of "dressed source", introduced long ago in quantum field theory [2]. In contrast with a real quantum, which is emitted by an energy- conserving process and which can leave the source in view of its infinite lifetime, a virtual quantum can only reach out at a finite distance

$$r \sim c\delta t \sim \hbar c/\delta E \tag{1.2}$$

from the source. Consequently one should expect the linear dimensions of the virtual cloud surrounding the atom to coincide roughly with r in (1.2), for virtual

finite period of time τ according to the Schrödinger equation. The set of processes taking place during this period is defined as the measurement and

$$| \Psi(t)\rangle \; = \; U(t) \, | \, \Psi(0)\rangle \; = \; \sum_k c_k(t) \, | \, v_k\rangle \, | \, \psi_k\rangle \qquad (2.3)$$

where c_k are c-number amplitudes and $U(t)$ is an appropriate unitary operator. The observation, which here is quite distinct from the measurement, takes place at time τ and it induces a nonSchrödinger collapse of $| \, \Psi(\tau)\rangle$ into one of the states $| \, v_k\rangle \, | \, \psi_k \, \rangle$. Thus in this more realistic description of the measurement process proposed by Peres [9] observation is separated from measurement and a new time scale is introduced, which is defined as the measurement time τ necessary in order to prepare the observation.

In preparation for application of the ideas above to the case of the dressed atom, we shall now discuss measurement of atomic population on the simple atomic model provided by a two-level atom. A two-level atom is most efficiently described in terms of the set of the three Pauli $S = \frac{1}{2}$ operators S_+, S_- and S_z. The latter possesses the eingenvalues $\pm\frac{1}{2}$, according to the atom being in the upper or lower state $| \pm\frac{1}{2}\rangle$, while S_+ and S_- are raising and lowering operators from the lower state $| -\frac{1}{2}\rangle$ (or $|\downarrow\rangle$) and from the upper state $| +\frac{1}{2}\rangle$ (or $|\uparrow\rangle$) respectively. Consider now the atom-pointer Hamiltonian [9]

$$H = \frac{1}{2M}p^2 + \hbar\omega_o S_z + g(t)p S_z \qquad (2.4)$$

where p is the momentum of a particle of (large) mass M, which we take as the pointer, and $\hbar\omega_o$ is the energy difference between the upper and lower states of the two-level atom. The atom-pointer coupling takes place through $g(t)$, which is a gate function of the form

$$g(t) = \frac{1}{\tau}g_o[\theta(t) - \theta(t - \tau)] \quad ; \quad \int_o^\infty g(t)dt = g_o \qquad (2.5)$$

Thus the atom-pointer interaction starts at $t = 0$, and it lasts for a time τ. During this measurement time the pointer gets correlated with the atom, and one can

404

evaluate in the usual way the Heisenberg time evolution of the position operator $q(t)$ of the pointer. If the initial state of the atom-pointer system is

$$| \Psi \rangle = | \psi \rangle \otimes | \Phi \rangle \tag{2.6}$$

where $| \psi \rangle$ is the pointer state and $| \Phi \rangle$ the atomic state of the form $\binom{a}{b}$, then $\langle \Psi | q(\tau) | \Psi \rangle \equiv \langle q(\tau) \rangle$ yields the average value of the distribution of the positions attained by the pointer at observation time τ. Moreover

$$\langle \Psi | q^2(\tau) | \Psi \rangle - (\langle \Psi | q(\tau) | \Psi \rangle)^2 \equiv \langle q^2(\tau) \rangle - \langle q(\tau) \rangle^2$$

is the variance of this distribution. The macroscopic nature of the pointer can be taken into account by assuming

$$\langle \psi | q(0) | \psi \rangle \sim \langle \psi | p(0) | \psi \rangle \sim 0 \tag{2.7}$$

A simple calculation yields

$$\langle q(\tau) \rangle = g_o \langle S_z(0) \rangle \quad ; \quad \langle q^2(\tau) \rangle - \langle q(\tau) \rangle^2 = g_o^2 \left(\frac{1}{4} - \langle S_z(0) \rangle^2 \right) \tag{2.8}$$

Thus if the atom is initially in the $| \pm \frac{1}{2} \rangle$ state, the pointer will always be found at $\langle q(\tau) \rangle = \pm \frac{g_o}{2}$ with zero variance (i. e. one peak only in the distribution of pointer's positions). On the other hand, if the atom is in a superposition $a | +\frac{1}{2} \rangle + b | -\frac{1}{2} \rangle$ at $t = 0$, two peaks in the pointer distribution are expected at $q = \pm \frac{g_o}{2}$ of intensities $| a |^2$ and $| b |^2$ in such a way that

$$\langle q(\tau) \rangle = \frac{1}{2} g_o (|a|^2 - |b|^2) \quad ; \quad \langle q^2(\tau) \rangle - \langle q(\tau) \rangle^2 = g_o^2 |a|^2 |b|^2 \tag{2.9}$$

It is also easy to realize that the position of the pointer at any observation at $\pm \frac{g_o}{2}$ is perfectly correlated with the atomic state $| \pm \frac{1}{2} \rangle$. In conclusion, the pointer is perfectly capable of measuring the atomic population, which is directly related to $\langle S_z(0) \rangle$ as

$$|a|^2 = \langle S_z(0) \rangle + \frac{1}{2} \quad ; \quad |b|^2 = -\langle S_z(0) \rangle + \frac{1}{2}$$

3. Measurement on atom dressed by ZP fluctuations

Take now the atom-pointer Hamiltonian to be of the form [10]

$$H = \frac{1}{2M}p^2 + \hbar\omega_o S_z + g(t)S_z p + \sum_k \hbar\omega_k a_k^\dagger a_k +$$
$$+ \sum_k \left(\epsilon_k a_k S_+ + \epsilon_k^* a_k^\dagger S_- \right) - \lambda \sum_k (\epsilon_k a_k^\dagger S_+ + \epsilon_k^* a_k S_-) \tag{3.1}$$

With respect to (2.4), a radiation field has been added, described by the Bose-like creation and annihilation operators for photons in mode k. This field interacts linearly with the atom through the coupling constant ϵ_k, which promotes photon absorption and emission processes together with changes of the atomic state. The term proportional to λ, which can be 1 or 0 (in the latter case one is working within the Rotating Wave Approximation or RWA), does not conserve the bare energy of the atom + field system and it gives rise to virtual processes. This term is responsible for the existence of the virtual photon cloud in the ground state of H. This ground state is the dressed atom ground state and it is described in perturbation theory at order ϵ^2 by [11]

$$|\tilde{O}\rangle = | \downarrow,\{O_k\}\rangle + (1 - P_o)\frac{1}{E_o - H_o}H_i| \downarrow,\{O_k\}\rangle +$$
$$+ (1 - P_o)\frac{1}{E_o - H_o}H_i(1 - P_o)\frac{1}{E_o - H_o}H_i| \downarrow,\{O_k\}\rangle - \tag{3.2}$$
$$- \frac{1}{2}\langle \downarrow,\{O_k\}|H_i(1 - P_o)\frac{1}{E_o - H_o}(1 - P_o)\frac{1}{E_o - H_o}H_i| \downarrow,\{O_k\}\rangle| \downarrow,\{O_k\}\rangle$$

where $| \downarrow,\{O_k\}\rangle$ is the bare atomic ground state with no photons, of energy $E_o = -\frac{\hbar\omega_o}{2}$, P_o is the corresponding projection operator, H_i contains the ϵ-dependent terms in (3.1) and

$$H_o = \hbar\omega_o S_z + \sum_k \hbar\omega_k a_k^\dagger a_k \tag{3.3}$$

The initial atom pointer state $|\Psi\rangle$ is taken to be of the same form as (2.6), except for the fact that $|\Phi\rangle$ is now given by the dressed state $|\tilde{O}\rangle$ in (3.2). Proceeding in the Heisenberg representation as in the previous section and taking quantum averages at $t = \tau$ on state $|\Psi\rangle$ one gets at order e^2, after a lengthy but straightforward procedure,

$$\langle q(\tau)\rangle = g_o \langle S_z(0)\rangle$$

$$\langle q^2(\tau)\rangle - \langle q(\tau)\rangle^2 = \frac{2}{\tau^2} g_o^2 \frac{\lambda^2}{\hbar^2} \sum_k \frac{|\epsilon_k|^2}{(\omega_o + \omega_k)^2} [1 - \cos(\omega_o + \omega_k)\tau] \quad (3.4)$$

where

$$\langle S_z(0)\rangle = -\frac{1}{2} + \frac{\lambda^2}{\hbar^2} \sum_k \frac{|\epsilon_k|^2}{(\omega_o + \omega_k)^2} \quad (3.5)$$

From (3.4) one can see that if the duration τ of the measurement is such that $(\omega_o + \omega_k)\tau \ll 1$ for any of the field modes, the variance in the pointer distribution takes the form

$$\langle q^2(\tau)\rangle - \langle q(\tau)\rangle^2 = g_o^2 \frac{\lambda^2}{\hbar^2} \sum_k \frac{|\epsilon_k|^2}{(\omega_o + \omega_k)^2} = g_o^2 \left(\frac{1}{4} - \langle S_z(0)\rangle^2\right) \quad (3.6)$$

which is of the same form as (2.8). In this case the pointer perceives the atom as completely bare and the vacuum fluctuations have no other role except determining the value of $\langle S_z(0)\rangle$ as in (3.5). If on the other hand τ is large enough to yield $(\omega_o + \omega_k)\tau \gg 1$ for any of the field modes, the variance of the pointer distribution is easily seen to vanish from (3.4). Thus there is only one peak in the position distribution of the pointer at $t = \tau$, as if the atom where in a ground state of some sort. This ground state, however, is not the bare one, since the position of the pointer is not $q = -\frac{g_o}{2}$, but rather $g_o\langle S_z(0)\rangle$, with $\langle S_z(0)\rangle$ given by (3.5). Thus the pointer perceives the atom as fully dressed by all of the field modes. For an intermediate value of τ, it is clear from (3.4) that some of the modes (the high frequency ones) will contribute to the variance whereas others (the low frequency ones) will not. Thus the pointer perceives the atom as half-dressed, or dressed only by the high-frequency modes [12].

4. Measurement on atom dressed by a strong field

Assume now that all modes in (3.1) have zero photons except one. This populated mode is taken to have $\omega_k \equiv \omega \approx \omega_o$ and to be so strongly populated that one is entitled to neglect the influence of all empty modes on the dynamics of the system. Thus the atom photon part of (3.1) can be approximated as $H_o + H_i$, where

$$H_o = \hbar\omega_o S_z + \hbar\omega a^\dagger a \quad ; \quad H_i = V_1 - \lambda V_2$$

$$V_1 = \epsilon a S_+ + \epsilon^* a^\dagger S_- \quad ; \quad V_2 = \epsilon a^\dagger S_+ + \epsilon^* a S_- \tag{4.1}$$

This is the so called Jaynes-Cummings Hamiltonian [13]. For $\omega \approx \omega_o$ the eigenvalue structure of H_o consists of a series of almost degenerate doublets $(|n, \downarrow\rangle, |n-1, \uparrow\rangle)$ of energy $\approx (n - \frac{1}{2})\hbar\omega$. V_1 in (4.1) connects states within the same doublet and it cannot be treated by perturbation theory in view of near degeneracy. V_2 connects different doublets whose energy differs by $\approx 2\hbar\omega$; hence its effects on the eigenstates of $H_o + V_1$ can be regarded as small. In these conditions one can use the RWA and set $\lambda = 0$. The eigenstates of $H_R = H_o + V_1$ are obtained by diagonalization within each doublet [5] as

$$|U_n^+\rangle = -\cos\frac{\theta}{2}|n, \downarrow\rangle + \frac{\epsilon}{|\epsilon|}\sin\frac{\theta}{2}|n-1, \uparrow\rangle;$$

$$|U_n^-\rangle = \frac{\epsilon}{|\epsilon|}\cos\frac{\theta}{2}|n-1, \uparrow\rangle - \sin\frac{\theta}{2}|n, \downarrow\rangle;$$

$$\sin\frac{\theta}{2} = \frac{1}{\sqrt{2}}\left[1 + \frac{\delta}{\Delta}\right]^{\frac{1}{2}} \quad ; \quad \cos\frac{\theta}{2} = -\frac{1}{\sqrt{2}}\left[1 - \frac{\delta}{\Delta}\right]^{\frac{1}{2}} \quad ;$$

$$\delta = \hbar(\omega_o - \omega) \quad ; \quad \Delta = \left[\delta^2 + 4|\epsilon|^2 n\right]^{\frac{1}{2}} \quad ; \quad E_n^\pm = (n - \frac{1}{2})\hbar\omega \pm \frac{1}{2}\Delta \tag{4.2}$$

Δ is the separation within each doublet and it is directly proportional to the Rabi frequency [6] discussed in sect. 1. The states $|U_n^\pm\rangle$ are states dressed by a strong field, clearly displaying correlations between bare atomic and photon occupation numbers. One would like to understand the physical consequences of these correlations, to investigate if it is meaningful to introduce concepts such as the occupation number of dressed states and what are the limits of this concept.

408

In order to answer these questions, one introduces a pointer as previously, leading to the Hamiltonian

$$H = H_R + \frac{1}{2M}p^2 + g(t)S_z p \tag{4.3}$$

and one chooses $|\Psi\rangle$ as in (2.6) with $|\Phi\rangle$ given by one of the dressed states (4.2). Taking for example $|\Psi\rangle = |u_n^+\rangle$ and proceeding as in the previous section, one has exactly [14]

$$\langle q(\tau)\rangle = g_o\langle S_z(0)\rangle;$$

$$\langle q^2(\tau)\rangle - \langle q(\tau)\rangle^2 = g_o^2 \left(\frac{1}{4} - \langle S_z(0)\rangle^2\right) \frac{1 - \cos\frac{\Delta\tau}{\hbar}}{\frac{\Delta^2\tau^2}{\hbar^2}} \tag{4.4}$$

where

$$\langle S_z(0)\rangle = \frac{1}{2}\frac{\delta}{\Delta} \tag{4.5}$$

Clearly for $\tau \ll \hbar/\Delta$, (4.4) reduce to the bare atom expressions. For $\tau \gg \hbar/\Delta$ however, the variance of the distribution of pointer's position vanishes and this distribution reduces to a single peak at $q = g_o\delta/2\Delta$. In this conditions one is observing a dressed atomic level rather than a bare one. It should be noted that here the time scale to distinguish between long and short measurements is the inverse Rabi frequency $\hbar/\Delta$. This is typically orders of magnitude longer than the time scale for dressing by ZP fluctuations [10,15], which is $\approx \omega_o^{-1}$. For times larger than $\hbar/\Delta$, observing $q(\tau)$ amounts to observing the dressed atomic population, whereas for times smaller than $\hbar/\Delta$, $q(\tau)$ indicates the bare atomic population. We see that the same apparatus can be used to observe quite different physical quantities exploiting measurements of different duration.

5. Conclusions.

Taking into account the finite time τ which is necessary in order to correlate the measuring apparatus and an atom, amount to introducing a new time scale which depends on the experimental conditions. This new time scale interferes with the time scale typical of the atom-field interactions which lead to the dressed atom. This is true both for an atom dressed by ZP fluctuations and for an atom dressed by a strong external field such as that provided by a laser, although the time scale of the processes leading to atomic dressing are very different in the two cases. As a result of this interference of the time scales, an experimental apparatus, designed to measure bare atomic quantities such as atomic populations, may in fact turn out to measure dressed or half-dressed quantities. Another way of stating the same concept is by saying that whether an atom should be considered as bare, as dressed or as half dressed depends very much on the experimental conditions under which a measurement is performed. Furthermore, with particular reference to the case of dressing by ZP fluctuations, the analysis developed in sect. 3 shows that the representation of a dressed atom in terms of a virtual photon cloud created by time-dependent fluctuations of bare quantities is physically sound.

Acknowledgements

It is a pleasure to dedicate this overview to Professor E.R. Caianiello and to acknowledge his significant contributions to the theory of quantized fields.

The research summarized in this paper was carried out under EEC contract $SCI - CT90/0572$ in the framework of the research project "Atoms in strong fields". Partial financial support by Comitato Regionale Ricerche Nucleari e Struttura della Materia, by Regione Siciliana Assessorato Beni Culturali e Ambientali and by Ministero Ricerca Scientifica is also acknowledged.

410

References

[1] G. Compagno, G.M. Palma. R. Passante, F. Persico, in *New Frontiers in QED and in Quantum Optics*, A.O. Barut ed (Plenum Press, New York 1990), p.129

[2] L. Van Hove, Physica 18, 145 (1952); 21, 910 (1955); 22, 343 (1956)

[3] R. Passante, G. Compagno, F. Persico, Phys. Rev. A31, 2827 (1985)

[4] R. Passante, E.A. Power, Phys. Rev. A35, 188 (1987)

[5] C. Cohen-Tannoudji, in Cargese Lectures in Physics vol 2, P. Levy ed. (Gordon and Breach, N.Y. 1968)

[6] P.L. Knight, P.W. Milonni, Phys. Rep. 66, 21 (1980)

[7] C. Cohen-Tannoudji, S. Reynaud, J. Phys. B10, 345 (1977)

[8] G. Rempe, H. Walther, Phys. Rev. A42, 1650 (1990)

[9] A Peres, W.K. Wootters, Phys. Rev. D32, 1968 (1985) A. Peres, Phys. Rev. D39, 2943 (1989)

[10] G. Compagno, R. Passante, F. Persico, Europhys. Lett. 12, 301 (1990)

[11] G. Compagno, R. Passante, F. Persico, Phys. Lett. A112, 215 (1985)

[12] E.L. Feinberg, Usp. Fiz. Nauk 132, 255 (1980) [Sov. Phys. Usp. 23, 629 (1981)] F. Persico, E.A. Power, Phys. Rev. A36, 475 (1987) G. Compagno, R. Passante, F. Persico, Phys. Rev. A38, 600 (1988)

[13] E.T. Jaynes, F.W. Cummings, Proc. IEEE 51, 89 (1965)

[14] L.Lo Cascio, F. Persico, to appear in J. Modern Opt.

[15] G. Compagno, R. Passante, F. Persico, Phys. Rev. A44, 1956 (1991)

EXPERIMENTS AND THEORIES ON CONVECTION.

G. Sonnino

Université libre de Bruxelles

Service de Chimie Physique

Bvd du Triomphe c.p.231, 1050 Bruxelles, Belgium.

M. De Paz

Dipartimento di Fisica

Via Dodecaneso, 33 16146 Genova, Italy.

M. Pilo

Dipartimento di Fisica Teorica

e Sue Metodologie per le Scienze Applicate

Università di Salerno, Baronissi, Italy.

During last years our group has concentrated experimental and theoretical efforts on the study of free convection starting from the observation of definite anomalies in the region of the density maximum of water (De Paz et al. 1987, De Paz and Sonnino 1988, 1989, Anselmi et al 1990, Sonnino and De Paz 1991).

Experiments on free convection in a fluid contained in a cylinder with its axis vertical were first performed by Mouton and De Röeck (M.D.R.) in 1977: they carried out temperature measurements versus time at various points of a cylindrical cell containing water and other liquids submitted to a fast initial temperature step at the lateral wall of the cylinder.

However, their experiments did not include the case of water in the region of its density maximum. Moreover, they advanced a theoretical model of convection, based on the results of their experiments, which subdivided the convecting cell in two regions, the boundary layer and the nucleus, moving in opposite directions to provide continuity. Their mathematical treatment of the model was approximated and the equations obtained by M.D.R. did not predict the inflection points of the temperature vs. time curves, contrary to the experimental evidence in small temperature intervals. Our experimental work consisted in measurements on water in temperature ranges not explored by M.D.R., mainly in the region of maximum density, where anomalous convection effects where observed. These effects where quite recently examined in water and heavy water mixtures and interpreted to provide new information about the density maximum position for these mixtures (Anselmi et al. 1990).

Theoretically, the convective model of M.D.R. has been improved firstly by stating a general theory of the phenomenon for a non-Boussinesq fluid in regions not including density maxima (De Paz and Sonnino 1988), then extending the treatment to include the anomalies (De Paz and Sonnino 1989).

This approach, if the central nucleus model is assumed to be valid, enabled us to solve analytically the set of hydrodynamical differential equations (Fourier, Navier-Stokes and continuity) to provide solutions in terms of only one variable and any boundary condition chosen at will. As pointed out by Sonnino and De Paz in 1991, the variable introduced by this approach to solve a free convection problem has a more general meaning to interpret non-equilibrium phenomena. It is a dimensionless "clock function" whose value tells us how far the convective motion has travelled towards equilibrium at a given instant.

In general, it is linked to a variable conjugated with the internal energy in the sense of Legendre's transformations (Weinreich 1968).

Here we report more recent free convection measurements in water in the region of the density maximum performed with the aim of investigating the limits within which the previously observed anomalies are reproduced. In fact, one of the outstanding results of our previous measurements obtained with only a thermocouple placed at the center of the cylinder, was the evidence of arrests in the temperature vs time curves, but these plateaux were considerably displaced with respect to the point of maximum density. This fact needs further experimental data to see the effect of width and absolute position of the temperature ranges on the motion of the fluid in the cell. This is done by placing in the cell three thermocouples and varying the temperature ranges in the vicinity of the density maximum. In this way we can understand better how convection works and also to detect the direction of the fluid motion in the nucleus.

Since we are going to interpret the experiments by means of a theoretical model of convection we first give the main information about this model.

Free convection at a wall having a different temperature from the surrounding fluid is usually described in terms of density gradients in the gravity field. The fluid close to the wall starts its motion shortly after a temperature difference is created between the wall and the fluid itself. In facts, due to thermal conduction, a thin layer of the fluid is heated or cooled and its density, if no anomalies are presented by this property, becomes correspondingly lower or higher than the surroundings.

Nearby a warmer wall a fluid starts to move upwards, while nearby a colder one the motion is dawnward. The thin layer moving at the wall is the well known boundary layer. If the fluid is contained in a hollow cylindrical cell with its axis vertical and well conducting lateral walls, in order to maintain continuity in the fluid, the motion at the walls must be counterbalanced by an apposite motion on the axis. This gives rise to the so called "nucleus", in the scheme of Fig. 1 observed and described by M.D.R..

In the case of water, which has a density maximum at 3.98°C, effects on the fluid motion are expected, since the density gradients in the neighbourhood of the maximum may generate turbulence and anomalous convection.

This implies a quite different approach to the mathematical problem when ranges including the density maximum are taken into consideration. In previous works (De Paz and Sonnino 1988, 1989), the hydrodynamical equations applied to the nucleus in both situations were solved: here we resume the more general treatment including the density maximum, since our new measurements are referred to

414

water in a temperature range including 4°C. The equations are:

$$\text{Fourier:} \qquad \partial_t T - w\partial_z T = \chi\partial_z^2 T + \nu^*(\partial_z w)^2 + \frac{p}{(\rho c)}\partial_z w$$

$$\text{Navier-Stokes:} \qquad \partial_t w - w\partial_z w = \nu\partial_z^2 w + \frac{1}{\rho}\partial_z p - g$$

Continuity:

$$\partial_t \rho - \partial_z(\rho w) = 0$$

where p and T are the pressure and temperature of the fluid and w and ρ are its vertical velocity and density respectively.

ν and ν^* are viscosity parameters, χ is the thermal diffusion coefficient, c is the constant volume heat capacity and g is the constant of gravity.

To solve these equations in the absence of a density maximum the temperature solution can be expanded in a complete basis set in terms of time and space (f_n, Φ_n) (De Paz and Sonnino 1988):

$$T(t, z) = T_S + \sum_n f_n(t)\Phi_n(z)$$

where:

$$f_n(t) = \exp(i\, n\, x(t))$$

$$\Phi_n(z) = \exp(i\, n\, z/h)$$

i.e. the variable $x(t)$ plays in the time the same role of z in the space.

The same simple role cannot be demonstrated in the case of a liquid with a density maximum, where the velocity field is perturbed in terms of time and space functions, but still the introduction of $x(t)$ seems to be important in view of a general description of the nucleus. In fact, in general, $x(t)$ satisfies a self-consistent equation, which we discuss briefly in what follows. In the absence of a density maximum the time derivative of x represents the velocity of the fluid in the nucleus.

For clarity, all the following graphs are obtained by averaging the data over ten, so that each point of the graph comes from the average of $1,000$.

In Fig. 2 the data of free convection between 7.8 $°C$ and -0.5 $°C$ are reported. The symmetry between the warming and cooling curve is evident for all the thermocouples in the cell, although the temperature interval is not symmetrical with respect to 4 $°C$. Moreover, the fluid velocity on the axis is upright either on cooling or in warming, as indicated by the relative positioning of the curves of $T1$, $T2$ and $T3$: $T1$ is always the first to show a temperature variation after the initial step at the walls.

This is consistent with the model of convection previously described. In facts, owing to the density maximum of water, when temperature is swiched, in both cases the density at the walls becomes larger than in the rest of the fluid and the velocity is upwards in the nucleus.

In Fig. 3 the data of free convection between 8.1$°C$ and 0.4$°C$ are reported . Also in this case the initial and final temperatures of the experiment are asymmetrical with respect to 4$°C$, but in the direction of a slightly larger gradient in the high temperature range. The asymmetry in convection is quite strong, amounting to a delay in the cooling curve of 20% in comparison with the warming curve.

Again, the fluid velocity in the nucleus is upward in both cases, as predicted by the model.

In order to interpret the strong asymmetry of convection, we note that the density curve of water as a function of temperature is slightly asymmetric with respect to the maximum at 3.98 $°C$, with a larger density variation (+8 ppm) on the lower temperatures side. This implies a slightly larger density variation at the cylinder walls when warming occurs. This phenomenon shows the great sensitivity of convection to extremely small density variations.

The existence of a density maximum sets qualitatively the condition for an upward motion of the nucleus both on cooling and in warming, in principle for all temperature ranges of the experiments which include 4 $°C$.

However, the typical temperature arrests of Figg. 2, 3 don't occur at 4 $°C$, but between about 5.8 $°C$ and 5.2 $°C$ on cooling and between 2.2 and 2.8 $°C$ on warming. This hystheresis is generated by a peculiar density distribution within the boundary layer, much different from one predicted by a linear temperature gradient between the walls and the nucleus. Resuming in few words the results of calculations performed with our model, we can tell that the plateaux indicate a sort of "egg cut" of the density function around the maximum, while the fast temperature fall after the arrests implies a cusp at the maximum. As a consequence of this picture, we expect some effects either on the shape of the curves or on the direction of the velocity when the width of the temperature range is reduced on one side of the maximum. This is argument of our present work, which is going to be published very soon.

Acknowledgements

This work is supported by the Italian Ministry of Education (M.P.I.) and by the National Research Council (C.N.R.). The work of G. Sonnino is financed by Economic European Communities.

References

Anselmi C., De Paz M., Pilo M., Sonnino G., 1990 *Int. J. Heat Mass Transfer*, vol. 33, p. 2519-2524.

De Paz M., Pilo M., Sonnino G., 1987 *Int. J. Heat Mass Transfer*, vol. 30, n. 2, P. 289 - 295.

De Paz M., Sonnino G., 1988 *Int. J. Heat Mass Transfer*, vol. 31, n. 6, P. 1167-1172

De Paz M., Sonnino G., 1989 *Phys. Rev. A*, vol. 39, p. 3031-3037.

Guckenheimeir J. and Holmes P., 1990 *Nonlinear Oscillations, Dynamical Systems, and Bifurcations of Vector Fields* Springer-Verlag.

Mouton H. and De Röeck H., *Int. J. Heat Mass Tranfer*, vol. 20, 627 (1977)

Sonnino G., De Paz M., 1991 *Taormina Conference on Thermodynamics*, Feb. 18-22, Taormina, Italy.

Weinreich G. 1968 *Fundamental Thermodynamics* Addition-Wesley Publishing Co.

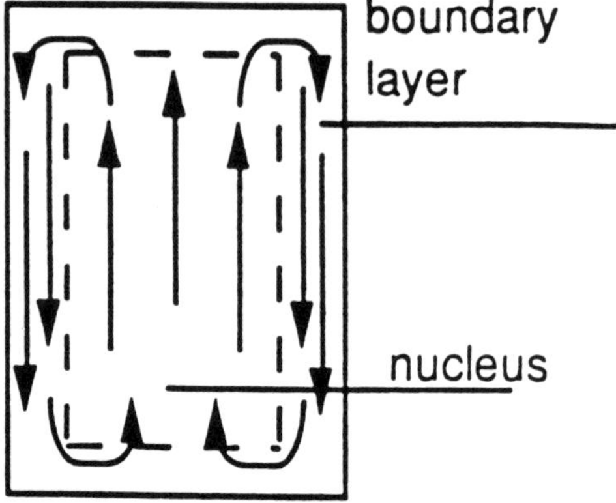

Figure 1.
Model of convection in a fluid contained in a hollow cylinder with conducting lateral walls. The arrows indicate the fluid motion in the boundary layer and in the nucleus in the case of initial larger density at the walls.

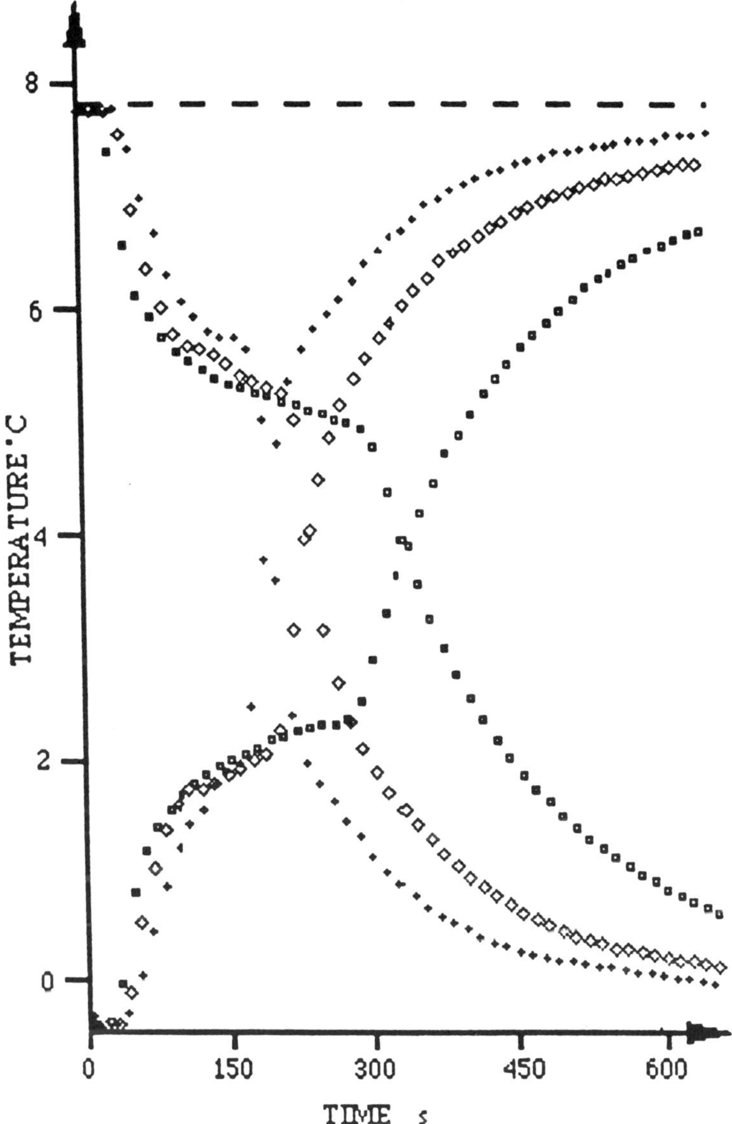

Figure 2.

Free convection in water between 7.8°C and −0.5°C (cooling and warming)
The dashed indicates the higher temperature limit, the lower being coincident with
the time axis.

q T1

◇ T2

+ T3

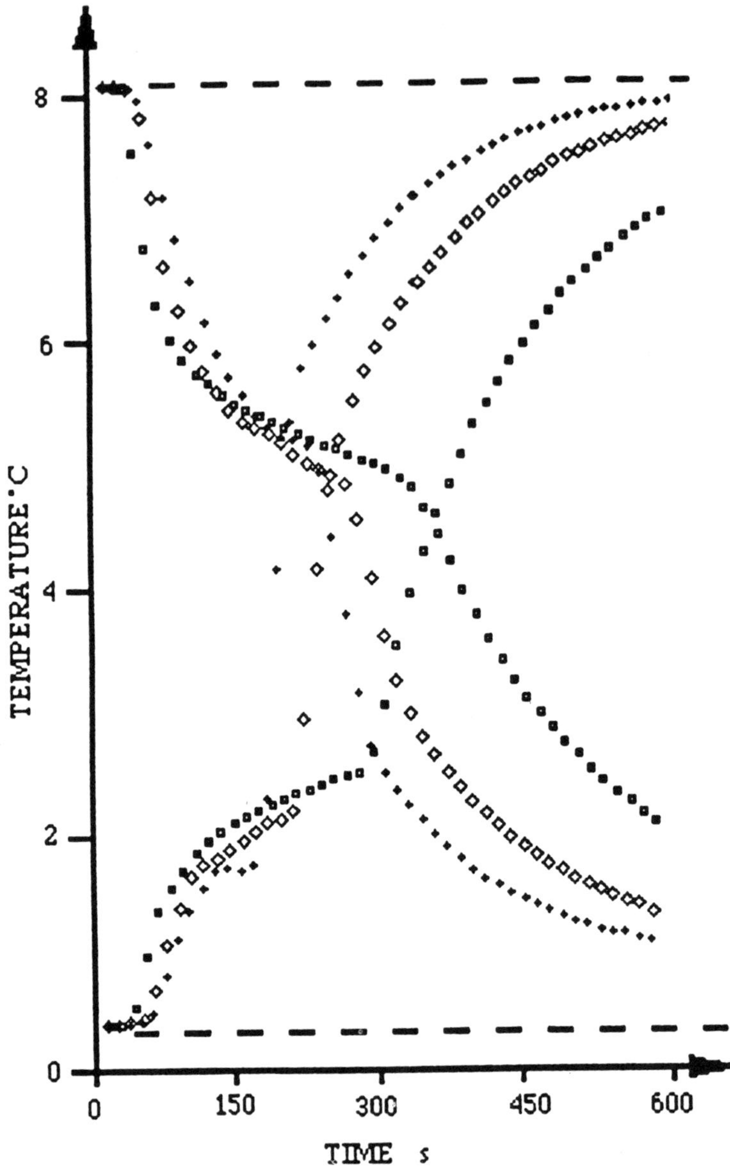

Figure 3.
Free convection in water between 8.1°C and 0.4°C. (cooling and warming)
- - - initial and final temperatures of the experiments.
q T1
◇ T2
+ T3

References

Anselmi C., De Paz M., Pilo M., Sonnino G., 1990 *Int. J. Heat Mass Transfer*, vol. 33, p. 2519-2524.

De Paz M., Pilo M., Sonnino G., 1987 *Int. J. Heat Mass Transfer*, vol. 30, n. 2, P. 289 - 295.

De Paz M., Sonnino G., 1988 *Int. J. Heat Mass Transfer*, vol. 31, n. 6, P. 1167-1172

De Paz M., Sonnino G., 1989 *Phys. Rev. A*, vol. 39, p. 3031-3037.

Guckenheimeir J. and Holmes P., 1990 *Nonlinear Oscillations, Dynamical Systems, and Bifurcations of Vector Fields* Springer-Verlag.

Mouton H. and De Röeck H., *Int. J. Heat Mass Tranfer*, vol. 20, 627 (1977)

Sonnino G., De Paz M., 1991 *Taormina Conference on Thermodynamics*, Feb. 18-22, Taormina, Italy.

Weinreich G. 1968 *Fundamental Thermodynamics* Addition-Wesley Publishing Co.

Contributors

M.A. Aizerman - Institute of Control Sciences of the Academy of Sciences of the USSR, Profsoyuznaya Str. 65, Moscow 117806, RUSSIA

T. Arecchi - Istituto Nazionale di Ottica, Largo E. Fermi, 6 50125 Firenze

T. Arimitsu - Institute of Physics, University of Tsukuba, Ibaraki 305, JAPAN

M. Ban - Advanced Research Laboratory, Hitachi, Ltd. Hatoyama, Saitama 350-03, Japan

A. Barone - Dip. di Scienze Fisiche, Facoltá di Ingegneria, Universitá di Napoli "Federico II", Napoli

V. Barone - Dip. di Fisica e Sezione I.N.F.N., Universitá di Perugia, Via Elce di Sotto, I-06100 Perugia

A.O. Barut - Department of Physics, University of Colorado, Boulder, CO 80309, U.S.A.

A. Bertoni - Dipartimento di Scienze dell'Informazione Universitá degli Studi di Milano, Via Comelico, 39, 20135, MILANO

C. Blundo - Dip. di Informatica e Applicazioni - Universitá di Salerno, Via S. Allende, 84081 Baronissi SALERNO

N.N. Bogolubov Jr. - Moscow Steclov Mathematical Institute, Vavilova 42, Moscow, Russia

C. Bohm - Dipartimento di Matematica, Istituto G. Castelnuovo, Universitá di Roma "La Sapienza", P.le Aldo Moro 5, I-00185 Roma

L. Borland - Institute for Theoretical Physics and Synergetics University of Stuttgart, Pfaffenwaldring 57/4, D-7000 Stuttgart 80 (Vaihingen) Germany

F. Brosens - Departement of Physics, University of Antwerp (UIA), Universiteitsplein 1, B-2610 Antwerpen, Belgium.

P. Budinich - SISSA, Strada Costiera, 11, TRIESTE

E. Burattini - Istituto di Cibernetica del CNR, 80072 Arco Felice, Napoli

G. Caglioti - Istituto di Fisica - Politecnico di Milano, Piazza Leonardo Da Vinci,
32, 20133 MILANO

P. Campadelli - C.N.R. , Istituto di Fisiologia dei Centri Nervosi, Via Mario
Bianco, 9 - 20131 Milano

A. Campolattaro - Depart. of Physics, University of Maryland Baltimore Country Campus, Baltimore, Maryland, 21228 U.S.A.

G. Carpenter - Center for Adaptive System, Boston University, 111 Cummington Street, Boston MA 02215 U.S.A.

E. Celeghini - Dipartimento di Fisica e Sezione I.N.F.N., Universitá di Firenze,
lg. E. Fermi, 2, 50125 Firenze.

S. Chillemi - Istituto di Biofisica del C.N.R. , Via S. Lorenzo 26, 56100 Pisa.

G. Cicuta - Dip. di Fisica, Universitá di Bari, Via Amendola, 173, 70126 Bari.

J.W. Clark - McDonnell Center for the Space Sciences and Dept. of Physics,
Washington University, St. Louis, MO 63130 USA

G. Compagno - Istituto di Fisica, Universitá di Palermo, Via Archirafi 36, 90123
PALERMO

R.M.J. Cotterill - Division of Molecular Biophysics, The Technical Univ. of
Denmark, Building 307, DK - 2800 Lyngby, Denmark

G. Cristofaro - Dip. di Scienze Fisiche Universitá di Napoli and I.N.F.N. Sezione
di Napoli, Mostra d'Oltremare Pad. 19 - 80125 Napoli

L. De Cesare - Dip. di Fisica Teorica, Universitá di Salerno, via S. Allende,
84081 Baronissi, SALERNO

S. De Filippo - Dip. di Fisica Teorica - Universitá di Salerno, Via S. Allende,
84081 Baronissi, SALERNO

G. Della Riccia - Dip di matematica e informatica, Universitá di Udine, Via
Zanon 6, 33100 UDINE

A. De Luca - Dip. di Matem., Universitá di Roma "La Sapienza", P. le Aldo Moro, 2, 00185 ROMA

S. De Martino - Dipartimento di Fisica, Universitá di Salerno, Via S. Allende, 84081 Baronissi, SALERNO

M. De Paz - Dip. di Fisica, Via Dodecaneso 33, 16146 Genova.

A. De Santis - Dip. di Informatica ed Applicazioni, Università di Salerno, Via S. Allende, 84081 Baronissi, SALERNO

S. De Siena - Dipartimento di Fisica, Universitá di Salerno, Via S. Allende, 84081 Baronissi, SALERNO.

J. T. Devreese - Departement of Physics, University of Antwerp (UIA), Universiteitsplein 1, B-2610 Antwerpen, Belgium.

A. Di Crescenzo - Dip. di Matem. ed Applicazioni, Università di Napoli "Federico II", Via Mezzocannone, 8, 80134 NAPOLI

A. Di Giacomo - Dipartimento di Fisica, Sezione dell'INFN, Universitá di Pisa, P.zza Torricelli 2, 56100 PISA

F. Esposito - Dip. di Scienze Fisiche, Universitá di Napoli "Federico II", P.le V. Tecchio 80 - 80125 NAPOLI

U. Esposito - Dip. di Scienze Fisiche, Universitá di Napoli "Federico II", P.le V. Tecchio 80 - 80125 NAPOLI

G. Falci - Istituto di Fisica, Facoltá di Ingegneria, V.le Doria 6, 95125 Catania

R. Fazio - Istituto di Fisica, Facoltá di Ingegneria, V.le Doria 6, 95125 Catania

A. Feoli - Dip. di Fisica Teorica, Università di Salerno, Via S. Allende, 84081 Baronissi, Salerno

F. Ferrucci - Dip. di Informatica, Univ. di Salerno, Via S. Allende, 84081 Baronissi, SALERNO

E.S. Fradkin - Lebedev Physical Institute, Moscow 117333 Russia

S. Fubini - Dip. di Fisica Teorica, Universitá di Torino, Via Pietro Giuria 8, 10125 TORINO

M. Fusco-Girard - Dip. di Fisica Teorica, Universitá di Salerno, Via S. Allende, 84081 Baronissi, SALERNO

G.M. Germano - Dip. di Informatica, Universitá di Pisa, C.so Italia 40, 56125 PISA

C. Gernoth - McDonnell Center for Space Sciences, Dept. of Physics, Washington University, St. Louis, MO 63130 USA

G. Giaquinta - Istituto di Fisica, Facoltá di Ingegneria, V.le Doria 6, 95125 Catania

V. Giorno - Dip. di Informatica ed Applicazioni, Universitá di Salerno, Via S. Allende, 84081 Baronissi, SALERNO

S. Grossberg - Center for Adaptive System, Cummington Street, Boston University, Massachussetts 02215 U.S.A.

S. Guccione - Dipartimento di Scienze Fisiche, Universitá di Napoli, Mostra d'Oltremare Pad. 19, 80125 NAPOLI

M. Guida - Dipartimento di Fisica, Universitá di Salerno, Via S. Allende, 84081 Baronissi, SALERNO

R. Haag - Istitut fur Theoretische Physik, Universitat Hamburg.

H. Haken - Institute for Theoretical Physics and Synergetics, University of Stuttgart, Pfaffenwaldring 57/4, D-7000 Stuttgart 80 (Vaihingen) GERMANY

F. Lauria - Dipartimento di Scienze Fisiche dell'Universitá di Napoli, Mostra d'Oltremare Pad. 19, I - 80125 NAPOLI

L. Lo Cascio - Istituto di Fisica, Universitá di Palermo, Via Archirafi 36, 90123 PALERMO

L. Lukierska-Walasek - Institute of Physics, Pedagocical University of Zielona Góra, Plac Slowianski 6, 65-069 Zielona Góra, Poland.

G. Maiella - Dip. di Scienze Fisiche Universitá di Napoli and I.N.F.N. Sezione di Napoli, Mostra d'Oltremare Pad. 19 - 80125 Napoli

O. Mannella - Dipartimento di Fisica, Universitá di Salerno, Via S. Allende, 84081 Baronissi, SALERNO

M. Marinaro - Dip. di Fisica Teorica, Universitá di Salerno, Via S. Allende, 84081 Baronissi, SALERNO.

G. Marmo - Dip. di Scienze Fisiche, Universitá di Napoli, Mostra d'Oltremare Pad. 19 - I-80125 Napoli

R.E. Marshak - Physics Dept. Virginia Polytecnic Institute, Blacksburg, Virginia 24061 U.S.A.

A. Martin - CERN - TH Division, CH 1211 Geneve 23, Switzerland

N. Markuzon - Center for Adaptive Systems, 111 Cummington Street, Boston University, Boston, Massachussetts 02215 U.S.A.

H. Matsumoto - Institute for Materials Research, Tohoku University, Katahira 2-1-1, Aoba-Ku, Sendai 980, JAPAN

G. Mauri - Dipartimento di Scienze dell'Informazione, Universitá degli Studi di Milano, Via Comelico 39, 20135 Milano

S. Mazzanti - Dip. di Matematica ed Informatica - Universitá di Udine, Via Zanon, 6-8, 33100 UDINE

L. Michel - Institut des Hautes, Etudes Scientifique - 35, Raute de Chartres F - 9111440 Bures sur Yvette - Francia

A. Molinari - Dip. di Fisica Teorica, Universitá di Torino, Via Pietro Giuria 1, 10125 TORINO

D. Montanari - Tema S.p.A. (ENI Group), Viale Aldo Moro 38, 40127 Bologna, Italy

C. Musio - Istituto di Cibernetica del CNR, Via Toiano 6, 80072 Arco Felice, Napoli

R. Musto - Dip. di Scienze Fisiche Universitá di Napoli and I.N.F.N. Sezione di Napoli, Mostra d'Oltremare Pad. 19 - 80125 Napoli

F. Nicodemi - Dip. di Scienze Fisiche Universitá di Napoli and I.N.F.N. Sezione di Napoli, Mostra d'Oltremare Pad. 19 - 80125 Napoli

C. Nielsen - Division of Molecular Biophysics, The Technical University of Denmark, Building 307, DK-2800 Lyngby, Denmark.

A.G. Nobile - Dip. di Informatica e Matematica Universitá di Udine, Via Zanon 6-8, 33100 UDINE

G. Pacini - Dip. di Informatica, Univ. di Salerno, Via S. Allende, 84081 Baronissi, SALERNO.

G. Papini - Dept. of Physics, University of Regina, Regina , Saskatchewan, S4S 0A2 CANADA

R. Passante - Ist. per le Applicazioni Interdisciplinari della Fisica, Universitá di Palermo, Via Archirafi 36, 90123 PALERMO

I.P. Pavlotsky - The Keldysh Institute of Applied Mathematics, Acad. of Sc. of the USSR, 125047 Moscow, RUSSIA.

V. Penna - Dip. Fisica e Unitá INFM , Politecnico di Torino, Via Corso Duca degli Abruzzi 24, 10129 TORINO

F. Persico - Istituto di Fisica e Ist. per le Applicazioni Interdisciplinari della Fisica, Universitá di Palermo, Via Archirafi 36, 90123 PALERMO

M. Pilo - Dipartimento di Fisica Teorica, Universitá di Salerno, Via S. Allende, 84081 Baronissi, SALERNO.

E. Predazzi - Dip. di Fisica Teorica - Universitá di Torino e INFN Sezione di Torino, Via Pietro Giuria 1, 10125 TORINO

Qi Xiang-Lin - Visual Information Processing Laboratory, Institute of Biophysics, Academia Sinica, BEIJING, CINA

I. Rabuffo - Dip. di Fisica Teorica Universitá di Salerno, Via S. Allende, 84081 Baronissi, SALERNO.

M. Rasetti - Dipartimento di Fisica e Unitá I.N.F.M., Politecnico di Torino, Corso duca degli Abruzzi, 24, 10129 TORINO.

A. Restivo - Dip. di Matematica ed Applicazioni, Universitá di Palermo Via Archirafi 34, 90123 PALERMO.

J.H. Reynolds - Center for Adaptive System, 111 Cummington Street, Boston University, Boston Massachussetts 02215 U.S.A.

L.M. Ricciardi - Dip. di Matem. ed Applicazioni, Universitá di Napoli "Federico II", Via Mezzocannone 8, 80134 NAPOLI.

A. Rimini - Dip. di Fisica Nucleare e Teorica, Universitá di Pavia, Via Bassi 6, 27100 Pavia

R. Rivers - Imperial College of Scienca and Technology, The Blackett Laboratory, Prince consort road, London SW7 2BZ, INGHILTERRA

D.B. Rosen - Center for Adaptive System, 111 Cummington Street, Boston University, Boston, Massachussetts 02215 U.S.A.

T. Saito - Institute of Physics, University of Tsukuba, Ibaraki 305, JAPAN

S. Salemi - Dip. di Matematica ed Applicazioni, Università di Palermo Via Archirafi 34, 90123 PALERMO.

N. Sanchez - Observatoire de Paris, Section de Meudon, Demirm 92195 Meudon Principal Cedex, FRANCE

G. Scarpetta - Dip. di Fisica Teorica, Università di Salerno, Via S. Allende, 84081 Baronissi, Salerno

R. Serra - Ferruzzi finanziaria (Ferruzzi Group), Via Ariani 1, 48100 Ravenna

P. Sodano - Dip. di Fisica e Sezione INFN, Università di Perugia, Via Elce di Sotto 10, I - 06100 PERUGIA

D. Somers - Center for Adaptive Systems, Boston University, 111 Cummington Street, Boston Massachussetts 02215 U.S.A.

G. Sonnino - Université Libre de Bruxelles, Service de Chimie Physique, Bvd du Triomphe, c.p. 231, 1050 Bruxelles, Belgium.

L. Stringa - IRST, Istituto per la ricerca Scientifica e Tecnologica, 38100 Trento, Italy.

V.M. Suslin - The Keldysh Institute of Applied Mathematics, Acad. of Sc. of the USSR, 125047 Moscow, Russia.

C. Taddei-Ferretti - Istituto di Cibernetica del CNR, Via Toiano 6, 80072 Arco Felice, Napoli

G. Tamburrini - Istituto di Cibernetica del CNR, 80072 Arco Felice, Napoli

G. Tortora - Tucci - Dip. di Informatica - Univ. di Salerno, Via S. Allende, 84081 Baronissi, SALERNO.

H. Umezawa - The Theoretical Physics Institute, University of Alberta, Edmonton, Alberta T6G 2J1 CANADA

R. Vaglio - Dip. di Fisica - Universitá di Salerno, Via S. Allende, 84081 Baronissi, SALERNO.

P. Vansant - Departement of Physics, University of Antwerp (UIA), Universiteitsplein 1, B-2610 Antwerpen, Belgium.

V. Varshavsky - R&D Coop. "Trassa", St. Peterburg, USSR

G. Vilasi - Dip. Fis. Teorica., Universitá di Salerno, Via S. Allende, 84081 Baronissi, SALERNO

G. Vitiello - Dipartimento di Fisica, Universitá di Salerno, Via S. Allende, 84081 Baronissi, SALERNO.

K. Walasek - Institute of Physics, Pedagocical University of Zielona Góra, Plac Slowianski 6, 65-069 Zielona Góra, Poland

Wang Yun-Jiu - Visual Information Processing Laboratory, Institute of Biophysics, Academia Sinica, BEIJING, CINA

W.R. Wood - Dept. of Physics, University of Regina, Regina, Saskatchewan, S4S 0A2 CANADA